Bayesian Philosophy of Science

Jan Sprenger Stephan Hartmann

Bayesian Philosophy of Science

Variations on a Theme by the
Reverend Thomas Bayes

$$p(\mathrm{H}|\mathrm{E}) = p(\mathrm{H})\,\frac{p(\mathrm{E}|\mathrm{H})}{p(\mathrm{E})}$$

OXFORD
UNIVERSITY PRESS

OXFORD
UNIVERSITY PRESS

Great Clarendon Street, Oxford, OX2 6DP,
United Kingdom

Oxford University Press is a department of the University of Oxford.
It furthers the University's objective of excellence in research, scholarship,
and education by publishing worldwide. Oxford is a registered trade mark of
Oxford University Press in the UK and in certain other countries

First Edition published in 2019

Published in the United States of America by Oxford University Press
198 Madison Avenue, New York, NY 10016, United States of America

British Library Cataloguing in Publication Data
Data available

Library of Congress Control Number: 2019941395

ISBN 978–0–19–967211–0

Ich bin ein Esel, und will getreu,
wie meine Väter, die Alten,
an der alten, lieben Eselei,
am Eseltume halten.

Und weil ich ein Esel, so rat ich euch,
den Esel zum König zu wählen;
wir stiften das große Eselreich,
wo nur die Esel befehlen.

Wir sind alle Esel! I-A! I-A!
Wir sind keine Pferdeknechte.
Fort mit den Rossen! Es lebe—Hurra!—
Der König vom Eselsgeschlechte!

—Heinrich Heine

First Author's Preface

Like many things in life that you do for the first time, writing a book is tempting and scary at the same time. It is scary because the stakes are high: A weaker journal article or edited volume chapter is easily pardoned by the academic community. It ends up not being cited and that's it. A book, however, is measured by different standards: it is assumed that you have written it without having to satisfy capricious referees, stressed editors and strict length and style constraints. It is supposed to show your highest level of philosophical thinking. (The analogue in music is perhaps the difference between writing a piano sonata and writing a symphony—especially when it is the first. We know well how much even the greatest composers have struggled with this step!) For personalities like me, who tend to be perfectionist and sloppy at the same time, this means that you spend an endless amount of time on revisions and corrections. At some point, then, the entire project feels like a never-ending story and you are just happy to complete it and to move on to new challenges.

At the same time, writing a book is a great temptation. Not only because having a bundle of casebound paper on your desk gives a different feeling of reward than being notified that your article has been published on the journal's homepage. Writing a book is something that allows you to reflect on your professional contributions and achievements. You can put them into a context that allows the readers to grasp the bigger picture and to perceive the links between the different parts of your work. While the process was at times tedious and frustrating, overall I enjoyed it very much. I enjoy even more that we have—finally!—brought the project to a successful conclusion.

Like with most young philosophers, the idea of publishing a book was born when I was about to complete my Ph.D. thesis in philosophy of science on the topic "Confirmation and Evidence". Different from the States, publishing their Ph.D. thesis as a book was a thing that young German *doctores philosophiae* were encouraged to do in order to launch their academic career—at least back in the days (2008). My supervisor, Professor Andreas Bartels at the University of Bonn, expressed a favorable opinion and suggested trying to publish the work with Mentis, a well-respected German publishing house. Although the thesis was certainly fine for obtaining the Ph.D., and although it contained several new ideas and results, I felt that it was not coherent and mature enough to survive as a book. Rather I hoped that the more interesting chapters could form the core of a different book. So the work disappeared in the proverbial file drawer, that is, the internet site of the *Universitäts- und Landesbibliothek Bonn* (ULB). (Making the contents publicly accessible, at least in electronic form, was a requirement for the award of the Ph.D.) Leaving the German academic system and taking up a post at Tilburg University in the Netherlands helped me to make that decision. I still think it was right. Too many pretentious phrases centered around points of marginal general interest, or as Stephan Hartmann likes to say in similar contexts, "calculation details for the umpteenth epicycle of the Ptolemaic system". If you are, like me, a young, callow and slightly overconfident philosopher of science, then I invite you to search for the work on the ULB's site and see whether there are parallels to your own Ph.D. thesis.

Years later Stephan and I started to work on this book—an outline and a résumé of philosophy of science conducted with Bayesian models. To my surprise, even the more original and significant parts of my thesis—the contributions to the theory of H-D confirmation and statistical inference—eventually dropped out of the picture. *"Ein schönes Lied, ein Meisterlied,/wie fass' ich da den Unterschied?"* asks Walther von Stolzing in what is perhaps the most inspired scene of Richard Wagner's *Meistersinger von Nürnberg*. With the above distinction, Walther's conversation partner Hans Sachs is alluding to the fact that creativity and inspiration ("ein schönes Lied") are not sufficient for making a lasting contribution that meets certain constraints and reaps widespread applause ("ein Meisterlied"). With the help of the experienced Sachs, Walther finally manages to condense his talent and enthusiasm into a piece that is also approved by the conservative Nuremberg master singers. I think the process of developing philosophical ideas and giving them form is similar. It is up to the reader's judgment whether we have succeeded at this task, but it was our ambition to balance accessibility for non-specialists and pleasing proportions with novel and inspiring contributions. It then dawned upon me that some of the contents which were dear to my heart and which I was planning to include in the book might rather be omitted.

The book is also a balance of my long-standing friendship and productive collaboration with the other author, Stephan Hartmann, who was my mentor and direct superior during five years in Tilburg (2008–2013). As director of the Tilburg Center for Logic and Philosophy of Science (TiLPS), Stephan managed to put a rather sleepy town on the global map of philosophy of science and to create a lively and stimulating environment where the central elements of this book were conceived and worked out—partly in collaboration with colleagues such as Matteo Colombo, visiting scholars such as Richard Dawid and Jonah Schupbach, and Ph.D. students such as Dominik Klein, Chiara Lisciandra, Carlo Martini and Soroush Rafiee Rad. We felt that our work in this period should be condensed into a monograph which does not only report our results, but shows the common theme in the individual pieces of research, and explains and defends our philosophical approach. Stephan left Tilburg in 2013 to become co-director of the Munich Center for Mathematical Philosophy (MCMP), and after succeeding Stephan as TiLPS director, I joined the University of Turin in 2017. The project and the friendship survived, however, and we are both very happy to have completed a manuscript which will—or so we hope—inspire future research in philosophy of science and beyond.

I would like to conclude with my words of thanks. In the first place, to Stephan for his continuous support and invaluable advice in becoming a better philosopher, navigating the academic world, and also in personal matters. Then to my mentors from my graduate student years at the University of Bonn, who introduced me to the world of contemporary philosophy and helped me to complete the transition from applied mathematics: my supervisor Andreas Bartels, his assistants Cord Friebe, Holger Lyre, Jacob Rosenthal (special thanks!) and Professor Rainer Stuhlmann-Laeisz. I would also like to mention Thomas Grundmann, Professor at the nearby University of Cologne (my hometown) and organizer of many terrific summer schools: for me, he is still the exemplary model for combining the German and the anglophone philosophy tradition. From the staff at the London School of Economics, where I spent a term as a junior visiting scholar during my Ph.D., I would like to mention Richard Bradley, Franz Dietrich and Matthew Parker for the inspiring example they set. Same for Mark Burgman, Mark Colyvan and Paul Griffiths, who were tremendous hosts and colleagues during my research stays in Sydney and Melbourne in 2009, 2011 and 2017.

Most credits go to my fantastic colleagues at TILPS, who created a unique atmosphere—cosy and dynamic at the same time—and who made me feel at home for nine long years. I fear I will never again play football on departmental corridors. Apart from my direct collaborators, I would like to thank Thomas Boyer-Kassem, Seamus Bradley, Colin Elliot, Silvia Ivani, Alessandra Marra, Felipe Romero and Naftali Weinberger for their tremendous feedback on the draft manuscript. From the not-yet-named visitors and colleagues, Claus Beisbart, Peter Brössel, Leandra Bucher, Filip Buekens, Luca Moretti, Reinhard Muskens and Marie Postma-Nilsenová. My student assistants Georgi Duev, Zhasmina Kostadinova and Sophia Shokuri for their corrections of earlier versions of the manuscript. Then, all those colleagues at different universities who helped us to improve this book through their generous feedback, and my new colleagues in Turin for their cordial welcome, which greatly helped me to finish the manuscript. Our copy editor and typesetter Christopher von Bülow, who found hundreds of mistakes and made literally thousands of valuable changes and suggestions. Peter Momtchiloff from Oxford University Press for his patience, support and belief in the project, and April Peake for guiding us through the production process. The Nederlandse Organisatie voor Wetenschappelijk Onderzoek (NWO), the Deutsche Forschungsgemeinschaft (DFG) and the European Research Council (ERC) for their generous financial support—especially through ERC Starting Grant No. 640638.

Finally, many thanks to all my friends: meeting you and spending time with you made life so enjoyable in the years that I was working on the book. Most of you will not read this book anyway, so I don't have to list you all! But please feel included, regardless of whether you are part of the Doppelkopf clique from high school or the (all-German) French connection *"Où est le vin?"*, whether we made friends at the TSVV Merlijn football club, whether we played chess together for Leidsche Schaakgenootschap or Šahovski Klub Brda, or whether you are just a good friend without any affiliations! Last, but certainly not least, I would like to thank my parents for their continuous support and belief in me. It is great to have a place that you can always come back to.

Turin, August 2018
Jan Sprenger

Second Author's Preface

In 2003, Luc Bovens and I published the book *Bayesian Epistemology*, which started with the sentence "Bayes is all the rage in philosophy". Now, some fifteen years later, Bayesianism is even more present in many subfields of philosophy, and a lot of exciting work has been done since our book appeared.

While *Bayesian Epistemology* showed how the powerful machinery of Bayesian networks can be used to address a number of rather general epistemological problems concerning coherence, confirmation and testimony, the present book focuses on selected normative questions raised by the practice of science, such as: How convincing are certain argumentative strategies? What makes one theory more explanatory than another? Why should we aim at reducing one theory to another? To address these questions, one cannot simply apply an existing normative framework. The details matter. At the same time, a purely descriptive or naturalistic approach to the philosophy of science does not suffice either. What is needed is a combination of the two approaches. The present book is our attempt to integrate details from the practice of science and to discuss them in a broader normative framework, informing philosophy and science at the same time.

For doing so, Bayesianism provides a suitable framework. It is well developed, it has a normative foundation, and the powerful machinery of Bayesian networks can be applied to take into account more and more details of scientific practice without getting lost in these details. Bayesianism also connects nicely to statistical practice, and it is probably the simplest formal approach to our questions one can think of. We leave it to our readers to judge how successful we have been and how convincing our analyses and models are, and we encourage our readers to challenge us by identifying aspects of the methodology of science that escape an illuminating Bayesian treatment.

Like *Bayesian Epistemology*, this book demonstrates by a detailed analysis of several examples how the Bayesian machinery can be productively applied. Many questions are still open, and we leave (for example) the Bayesian analysis of social aspects of science and the exploration of closer connections between Bayesian philosophy of science and the psychology of reasoning for another occasion.

My research was supported by the Nederlandse Organisatie voor Wetenschappelijk Onderzoek (NWO), the Deutsche Forschungsgemeinschaft (DFG, Priority Program "New Frameworks of Rationality"), the Center of Advanced Studies (CAS) at Ludwig Maximilian University (LMU) Munich (Senior Researcher in Residence Program) and the Alexander von Humboldt Foundation (Alexander von Humboldt Professorship). The Tilburg Center for Logic and Philosophy of Science (TiLPS), the Munich Center for Mathematical Philosophy (MCMP) and the CAS at LMU Munich provided excellent working environments for conducting this research.

Over the years, I have profited from many discussions with friends and colleagues on various aspects of this book, including Claus Beisbart, Cristina Bicchieri, Lorenzo Casini, Chiara Lisciandra, Matteo Colombo, Mark Colyvan, Eleonora Cresto, Erik Curiel, David Danks, Radin Dardashti, Richard Dawid, Franz Dietrich, Igor Douven, Frederick Eberhardt, Michael Esfeld, Benjamin Eva, Christian Feldbacher-Escamilla, Branden Fitelson, Roman Frigg, Mathias Frisch, Alexander Gebharter, Mario Günther, Clark Glymour, Paul Griffiths, Ulrike Hahn, Alan Hájek, Bill Harper, Remco Heesen, Colin Howson, Paul Hoyningen-Huene, Franz Huber, Jim Joyce, Jürgen Landes, Hannes Leitgeb, Kristina Liefke, Christian List, Sebastian Lutz, Aidan Lyon, Conor Mayo-Wilson, Ryan Muldoon, Barbara Osimani, Huw Price, Soroush Rafiee Rad, Miklós Rédei, Raphael van Riel, Jan-Willem Romeijn, Jonah Schupbach, Gerhard Schurz, Wolfgang Spohn, Reuben Stern, Manfred Stöckler, Michael Stöltzner, Patrick Suppes, Marko Tesic, Karim Thébault, Johanna Thoma, Jos Uffink, Greg Wheeler and Eric Winsberg. A special thanks goes to my secretary Maria Csauscher for her excellent work.

The late Patrick Suppes was my academic mentor for more than twenty years, and I owe him more than I can say in words. I am also tremendously thankful for the constant support and encouragement of Jeremy Butterfield, Roman Frigg and Margie Morrison. Luc Bovens got me excited about Bayesianism in the late 1990s, for which I am very grateful. I would also like to thank him for our always enjoyable and productive collaboration. It turned my academic life around and opened a whole new field for me. Peter Momtchiloff has been an excellent and patient editor and it was a pleasure to work again with him. Several chapters of this book contain material which appeared already in joint papers with Benjamin Eva, Foad Dizadji-Bahmani, Richard Dawid, Branden Fitelson, Roman Frigg and Soroush Rafiee Rad, and I would like to thank them for their input and the wonderful collaborations. Erik Curiel provided detailed written feedback on an earlier version of the manuscript and Christopher von Bülow did a fantastic job as a copy editor. Thank you both! My course "Central Topics in the Philosophy of Science" at LMU Munich in the academic years 2015/2016 and 2017/2018 was partly based of the manuscript of this book, and I would like to thank my students for their very helpful feedback.

My biggest thanks goes to my friend and co-author Jan Sprenger who did the lion's share of this work and who also deserves the lion's share of the credit for it. It has been a lot of fun to discuss and work out with him the ideas presented in this book. I would also like to thank my friends and family for their support over the years. It means a lot to me.

Munich, September 2018
Stephan Hartmann

Contents

List of Figures

List of Tables

Book Overview

The plan for this book emerged in 2011, when we noticed that we were both using Bayesian models for explicating central scientific concepts and capturing arguments in scientific reasoning, sometimes individually, sometimes together. We also discovered a lot of common themes in our work; however, these parallels were difficult to bring out in the compressed form of a journal article or contribution to an edited volume. So we decided to systematize and synthesize our work in a monograph with the label "Bayesian Philosophy of Science", allowing the reader to appreciate the manifold use of Bayesian models in philosophy of science—philosophical foundations, conceptual advances and practical applications in statistical inference.

In some sense, this book is a summary of the work that we have done over the past ten years. The idea that rational degrees of belief can be described by the probability calculus and changed via a specific formula—Bayesian Conditionalization—is a recurring motive in the book and we conjoin it with central issues of philosophy of science. We do not provide new philosophical foundations for rational belief, nor do we defend a particular variety of Bayesian reasoning. In this respect, the book is different from monographs that aim at rethinking the foundations of rational belief, knowledge or rational decision-making (Levi 1980; Buchak 2013; Pettigrew 2016; Moss 2018). Compared to those publications, our effort devotes more space to problems that are typical of scientific reasoning, and to concepts that are central in (general) philosophy of science, such as causation, explanation and confirmation. The same holds with respect to Henry Kyburg's (1961; 1974) classics on probability and inference, Richard Jeffrey's (2004) primer on subjective probability and Franz Huber's (2019) recent textbook on probability and induction.

Brössel forthcoming is closer to our project, but while that author shares our interest in Bayesian reasoning and confirmatory relationships, his work is more firmly anchored in formal epistemology. From Williamson 2017, we adopt a part of the formal calculus (see also Hailperin 1996), but his focus is, unlike ours, on inductive logic in the tradition of Rudolf Carnap and Jeff Paris.

Finally, there are books which aim at evaluating the scope and limits of Bayesian reasoning as a logic of scientific inference. Earman 1992 is perhaps the most notable example. There is also Colin Howson and Peter Urbach's (2006) spirited defense of subjective Bayesian inference vis-à-vis other schools of statistical reasoning, and frequentist inference in particular. Or Jon Williamson's (2010) monograph on Objective Bayesianism and its philosophical foundations. Finally, there are various manuals for applying Bayesian inference to data analysis in science (e.g., Bernardo and Smith 1994; Lee and Wagenmakers 2014). However, apart from Variations 9–11, we do not engage in comparisons of Bayesian and frequentist statistics. Rather than proving an alleged superiority of Bayesian inference, or giving a critical appraisal, we take a *constructive approach*: we show how to use Bayesian models for explicating important concepts in philosophy of science, such as explanatory power, causal strength, intertheoretic reduction or degree of corroboration, and for reconstructing prominent elements of scientific argumentation (e.g., the No Alternatives or the No Miracles Argument). It remarkable that a theory as simple as Bayesian inference, which builds only on three axioms of probability plus a single updating rule, has such a broad scope of applications, and that the results can be quite surprising and illuminating. Since the individual chapters are unified rather by their methods than by their (quite diverse) topics, we have decided to structure our book as variations on a common theme: uncertain reasoning in science, modeled by subjective Bayesian inference.

The biggest part of the book is based on research articles that the authors wrote in the years 2009–2018. Partly they were published jointly, partly they were published individually or with other authors. Typically, we have merged one or two research articles with new material and modified the writing style for the purpose of this book. We hope that our presentation outlines the common elements in what is often published as contributions to specialized debates, and to have convinced the reader that there is a unified and promising research program in Bayesian philosophy of science.

We conclude this brief overview by relating each chapter to previous publications. Variation 6 and 11 are suitably amended republications of recent research articles. The same can be said of Variation 2, but here the revisions were more substantial. Variation 3, 5 and 8 are each based on two journal articles; considerable rewriting was necessary in order to give an organic synthesis of our insights. The other Variations (1, 4, 7, 9, 10 and 12) present either original research, or a mixture of novel and published material. The contents of the Variations are described in detail at the end of the introductory chapter; here we just point the reader to the relevant previous publications and give credit where appropriate.

Theme: Bayesian Philosophy of Science This is an original exposition of static and dynamic principles of Bayesian inference, including a short section on causal Bayesian networks.

Variation 1: Confirmation This chapter is original work based loosely on Hartmann and Sprenger 2010, Sprenger 2010a and Sprenger 2016c. Our presentation of the material was also influenced by Vincenzo Crupi's and Branden Fitelson's work on confirmation theory (e.g., Fitelson 2001b; Crupi 2015).

Variation 2: The No Alternatives Argument This chapter is essentially a simplified and compressed version of "The No Alternatives Argument" (*British Journal for the Philosophy of Science*, Volume 66, No. 1, pp. 213–234; 2015) by Richard Dawid, Stephan Hartmann, and Jan Sprenger, with changes to the presentation. Adapted under the terms of the Creative Commons Attribution 4.0 International (CC BY 4.0) license (http://creativecommons.org/licenses/by/4.0/).

Variation 3: Scientific Realism and the No Miracles Argument This chapter amalgamates two stand-alone papers on Bayesian reconstructions of the No Miracles Argument. The first part of the chapter has been adapted from Jan Sprenger's "The Probabilistic No Miracles Argument" (*European Journal for Philosophy of Science*, Volume 6, No. 2, pp. 173–189; 2016) under the terms of the Creative Commons Attribution 4.0 International (CC BY 4.0) license. The argument in the second part of the paper is presented in more detail in Richard Dawid and Stephan Hartmann's "The No Miracles Argument Without the Base Rate Fallacy" (*Synthese*, Volume 195, No. 9, 4063–4079).

Variation 4: Learning Conditional Evidence This chapter is original work; the results are also discussed in the forthcoming paper "Updating on Conditionals" (*Mind*) by Benjamin Eva, Stephan Hartmann, and Soroush Rafiee Rad.

Variation 5: The Problem of Old Evidence This chapter merges two papers on the Problem of Old Evidence: "A Novel Solution to the

Problem of Old Evidence", by Jan Sprenger (*Philosophy of Science*, Volume 82, No. 3, pp. 383–401; 2015), and "A New Garber-Style Solution to the Problem of Old Evidence" by Branden Fitelson and Stephan Hartmann (*Philosophy of Science*, Volume 82, No. 4, pp. 712–717; 2015).

Variation 6: Causal Strength This chapter is essentially a republication of Jan Sprenger's "Foundations of a Probabilistic Theory of Causal Strength" (*Philosophical Review*, Volume 127, No. 3, pp. 371–398; 2018), integrating material from Sprenger and Stegenga 2017.

Variation 7: Explanatory Power This chapter draws from the formal and empirical work that Jan Sprenger conducted with different co-authors, as well as from recent contributions by Michael Cohen, Vincenzo Crupi, Benjamin Eva and Reuben Stern. Schupbach and Sprenger 2011 and Crupi and Tentori 2012 are the most important reference papers.

Variation 8: Intertheoretic Reduction This chapter synthesizes and rearranges two papers that Stephan Hartmann wrote on intertheoretic reduction: "Who is Afraid of Nagelian Reduction?" (*Erkenntnis*, Volume 73, No. 3, pp. 393–412; 2010) and "Confirmation and Reduction: A Bayesian Account" (*Synthese*, Volume 179, No. 2, pp. 321–338; 2011). Both papers have been co-authored with Foad Dizadji-Bahmani and Roman Frigg.

Variation 9: Hypothesis Tests and Corroboration The negative part of the chapter presents the arguments from Jan Sprenger's "Two Impossibility Results for Popperian Corroboration" (*British Journal for the Philosophy of Science*, Volume 69, No. 1, pp. 139–159; 2018). Adapted with permission of Oxford University Press. The positive part is unpublished material presented by Jan Sprenger at a symposium on statistical inference and the replication crisis at PSA2018 in Seattle.

Variation 10: Simplicity This chapter combines a general discussion of simplicity as a cognitive value with a more specialized investigation of the role of simplicity in Bayesian model selection (Sprenger 2009b). The latter part of the chapter is based on Jan Sprenger's paper "The Role of Bayesian Philosophy within Bayesian Model Selection" (*European*

Journal for the Philosophy of Science, Volume 3, No. 1, pp. 101–114; 2013). © 2012 by Springer Nature, used with permission.

Variation 11: Objectivity This chapter is essentially a republication of Jan Sprenger's "The Objectivity of Subjective Bayesian Inference" (*European Journal for Philosophy of Science*, Volume 8, No. 3, pp. 539–558; 2018). © 2018 by Springer Nature, reproduced with permission. The conceptual background was provided by Heather Douglas's book "Science, Policy and the Value-Free Ideal" (Douglas 2009b) and the SEP article "Scientific Objectivity" by Julian Reiss and Jan Sprenger (Reiss and Sprenger 2014).

Variation 12: Models, Idealizations and Objective Chance This chapter is mainly based on Jan Sprenger's work in progress "Conditional Degrees of Belief" (Sprenger 2018a). The application to reasoning with idealized statistical models is original and unique to this book.

Conclusion: The Theme Revisited The final chapter looks back on the achievements of Variation 1–12 and sketches future research projects. Like the introductory chapter, it has been written exclusively for the purpose of this book.

At this point, it only remains to hope that you, the reader, take a lot of inspiration from the book, and that you read it with pleasure!

Theme: Bayesian Philosophy of Science

Subjective Bayesian inference is a theory of uncertain reasoning that represents an agent's epistemic attitudes by degrees of belief that follow the laws of probability. They are changed by a particular mechanism for learning from evidence: Bayesian Conditionalization. The evaluation of a hypothesis in the light of observed evidence, and decisions to act in a particular way, are based on these updated degrees of belief.

Bayesian inference is applied in a large variety of domains in science and philosophy. It is rapidly gaining popularity as an alternative to frequentist statistics and it is used in almost all empirically oriented parts of science. This includes areas as diverse as behavioral psychology, clinical trials and high-energy physics (Bernardo and Smith 1994; Howson and Urbach 2006; US Food and Drug Administration 2010). The Bayesian framework also provides the foundations for the most popular theory of rational choice, Expected Utility Theory (Jeffrey 1965; Savage 1972). That theory amalgamates utility assessments with judgments of (subjective) probability and enjoys widespread popularity in decision theory, economics and beyond. Finally, Bayesian models are frequently used in theories of rationality in cognitive psychology (Oaksford and Chater 2000).

There is a comparably rich number of applications of Bayesian inference in philosophy, particularly in epistemology. It provides the formal basis for responses to the notorious Lottery and Preface paradoxes (Kyburg 1961; Makinson 1965), and for theories that connect the all-or-nothing concept of belief with the concept of degree of belief (Leitgeb 2014, 2017; Easwaran 2016). It stands behind the explication of epistemic rationality as minimizing the divergence between one's epistemic attitudes and the truth (Pettigrew 2016; Fitelson 2018). It serves as a tool for describing how information spreads in social networks, and it assigns a precise meaning to central epistemic concepts such as coherence, reliability and testimony (Bovens and Hartmann 2003; Fitelson 2003; Olsson 2011).

This book proposes Bayesian inference as a normative theory of scientific reasoning that is anchored in general principles of rationality. At the same time, we argue that Bayesian models are, on the descriptive level, reasonably in sync with specific judgments scientists make (e.g., regarding the confirmatory power of certain pieces of evidence, or the value of intertheoretic reduction). Thus, Bayesian reasoning can also provide a rational reconstruction of scientific practice. Both aspects—outlining a normative theory and providing a rational reconstruction—are part of Bayesian philosophy of science. While we do not want to claim that scientific reasoning is essentially reasoning according to the laws of probability, we claim that Bayesian models can elucidate important aspects of scientific inference, increasing our understanding of how science works and why it is so successful. The book is written as a cycle of variations on this theme.

Of course, there are other frameworks for uncertain reasoning such as evidential probability (Kyburg 1961), Dempster–Shafer theory (Shafer 1976) or the theory of ranking functions (Spohn 1988, 2012; Huber 2006). Each of them has their specific merits (Haenni et al. 2011), but we feel that Bayesian inference is distinguished by the simplicity of the formal apparatus, its closeness to everyday reasoning, its proven fruitfulness and its coherence with probabilistic inference in science.

We understand Bayesian philosophy of science as the **use of Bayesian principles and methods for modeling scientific reasoning.** We set up a spectrum of Bayesian models across the book, and we do so in a candidly eclectic fashion. Some of our models are based on Bayesian Conditionalization and others on more general forms of belief change (e.g., the theory of *f*-divergences in Variation 4). Sometimes we import models from Bayesian statistics, sometimes from the philosophical literature on Bayesian inductive logic. To our mind, this eclectic approach is not a problem. No single Bayesian model will be able to succeed at modeling concepts as diverse as scientific confirmation, explanatory power and simplicity. Diverse targets ask for a diversity of models. In this respect, our book differs from previous, more unified monographs on Bayesian reasoning in science, such as Earman 1992 or Howson and Urbach 2006. These authors focus on the scope and limits of Bayesian inductive reasoning, and on comparisons to rivalling approaches such as frequentist statistics. Our approach may be less unified and more eclectic, but this is fully compensated by the strength and diversity of the results we present. By applying Bayesian models to a wide variety of concepts and reasoning patterns in science, we expand their scope considerably. Moreover, we are (unlike Howson and Urbach in an earlier edition of their book) not interested in defending Bayesian inference as the uniquely correct theory of scientific reasoning, but rather in showing that it is a *fruitful* one. In other words, we make a case for affirmative answers to questions such as, *Does Bayesian inference provide unexpected insights into scientific reasoning? Does it solve problems that other models struggle with? Does it suggest interesting experiments or questions for future research?* Such criteria are also standardly used for the evaluation of scientific models (Weisberg 2007, 2012; Frigg and Hartmann 2012), and we subject our Bayesian models of scientific reasoning to the same standards.

In this book, we construct two different types of Bayesian models. First, there are **explicative Bayesian models,** which serve to replace a vague concept, the *explicandum*, by an exact one, the *explicatum*. Rudolf Carnap (1950) introduced this method in his book *Logical Foundations of Probability*:

> If a concept is given as explicandum, the task consists in finding another concept as its explicatum which fulfils the following requirements to a sufficient degree.
>
> (1) The explicatum is to be *similar* to the explicandum in such a way that, in most cases in which the explicandum has so far been used, the explicatum can be used; however, close similarity is not required, and considerable differences are permitted.
>
> (2) The characterization of the explicatum, that is, the rules of its use (for instance, in the form of a definition), is to be given in an *exact* form, so as to introduce the explicatum into a well-connected system of scientific concepts.
>
> (3) The explicatum is to be a *fruitful* concept, that is, useful for the formulation of many universal statements (empirical laws in the case of a non-logical concept, logical theorems in the case of a logical concept).
>
> (4) The explicatum should be as *simple* as possible; this means as simple as the more important requirements (1), (2), and (3) permit. (Carnap 1950, 7)

For our project, explication means providing a quantitative dimension for central concepts in scientific reasoning, such as confirmation, explanatory power, simplicity and causal strength. Explication involves a tight interconnection of conceptual analysis and formal methods: conceptual analysis identifies adequacy conditions that the explicatum has to satisfy, while formal reasoning leads us to an explicatum that satisfies these constraints. Strawson (1963) and, more recently, Boniolo (2003) and Eagle (2004), voice scepticism about Carnapian explication as a philosophical method; Maher (2007) responds to their objections and defends the philosophical merits of the explicative program.

Next to the explicative models, there are **Bayesian models of scientific argumentation.** In particular, we reconstruct the No Alternatives Argument (Variation 2) and the No Miracles Argument (Variation 3) from a Bayesian point of view. To this end, we identify sets of variables that matter for a particular argument scheme and we define conditions on an agent's degrees of belief that validate these arguments from a probabilistic point of view. As before, this type of research demands a tight interplay between conceptual analysis and formal modeling. In both the explicative and the argumentative models, we use empirical and computational methods where appropriate: experimental findings are evaluated in order to judge the adequacy of an explicatum with respect to the concept that it targets, and computational methods are used for exploring the consequences of our models in cases where we cannot find analytical solutions. Finally, case studies are used to check our results against scientific practice.

We now explain the constitutive principles of Bayesian inference (for recent primers, see Easwaran 2011a,b; Briggs 2015; Titelbaum forthcoming). The level is introductory; no knowledge of calculus or higher mathematics is required. We begin with the **statics of Bayesian inference:** the representation of degrees of belief by probabilities, the concept of conditional degrees of belief, and Bayes' Theorem. Then we turn to the **dynamics of Bayesian inference:** the principle of Bayesian Conditionalization and its generalizations. Finally, we introduce causal Bayesian networks and outline the structure of the book.

Probability and Degrees of Belief

In science as well as in ordinary life, we routinely make a distinction between more and less credible hypotheses. Consider the potential winners of the 2020 European Football Cup. Albania is a less plausible candidate than Belgium, and Belgium is again less plausible than France. Sill, few people would say that France is likely to win. This example illustrates that the epistemic standing of an empirical hypothesis is no all-or-nothing affair.

Traditional descriptions of epistemic attitudes that just distinguish between belief, disbelief and suspension of judgment struggle to account for the fine gradations between different epistemic attitudes that we have toward scientific hypotheses, and toward propositions in general.

Here the Bayesians step in: they use the concept of **degrees of belief** to describe epistemic attitudes, and they represent these degrees of belief by a particular mathematical structure: probability functions. In other words, Bayesians regard probabilities as expressions of subjective uncertainty. This interpretation goes back to philosophers and scientists as prominent as Blaise Pascal (1623–1662) and Pierre-Simon Laplace (1749–1827) and is named after the English mathematician, philosopher and Presbyterian minister Thomas Bayes (1701–1761), whose "Essay towards Solving a Problem in the Doctrine of Chances" (Bayes 1763) contains the mathematical basis for what later became known as "Bayes' Theorem". Famous champions of Bayesian inference in the twentieth century have been Frank P. Ramsey, Bruno de Finetti and Richard Jeffrey.

Following Hailperin (1984, 1996), Popper (1959/2002) and Williamson (2017), we conceptualize probability functions as operating on sentences of a propositional language L, consisting of a (finite or infinite) set of propositional variables $\{A, B, C, \ldots\}$. While these variables are written in uppercase italic letters, sentences about their values are abbreviated by regular Roman letters (Bovens and Hartmann 2003):

- A: "The variable A takes the value 'true'."
- $\neg$B: "The variable B takes the value 'false'."
- etc.

These are the **atomic sentences** of L. More complex sentences can be generated with the sentential connectives "$\wedge$" (conjunction, "A and B"), "$\vee$" (disjunction, "A or B") and "$\neg$" (negation, "not A"). $\mathcal{L}$ denotes the set of all sentences of L, including the special symbols "$\top$" and "$\bot$" for the tautology and the contradiction. In addition, we assume that $\mathcal{L}$ is closed under countable disjunction of its elements so that it has the structure of a σ-algebra.

Probability functions express our degrees of belief in the truth of these sentences numerically. However, since epistemologists traditionally view *propositions* rather than sentences as objects of degrees of belief, we allow for some imprecision and identify sentences with the propositions they express, such as scientific hypotheses. This allows us to stick with common parlance and to speak about the probability of a proposition. Adopting this convention is innocuous: as long as logically equivalent sentences are assigned the same degree of belief, algebras of sentences and propositions have isomorphic probabilistic models (Huber 2016). In particular, different formulations of one and the same hypothesis will have the same probability.[1]

The axioms of probability specify how our degrees of belief should be structured. First, the axioms specify their range: 0 denotes minimal and 1 denotes maximal degree of belief. Moreover, sentential connectives such as negation and disjunction impose precise constraints on the degree of belief that we should have in the resulting sentences or propositions (Kolmogorov 1933):

Probability Function *For a propositional language L with set of sentences $\mathcal{L}$, a mapping $p\colon \mathcal{L} \to [0;1]$ is called a* probability function *if and only if it satisfies the following three conditions:*

1. $p(\top) = 1$.
2. *For any* $\mathrm{A} \in \mathcal{L}$, $p(\neg \mathrm{A}) = 1 - p(\mathrm{A})$.
3. *For sentences* $\mathrm{A}_1, \mathrm{A}_2, \mathrm{A}_3, \ldots \in \mathcal{L}$ *with mutually exclusive truth conditions (i.e., for any $i \neq j$, A_i and A_j cannot be jointly true):*

$$p\left(\bigvee_{n=1}^{\infty} \mathrm{A}_n\right) = \sum_{n=1}^{\infty} p(\mathrm{A}_n). \tag{T.1}$$

It is not hard to motivate the three above constraints: The tautology is assigned maximal degree of belief. If A is strongly believed, its negation $\neg$A is weakly believed, and vice versa. In particular, the degrees of belief in A and in $\neg$A add up to unity because by the Laws of Bivalence and Non-Contradiction for propositional logic, exactly one of these sentences has to be true. Similarly, the degree of belief in a disjunction of mutually exclusive

[1]The alternative to this sentential approach consists in Kolmogorov's measure-theoretic approach: probabilities operate on σ-algebras of sets, and the objects of degrees of belief correspond to the epistemic possibilities that an agent considers—for example, sets of possible worlds (e.g., Easwaran 2011a; Huber 2019). Compare also the following footnote. In this context, one usually speaks of "probability measures" and the set-theoretic operations of intersection, union and complement replace the sentential connectives of conjunction, disjunction and negation. Our use of the term "probability function" indicates that we prefer the sentential interpretation. It strikes us as more natural, simpler and closer to the purpose of this book, namely to model scientific reasoning and the assessment of logically interconnected scientific hypotheses. Naturally, our results can also be interpreted from a measure-theoretic point of view.

sentences corresponds to the sum of the degrees of belief in the individual sentences (or propositions).[2]

All this is in line with our everyday use of the word "probable". Bayesians of all kinds and shades agree that rational degrees of belief over a set of logically interconnected propositions should satisfy the three above axioms. While there may not be, or may never have been, an *actual* agent with fully probabilistic degrees of belief, Bayesians find the axioms plausible enough to defend probabilistic degrees of belief as a normative epistemic ideal.

It is notable that the third condition—Equation (T.1)—uses an infinite instead of a finite sum. Indeed, there is a substantial debate about whether probabilistic degrees of belief should satisfy this condition of **countable additivity** instead of the weaker requirement of **finite additivity:** $p(\mathrm{A})+p(\mathrm{B}) = p(\mathrm{A}\vee\mathrm{B})$ for two mutually exclusive sentences A and B.[3] However, since countable additivity is standardly assumed in most mathematical textbooks on probability theory and in applications of Bayesian inference in statistics, we take all probability functions to be countably additive.

While the qualitative motivation of the probability axioms is highly plausible, their precise quantitative statement is harder to justify. Why should rational degrees of belief satisfy the axioms of probability rather than another set of axioms with similar qualitative properties? Four types of arguments for an affirmative answer—**Probabilism**—have been proposed (see also Briggs 2015; Easwaran and Fitelson 2016):

1. Dutch book arguments relating degrees of belief to betting behavior (Ramsey, de Finetti, Jeffrey, etc.);

2. representation theorems relating degrees of belief to rational preferences and choices under uncertainty (Savage, Maher, etc.);

[2] As mentioned in the previous footnote, a possible semantics for degrees of belief maps propositions—the objects of degrees of belief—to the sets of the possible worlds where they are true. The probability of a proposition is then equal to the weight of those worlds which make it true. This semantics gives a particularly intuitive basis for the three axioms; for example, the third axiom just sums up the weights of the possible worlds where A_1, A_2, A_3, etc. obtain.

[3] Several authors argue that accepting countable additivity amounts to making substantial and unwarranted epistemological assumptions (de Finetti 1972, 1974; Kelly 1996; Howson 2008). Jaynes (2003) responds that countable additivity naturally follows from a proper mathematical modeling of uncertainty. Kadane, Schervish and Seidenfeld (1999) discuss further consequences of choosing finitely instead of countably additive probability functions. Fortunately, this choice does not make a difference for most applications in this book.

3. representation theorems relating degrees of belief to qualitative constraints on a plausibility measure (Cox, Villegas, Jaynes, etc.);
4. arguments from minimizing the total inaccuracy of one's system of degrees of belief (de Finetti, Joyce, Pettigrew, etc.).

These arguments differ in the type of rationality on which they build—pragmatic vs. epistemic—as well as in their methodological approach. Some strategies contend that any (epistemically or pragmatically) rational agent can be represented *as if* her (actual or hypothetical) degrees of belief obeyed the degrees of probability: this is the **method of representation theorems.** Savage and others in the decision-theoretic camp (e.g., Maher 1993) connect degrees of belief to rational preferences and axioms of rational choice, whereas Cox (1946) and other proponents of the epistemic approach characterize a pre-theoretic notion of plausibility such that agreement with these qualitative constraints ultimately yields the laws of probability.

By contrast, both the popular Dutch book justification and the justification from minimizing inaccuracy contend that any agent whose degrees of belief violate the laws of probability exposes herself to an act of irrationality. More precisely, these approaches establish **dominance arguments for Probabilism:** regardless of the actual state of the world, non-probabilistic degrees of belief yield a lesser (epistemic or monetary) utility than their probabilistic counterparts. In the case of **Dutch book arguments,** non-probabilistic degrees of belief lead to a system of bets which appear individually rational, but collectively lead to a sure loss. By contrast, accuracy-based arguments show that for each non-probabilistic credence function, there is a probabilistic one which receives a lower inaccuracy score in all possible worlds. Table T.1 places the four strategies in a simple 2×2-matrix.

Argument type / Notion of rationality	Dominance arguments	Representation theorems
Epistemic rationality	Minimizing inaccuracy (Joyce, Pettigrew)	Axioms on plausibility (Cox, Villegas)
Pragmatic rationality	Dutch book arguments (Ramsey, de Finetti)	Axioms on preferences (Savage, Maher)

Table T.1: A schematic overview of the arguments for Probabilism.

It is important to mention up front that all these arguments are normative and involve some form of idealization: they do not establish that real agents have probabilistically coherent degrees of belief. Rather, they intend to show that *ideally rational* agents follow the axioms of probability. To this end, positivist-minded Bayesians connect degrees of belief to (real or hypothetical) betting behavior, based on an observation by Ramsey (1926) that human action is often akin to accepting a bet on the occurrence of an event:

> [A]ll our lives we are in a sense betting. Whenever we go to the station we are betting that a train will really run, and if we had not a sufficient degree of belief in this we should decline the bet and stay at home. (Ramsey 1926, 85)

Acting in a certain way corresponds to an implicit bet, and these bets are informed by degrees of belief. The most pervasive example are perhaps transactions on financial markets—traders buy and sell stocks, certificates and options according to their degrees of belief that these will rise or fall, respectively. Someone with a high degree of belief that an option will become worthless will sell it eagerly, while someone who is convinced that it will increase in value will keep it in her portfolio. This approach naturally leads to a fully dispositional, behaviorist definition of degrees of belief: agent *S*'s degree of belief in proposition A is equal to $x \in [0;1]$ if for all stakes smaller than €x, *S* would buy a bet on A that pays €1 if A occurs, and for all stakes greater than x, *S* would sell this bet if she were a bookie (e.g., de Finetti 1937).

For example, if *S* has degree of belief 1/10 in Germany being the 2020 European football champion, then she would at most spend 10 cent on a bet on this proposition that pays €1 if successful. The proportion between return and stakes is called the **betting odds.** In this case, *S*'s degrees of belief correspond to $10:1$ odds; in general, the odds are just the inverse of the degrees of belief. A variant of this approach defines *S*'s degree of belief in proposition A as the stake €x at which *S* is indifferent between taking the role of the bettor and the bookie in a bet that pays €1 if A occurs. In other words, *S* is equally willing to pay €x for a bet on A with payoff €1 as to sell that bet for €x. This technique resembles the famous veil of ignorance for disclosing judgments about the fair distribution of goods in a society (Rawls 1971): the agent has to specify a fair price for the bet, not knowing whether she will end up as the bettor or as the bookie.

Ramsey argues that no system of degrees of belief can be reasonable if it allows for constructing a system of bets that implies a sure loss for the bettor—or the bookie. Such a system of bets is called a **Dutch book.** By the above mapping between degrees of belief and betting odds, Ramsey grounds the famous **Dutch Book Theorem:** degrees of belief which violate the axioms of probability will give rise to Dutch books and imply a sure loss for one side. Since the proof is quite instructive, we give it below in the main text (less mathematically minded readers may skip it).

Proof of the Dutch Book Theorem: For each axiom that is violated, we construct a Dutch book. It is assumed that p denotes the degrees of belief of an agent, and that these degrees of belief conflict with one (or more) of the axioms.

First Axiom: $p(\top) = 1$. Suppose that $p(\top) < 1$, in violation of the first axiom. If a bookie offers a bet on the tautology with odds greater than 1:1 (e.g., a stake of 80 cents for a return of €1), then he will lose money for sure since $\top$ is always the case. Thus we have constructed a Dutch book.

Second Axiom: $p(\mathrm{A}) + p(\neg\mathrm{A}) = 1$ for any $\mathrm{A} \in \mathcal{L}$. Suppose that $p(\mathrm{A}) + p(\neg\mathrm{A}) > 1$, in violation of the second axiom, and let $x := p(\mathrm{A})$, $y := p(\neg\mathrm{A})$. The agent's degrees of belief in A and ¬A correspond to bets on both propositions, with a return of €1 for each bet and respective stakes of €x and €y. Since either A or ¬A must occur, the agent's total payoff in such a system of bets is fixed at €1. However, since her stakes exceed that amount, she will always lose money. Similarly, degrees of belief where $p(\neg\mathrm{A}) + p(\mathrm{A}) < 1$ would give rise to a system of bets where the bookie would always lose money. Either way, one side in this system of bets can be Dutch booked.

Third Axiom: $p(\mathrm{A}) + p(\mathrm{B}) = p(\mathrm{A} \vee \mathrm{B})$ for mutually exclusive $\mathrm{A}, \mathrm{B} \in \mathcal{L}$. Suppose that $p(\mathrm{A}) + p(\mathrm{B}) > p(\mathrm{A} \vee \mathrm{B})$, in violation of the third axiom, and let $x := p(\mathrm{A})$, $y := p(\mathrm{B})$ and $z := p(\mathrm{A} \vee \mathrm{B})$. It follows that $p(\neg(\mathrm{A} \vee \mathrm{B})) = 1 - z$ because otherwise the second axiom would be violated and we could construct a Dutch book against such degrees of belief, as shown above. Now consider a system of bets on A, B and $\neg(\mathrm{A} \vee \mathrm{B})$ with return €1 and stakes €x, €y and €$1-z$, respectively, as shown in Table T.2. Exactly one of these propositions must be true, and the stakes are chosen such that the winning bet always returns €1. On the other hand, the total stake, €$x+y+1-z$, will be greater than €1 because, by assumption, $x + y > z$. So this system of bets implies a sure loss for the bettor. For the case of $p(\mathrm{A}) + p(\mathrm{B}) < p(\mathrm{A} \vee \mathrm{B})$, the same system will lead to a sure win for the bettor, and a sure loss for the bookie. Again, we have constructed a Dutch book. (The case of countably many propositions is proved analogously.)

Hence non-probabilistic degrees of belief fail the rationality criterion of avoiding Dutch books. The converse can be shown, too: All probabilistic systems of degrees of belief are immune to Dutch books (see also Kemeny 1955; de Finetti 1974; Skyrms 2000). □

Bet on Proposition	Stake	State of the World A	B	$\neg(A \vee B)$
A	€x	€1	€0	€0
B	€y	€0	€1	€0
$\neg(A \vee B)$	€$1-z$	€0	€0	€1
Total Stake and Return	€$x+y+1-z$	€1	€1	€1

Table T.2: The Dutch book argument in defense of the third axiom of probability. Each line represents an individual bet with a fixed stake, and variable returns dependent on which state of the world obtains. The entire system of bets has a fixed total return, which for non-probabilistic degrees of belief differs from the total stake.

The cogency of Dutch book arguments has been debated in various places (e.g., de Finetti 1972; Howson 2008; Hájek and Hartmann 2010; Hartmann and Sprenger 2010; Easwaran 2011a,b; Vineberg 2016). For the sake of simplicity, suppose a fully behaviorist interpretation of degrees of belief. The Dutch Book Theorem assumes that the agent accepts all bets where the proposed odds are higher than her personal odds (viz., degrees of belief), and is ready to act as bookie on all bets where the proposed odds are lower than her personal odds. Real agents, however, are often risk-averse: the size of the stake may influence their willingness to take a side in the bet. When degrees of belief are identified with betting odds, an agent's degree of belief in a proposition may thus depend on the size of the stake in the elicitation procedure. In particular, the agent may be unwilling to engage in any bet if the stakes are high enough, and may be willing to suffer a Dutch book if the stakes are low enough (e.g., 1 cent). None of these behaviors strikes us as blatantly irrational unless we presuppose what is to be shown: that rationality equals immunity to Dutch books.

These objections suggest that a straightforward operationalization of degrees of belief in terms of betting behavior, or in terms of fairness judgments about bets, is problematic (see also Eriksson and Hájek 2007). However, when the link between degrees of belief and fair betting odds crumbles, a Dutch book argument for Probabilism loses part of its normative force. Sure, in many situations we can still argue for a strong dependency between degrees of belief and betting odds, but it might go too far to identify both concepts with each other. For this reason, we now move to the second argument in favor of Probabilism: the **representation of rational preferences in terms of subjective degrees of belief.**

The idea of this argument is that probabilistic degrees of belief can represent the epistemic state of an agent who bases her choices on rational preferences. First, a number of axioms are imposed on rational preference, represented by the binary relation $\preceq$. For example, it is usually assumed that such preferences are *transitive*: if an agent prefers apples to bananas and bananas to cherries, then she will also prefer apples to cherries. Similarly, it is often assumed that such preferences are *complete*: either the agent strictly prefers one of two options or is indifferent between them.

In his 1954 landmark book *The Foundations of Statistics*, Leonard J. Savage sets up an entire system of such axioms, called P1–P7 (for an accessible overview, see Ellsberg 1961). They contain transitivity and completeness as well as more demanding axioms, such as the **Sure-Thing Principle** (or Axiom P2): Preference between two actions cannot depend on states of the world where they yield the same payoff. Suppose, for example, that in Table T.3, the agent prefers Gamble I to Gamble II. Apparently, her preference for Gamble I can only be explained by the fact that she considers A more likely than B. After all, if neither A nor B is the case, her choice does not make a difference. Likewise, there seems to be no rational explanation of preferring Gamble IV to Gamble III unless she considers B more likely than A. Taken together, both preferences lead to inconsistent (implicit) beliefs, however. Therefore, Savage postulates that anybody who prefers Gamble I to Gamble II should also prefer Gamble III to Gamble IV (see also Ellsberg 1961, 648–649).

Savage then proceeds to proving his famous representation theorem. If the preferences of an agent X satisfy the axioms P1–P7, then there is

1. a probability function p, representing the agent's subjective degrees of belief over (future) states of the world, and

2. a real-valued utility function u (unique up to affine transformation)

such that for any two acts f and g, act g is preferred to act f if and only if the expected utility of g relative to p exceeds the expected utility of f. In

	State of the World		
Choice Problem 1	A	B	$\neg(A \wedge B)$
Gamble I	€100	€0	€0
Gamble II	€0	€100	€0
Choice Problem 2	A	B	$\neg(A \wedge B)$
Gamble III	€100	€0	€50
Gamble IV	€0	€100	€50

Table T.3: Motivation of the Sure-Thing Principle. Preferring Gamble I to Gamble II implies a judgment on the relative probability of A and B (assumed to be mutually exclusive); therefore, the agent should also prefer Gamble III to Gamble IV.

formulas:

$$f \preceq g \quad \Leftrightarrow \quad \mathbb{E}_p[u(f)] \leq \mathbb{E}_p[u(g)].$$

In other words, we can represent a rational agent as maximizing the utility of her actions relative to her personal degrees of belief.

Savage's approach has been very influential in economics under the label of Subjective Expected Utility Theory, and it bridges the gap between epistemology and decision theory in an attractive and elegant way. However, it is not without drawbacks. First, the probability function p describing the agent's degrees of belief is not unique by itself: it is only unique jointly with the utility function u. This weakens the appeal of Savage's results for models of scientific reasoning, where pragmatic utility considerations are often secondary to pursuit of truth. Second, Savage's axioms on rational preferences are not all equally compelling. The Sure-Thing Principle, for instance, fails to take into account considerations of uncertainty aversion and possible regret. Not following the principle can be explained rationally in a variety of circumstances.

The economists Maurice Allais (1953) and Daniel Ellsberg (1961, 2001) have conducted influential experiments that show mass defections from the Sure-Thing Principle when choosing between two different lotteries (see also Allais and Hagen 1979). Table T.4 presents the **Ellsberg Paradox**—a variation of the choice problem that has motivated the Sure-Thing Principle. An urn contains ninety balls in total, thirty of which are red. The other sixty balls are either black or yellow, but it is unknown in which proportion. A ball is drawn at random from the urn. The participant is asked to choose between Gambles A (red wins) and B (black wins), with equal payoff. Then she is asked to choose between Gambles C (red or yellow wins) and D (black or yellow wins). The most frequently observed response pattern is A and D—the participant avoids lotteries with unknown probabilities—but this finding contradicts Savage's Sure-Thing Principle that preferring A over B should also entail preferring C over D. After all, if I prefer A to B, then it must be because of the different payoffs in the Red and Black states. Same for C versus D. However, as far as payoffs in these states are concerned, A is identical to C, and B is identical to D.

Choice Problem 1	Red (30)	Black (?)	Yellow (?)
Gamble A	€100	€0	€0
Gamble B	€0	€100	€0
Choice Problem 2	**Red (30)**	**Black (?)**	**Yellow (?)**
Gamble C	€100	€0	€100
Gamble D	€0	€100	€100

Table T.4: Payoff table for a classical instance of the Ellsberg paradox. There are thirty red balls in an urn, and an unknown proportion of black and yellow balls. The total number of balls is ninety.

Ellsberg's experiment has been replicated many times and participants keep violating the Sure-Thing Principle. What is more, they stick to their guns when confronted with the fact that their behavior is in conflict with an axiom of rational choice. This suggests that they are not committing a cognitive error, such as in the base rate fallacy explained on page 22. Instead, they deny the universal scope of Savage's axioms—perhaps because they feel uncomfortable with unknown probabilities. Hence axiomatic justifications of probabilistic degrees of belief from principles of rational choice are based on a very restrictive and perhaps unrealistic interpretation of rationality.

Finally, there are purely **epistemic arguments for the probabilistic nature of degrees of belief.** They fall into two categories: those that postulate a set of qualitative rules for plausible reasoning, such as the arguments by Cox, Villegas and Jaynes, and those that evaluate the rationality of a system of degrees of belief by its inaccuracy, that is, its distance from the truth. We begin with the first family of arguments and expound one of its representatives.

A tradition going back to an early article of Bruno de Finetti (1931) tries to connect qualitative or comparative reasoning about degrees of plausibility to the axioms of probability. A particularly simple and intuitive attempt along the former lines was made by the physicist Richard Cox in 1946. He demonstrated that any real-valued function $q\colon \mathcal{L} \rightarrow [0;1]$ representing the plausibility of a proposition is isomorphic to a probability function if the following two assumptions (plus minor technical requirements) are satisfied:

Complementarity There is a monotonically decreasing function $f\colon \mathbb{R} \rightarrow \mathbb{R}$ such that $q(\neg A) = f(q(A))$.

Compositionality There is a function $g\colon \mathbb{R} \times \mathbb{R} \rightarrow \mathbb{R}$ such that $q(A \wedge B) = g(q(A), q(B|A))$ where $q(B|A)$ denotes the plausibility B has for us if we suppose that A is true.

In other words, if (i) the plausibility of the negation of a proposition is a decreasing function of the plausibility of the proposition itself, and (ii) the plausibility of $A \wedge B$ is determined by the plausibility of A and the plausibility of B given A, then plausibility measures are structurally identical to probability functions and differ only with respect to scaling properties (see also de Finetti 1936). However, Compositionality is a strong assumption: it is not clear beyond reasonable doubt that the plausibility of $A \wedge B$ should depend only on the plausibility of A and the plausibility of B given A. Similar questions can be asked for the like-minded and technically more complex proposals by Koopman (1940), Scott (1964), Villegas (1964) and Jaynes (2003). Nevertheless, the method of epistemic representation theorems remains a highly principled and respectable way of motivating probabilistic models of degrees of belief.

In more recent times, the epistemic approach to justifying probabilistic degrees of belief has been resuscitated from a different perspective: as an **epistemic utility argument** (de Finetti 1974; Rosenkrantz 1981; Joyce 1998, 2009; Pettigrew 2016). Degrees of belief are evaluated in terms of their **inaccuracy** in a possible world ω, that is, the degree of belief $p(A)$ in a proposition A is compared to the truth value $v(A, \omega) \in \{0, 1\}$ of A in ω, where 0 denotes falsehood and 1 denotes truth. Extending this evaluation to an entire algebra $\mathcal{L}$ (e.g., the elements of a finite propositional language), we can define measures of inaccuracy for credence functions p on $\mathcal{L}$, for example:

$$B(p, \omega) = \sum_{A \in \mathcal{L}} (p(A) - v(A, \omega))^2. \qquad \text{(Brier score)}$$

This evaluation is relative to the truth values of the A's in a possible world ω, and it uses the **Brier score** (Brier 1950), that is, the quadratic **scoring rule** $s_q\colon [0;1]^2 \rightarrow [0;\infty];\ (x, y) \mapsto (x - y)^2$, for penalizing deviations of the credence function from the actual state of affairs. The Brier score is a representative of a very general class of scoring rules, namely **Bregman divergences,** which play an important role in epistemic utility theory (Pettigrew 2016). In a nutshell, for any Bregman divergence s_B, there is a convex and continuous function F: $[0;1] \to \mathbb{R}$, differentiable on $(0;1)$, such that $s_B(x,y) = F(y) - F(x) - F'(x)(y-x)$. By choosing $F(x) := x^2$, we can represent the Brier score as a Bregman divergence.

Generally, a credence function p has a high degree of accuracy in those possible worlds where subjectively probable propositions are true, and a low degree of accuracy in worlds where subjectively probable propositions are false. Hence, some credence functions will be more accurate in some worlds, and other credence functions in other worlds. However, de Finetti (1974) showed that if a credence function p does not satisfy the probability axioms, then there is a probabilistic credence function that is more accurate *in all possible worlds*, as measured by the Brier score. That is, credence functions that violate the axioms of probability are *dominated* by probabilistic credence functions. Conversely, no probabilistic credence function is dominated by any other credence function.

Joyce (1998) extended de Finetti's dominance argument to a broader class of inaccuracy measures characterized by a parsimonious set of axioms. Maher (2002) disputes the motivation and plausibility of some of Joyce's axioms, while Hájek (2008) questions whether Joyce's results really make a convincing case for Probabilism.

That said, there are other dominance results which the epistemic inaccuracy argument can use. Imagine that q expresses a probabilistic prediction of an event A and that $p(\mathrm{A})$ denotes our credence in the occurence of A. The function s: $[0;1]^2 \to [0;\infty]$ is a **strictly proper scoring rule** if and only if

$$\mathbb{E}_p^{\mathrm{A}}[s(\cdot, q)] = p(\mathrm{A})\, s(1,q) + (1 - p(\mathrm{A}))\, s(0,q)$$

takes its minimum (as a function of q) for $q = p(\mathrm{A})$. In other words, if we believe that A will occur with probability α, then the prediction $q = \alpha$ minimizes expected inaccuracy. Predictions should track our degrees of belief. Based on this definition, Predd et al. (2009, 4788) have shown the following generalization of de Finetti's 1974 result: For all inaccuracy measures generated by a continuous strictly proper scoring rule, non-probabilistic credence functions are accuracy-wise dominated by probabilistic ones. Conversely, no probabilistic credence function is ever dominated by a non-probabilistic one. For recent research in this program and an overview of the current state of the art, see Pettigrew 2016. In general, the recent inaccuracy-based arguments for Probabilism have a clear motivation and a rigorous rationale (dominance), but they are not immune to criticism either, be it about the choice of the scoring rule or about their philosophical implications (see, e.g., Pettigrew 2015).

There are also interesting formal connections between the four arguments for Probabilism. Williams (2012) establishes that Dutch book arguments and arguments from epistemic inaccuracy are both consequences of the same mathematical result (the Separating Hyperplane Theorem). Nonetheless, it is notable that the same result—rational degrees of belief should satisfy the axioms of probability—is reached from completely different approaches and perspectives: an operationalist perspective that explicates probability as betting behavior, a decision-theoretic perspective, and two different epistemic perspectives (qualitative characterization of plausibility and minimizing inaccuracy). While none of the four arguments is immune to objections, the proposed justifications are by all means reasonable and make a good cumulative case for Probabilism (see also De Bona and Staffel 2018). Moreover, given that probability is a very specific mathematical structure, it is perhaps not surprising that we have to make substantial assumptions in order to obtain a unique mathematical representation of degrees of belief.

Conditional Degrees of Belief and Bayes' Theorem

The previous section motivated why the degrees of belief of a rational agent at a particular time should satisfy the axioms of probability. What about the **dynamics of degrees of belief?** How should they change in the light of incoming evidence? To answer this question, Bayesians make use of **conditional degrees of belief.** These capture the idea that we sometimes judge the plausibility of a proposition A in the light of another proposition B. The notation for this is $p(\mathrm{A}\,|\,\mathrm{B})$. For example, what is our degree of belief that Juventus Turin will win the next Champions League if we suppose that Cristiano Ronaldo is injured for a period of six months? But also less profane examples abound. Especially in science, we often judge the plausibility of an event conditional on assuming the truth of a particular hypothesis. What is our degree of belief that crop yields will decrease by more than 50 % if we suppose that there will be a draught this summer? What is our degree of belief that infections with sexually transmitted diseases will drop substantially if we conduct a large-scale sexual education program? What is our degree of belief that there is an intelligent form of life outside the solar system if we make a particular assumption about the number of Earthlike planets in the galaxy? And so on.

In these examples, we have suggested a **suppositional analysis of conditional degrees of belief:** they should be evaluated in the light of supposing the proposition we condition on. This account has its roots in Charles S. Peirce's philosophy (Peirce 1931–1935) and has been articulated explicitly by Frank P. Ramsey (1926) and Bruno de Finetti (1937). Here is Ramsey's famous analysis describing how hypothetically assuming B determines one's conditional degrees of belief in A, given B (see also de Finetti 1972, 2008):

> If two people are arguing "if B will A?" and both are in doubt as to B, they are adding B hypothetically to their stock of knowledge and arguing on that basis about A. We can say that they are *fixing their degrees of belief in* A *given* B. (Ramsey 1926, our emphasis, notation changed)

Due to the suppositional and sometimes outright counterfactual nature of these statements, the concept of conditional degrees of belief is qualitatively different from unconditional degree of belief. This means that we have to answer two questions:

(1) How do conditional degrees of belief relate to unconditional degrees of belief?

(2) Why should they satisfy the axioms of probability?

A key to answering the first question is the observation that the conditional degree of belief in A given B, described by $p(\mathrm{A}\,|\,\mathrm{B})$, has to be equal to the ratio of the degrees of belief in $\mathrm{A} \wedge \mathrm{B}$ and B (assuming $p(\mathrm{B}) > 0$):

$$p(\mathrm{A}\,|\,\mathrm{B}) = \frac{p(\mathrm{A} \wedge \mathrm{B})}{p(\mathrm{B})}. \qquad \text{(RATIO ANALYSIS)}$$

This move is primarily motivated by Kolmogorov's (1933) mathematical development of probability theory, but we can also justify it philosophically, by means of a Dutch book argument. Like unconditional degrees of belief, conditional degrees of belief express dispositions to bet, or judgments on the fairness of bets. More precisely, the conditional degree of belief in A given B induces a **conditional bet:** if B is true, it is a regular bet on A that wins when A is true and loses when A is false, with odds given by $1/p(\mathrm{A}\,|\,\mathrm{B})$, whereas the bet is *called off* if B is false (de Finetti 1936, 1937). In that case, the stake is returned to the bettor with no further consequences. Failure to comply with RATIO ANALYSIS leads to a Dutch book against the agent, involving a combination of conditional and unconditional bets. For $x := p(\mathrm{A} \wedge \mathrm{B})$, $y := p(\mathrm{B})$ and $z := p(\mathrm{A}\,|\,\mathrm{B})$, consider the system of bets in Table T.5. It leads to a fixed return of €1, but the stakes are only equal to €1 if $x = yz$, that is, if RATIO ANALYSIS is respected. Otherwise the system of bets leads to a sure loss for either the bettor or the bookie. Therefore RATIO ANALYSIS is broadly accepted as a rationality constraint on conditional degrees of belief.

Bet on Proposition	Stake	State of the World $\mathrm{A} \wedge \mathrm{B}$	$\neg\mathrm{B}$	$\neg\mathrm{A} \wedge \mathrm{B}$
$\mathrm{A} \wedge \mathrm{B}$	€x	€1	€0	€0
$\neg\mathrm{B}$	€$z \cdot (1-y)$	€0	€z	€0
$\neg\mathrm{A}\,|\,\mathrm{B}$	€$1-z$	€0	€$1-z$	€1
Total Stake and Return	€$1+x-yz$	€1	€1	€1

Table T.5: The Dutch Book argument in defense of RATIO ANALYSIS as a constraint on conditional degrees of belief. Each line represents an individual bet with a fixed stake and variable returns dependent on which state of the world obtains. The entire system of bets has a fixed total return, which for degrees of belief that violate RATIO ANALYSIS differs from the total stake.

Now we turn to the second question. Consider degrees of belief conditional on a proposition B. As stated above, a conditional degree of belief $p(\mathrm{A}\,|\,\mathrm{B})$ corresponds to a conditional bet on A which is winning if A and B both occur, losing if ¬A and B occur, and called off if ¬B is the case. Again, Dutch book arguments for probabilistic coherence apply equally well to $p(\cdot\,|\,\mathrm{B})$ as to p. If there is a probabilistic incoherence in $p(\cdot\,|\,\mathrm{B})$, involving the propositions A_1, A_2, etc., RATIO ANALYSIS implies that there is also a probabilistic incoherence in p, based on the propositions $\mathrm{A}_1 \wedge \mathrm{B}$, $\mathrm{A}_2 \wedge \mathrm{B}$, etc. So we can construct a Dutch book for the latter set of propositions, and define a corresponding set of conditional bets that will lead to a sure loss (unless ¬B occurs and all bets are called off). The arguments for the decision-theoretic and the epistemic justification of probabilistic degrees of belief transfer in a similar way.

It is an exciting philosophical question whether conditional degrees of belief can be reduced to unconditional degrees of belief. Kolmogorov (1933) uses RATIO ANALYSIS as a *definition* of conditional probability, and most textbooks in (mathematical) probability theory and statistics reproduce this definition. Notably, also philosophical primers on Bayesian inference (e.g., Earman 1992; Skyrms 2000; Howson and Urbach 2006; Williamson 2017) follow Kolmogorov's route and define conditional probability in terms of unconditional probability. Since these probabilities are interpreted as (conditional) degrees of belief, RATIO ANALYSIS naturally leads us to a definition of conditional degrees of belief in terms of unconditional degrees of belief. That is, a given conditional degree of belief in A given B is equivalent to having particular unconditional degrees of belief in $A \wedge B$ and B—namely those whose ratio equals the stipulated value of $p(A \mid B)$.

We do not think that such a reductive analysis does justice to the concept of unconditional degrees of belief. Neither does it give a good description of how people reason with conditional degrees of belief and conditional probabilities (e.g., Zhao, Shah and Osherson 2009), nor does it grasp the normative force of statements like "the probability of tossing heads in a single toss, given that the coin is fair, is 1/2". In the literature, Hájek (2003) develops and synthesizes various criticisms of RATIO ANALYSIS as a definition of conditional probability (see also Fitelson and Hájek 2017) while Easwaran (2011c) and Myrvold (2015) explore avenues for parrying Hájek's criticism.

A more detailed investigation of the topic would go beyond this introductory chapter and therefore we postpone it to Variation 12. There, we will defend a fully suppositional analysis of conditional degrees of belief in the Ramsey–de Finetti tradition, and show its implications for chance–credence coordination and reasoning with complex, highly idealized Bayesian models. For those who—like us—think that conditional degrees of belief are at least equally fundamental to Bayesian inference as unconditional degrees of belief, there is an alternative axiomatization of probability supported by numerous mathematicians, epistemologists and philosophers of science. They propose to define conditional probability as a primitive concept (Rényi 1970; Popper 1959/2002; Hájek 2003; Maher 2010). For a σ-algebra $\mathcal{L}$ of sentences of a propositional language, the function $p\colon \mathcal{L} \times (\mathcal{L} \setminus \{\bot\}) \to [0;1]$ is called a **Popper–Rényi function** or simply a "Popper function" (see also Weisberg 2009) if and only if for all A, B and $C \in \mathcal{L}$:

1. $p(\cdot \mid A)$ is a (standard) probability function;
2. $p(A \mid A) = 1$;
3. $p(B \wedge C \mid A) = p(B \mid A \wedge C)\, p(C \mid A)$.

On this account, the unconditional probability of A can be defined as the probability of A conditional on a tautological proposition. In other words, unconditional probability becomes a special case of conditional probability. We will make use of this interpretation in several parts of the book, mainly in order to avoid that $p(\cdot\,|\mathrm{A})$ is undefined when $p(\mathrm{A}) = 0$.[4] This case occurs in several applications of Bayesian inference, especially in the context of statistical reasoning.

RATIO ANALYSIS remains important, though, as a coherence requirement for conditional and unconditional degrees of belief in the case that the proposition conditioned on has a probability greater than zero. One of its consequences is Bayes' famous theorem. Assume, for some scientific hypothesis H and some experimental evidence E, that $p(\mathrm{E}), p(\mathrm{H}) > 0$. If we combine

$$p(\mathrm{E}\,|\,\mathrm{H}) = \frac{p(\mathrm{E} \wedge \mathrm{H})}{p(\mathrm{H})}$$

with the cognate equation

$$p(\mathrm{H}\,|\,\mathrm{E}) = \frac{p(\mathrm{E} \wedge \mathrm{H})}{p(\mathrm{E})}$$

then we obtain, by a simple substitution, **Bayes' Theorem:**

$$p(\mathrm{H}\,|\,\mathrm{E}) = p(\mathrm{H})\frac{p(\mathrm{E}\,|\,\mathrm{H})}{p(\mathrm{E})}. \qquad \text{(Bayes' Theorem)}$$

This equation will accompany us throughout the book—it describes how the degree of belief in hypothesis H given evidence E relates to the unconditional degrees of belief in H and E, and to the conditional degree of belief in E given H. Note that Bayes' Theorem does *not* (yet) describe how agents should change their degree of belief in H upon learning E. We are still concerned with the *statics* of Bayesian inference, with the question of how conditional and unconditional degrees of belief cohere. The dynamic dimension is introduced in what follows.

Conditionalization and Varieties of Bayesian Inference

The previous sections have explained the static dimensions of Bayesian inference: representing degrees of belief in terms of probabilities and coordinating unconditional with conditional degrees of belief. But how should we change our degrees of belief in the light of incoming evidence? Fortunately, the

[4] Another option consists in unifying conditional and unconditional probabilities under the umbrella of the concept of conditional expectations, and a σ-algebra that is conditional on a random variable. For this rather technical route, see Rédei and Gyenis 2016; Gyenis, Hofer-Szabó and Rédei 2017.

Bayesian answer is very simple: the rational degree of belief in hypothesis H after learning evidence E is expressed by the conditional degree of belief in H given E.

Bayesian Conditionalization The rational degree of belief in a proposition H after learning evidence E equals the conditional probability of H given E according to the agent's original degrees of belief. In other words, $p'(\mathrm{H}) = p(\mathrm{H}\,|\,\mathrm{E})$.

Conventionally, $p(\mathrm{H})$—the degree of belief in H before learning E—is called the **prior probability** of H, while $p'(\mathrm{H}) = p(\mathrm{H}\,|\,\mathrm{E})$—the degree of belief in H after learning E—is called the **posterior probability** of H.

Bayesian Conditionalization is inspired by the same idea that Ramsey proposed for conditional degrees of belief: when we learn a piece of evidence E, we add it to our background knowledge ($p'(\mathrm{E}) = 1$) and infer the consequences of this move for the rest our epistemic attitudes. This is why the new degree of belief in H is set equal to the conditional probability of H given E. Due to its generality and simplicity, Bayesian Conditionalization is regarded as a defining principle of Bayesian inference (e.g., Earman 1992, 34). Sometimes it also figures under the name of **Bayes' Rule.**

Combining Bayesian Conditionalization with Bayes' Theorem, we can derive the following equality:

$$p'(\mathrm{H}) = p(\mathrm{H}\,|\,\mathrm{E}) = \left(1 + \frac{p(\neg\mathrm{H})}{p(\mathrm{H})} \cdot \frac{p(\mathrm{E}\,|\,\neg\mathrm{H})}{p(\mathrm{E}\,|\,\mathrm{H})}\right)^{-1}. \tag{T.2}$$

In (T.2), $p(\mathrm{E}\,|\,\mathrm{H})$ and $p(\mathrm{E}\,|\,\neg\mathrm{H})$—the probabilities of the observed evidence E under specific hypotheses—are called the **likelihoods** of H and ¬H on E. This version of Bayesian Conditionalization is especially useful for applications in statistical inference, where the statistical model often provides us with the likelihood function $p(\mathrm{E}\,|\,\mathrm{H}_\theta)$, with the competing hypotheses indexed by a real-valued parameter θ. Bayesian Conditionalization provides a way of learning novel evidence by means of trading off the prior probability of hypothesis H with the likelihoods of H and ¬H on the observed evidence.

Equation (T.2) allows us to detect numerous reasoning errors. The classical example is the **base rate fallacy.** Consider a medical test for a rare disease which gives a correct result in 95 % of all cases, regardless of whether you have the disease or not. Thus it is quite reliable. Imagine that you undergo a routine blood screening (e.g., because you work in a hospital) and that the test is positive. In that case, one might draw the conclusion that you probably have the disease and should be treated accordingly.

A Bayesian statistical model shows that this conclusion is premature. If you were not exposed to a particular risk of contracting the disease, then your prior probability of being infected (H) may, before the blood screening, be as low as 0.1 % ($p(\text{H}) = 0.001$). After all, the disease is rare. Learning the result of the blood test (E) implies that being infected is still very unlikely:

$$p(\text{H}\,|\,\text{E}) = \left(1 + \frac{1-0.001}{0.001} \cdot \frac{0.05}{0.95}\right)^{-1} \approx 0.02.$$

It is easy to see that for a less frequent disease, the posterior probability of H would be even lower. People are, however, inclined to discount the prior implausibility of a hypothesis and to accept it on the basis of apparently striking evidence (Kahneman and Tversky 1973; Kahneman, Slovic and Tversky 1982; Goodman 1999a). Even most students at the renowned Harvard Medical School failed to reach the correct conclusion (Casscells, Schoenberger and Graboys 1978). With Bayesian inference, we have a tool for detecting cognitive fallacies and for correcting uncertain reasoning where appropriate.

What are the reasons for accepting Bayesian Conditionalization in the first place? Essentially it identifies learning E with *supposing* E: $p'(\text{H}) = p(\text{H}\,|\,\text{E})$. However, in spite of an intuitive similarity between learning and supposing, this equation is not easy to justify. After all, there are non-trivial psychological differences: Zhao et al. (2012) found that participants who *learned* evidence E (e.g., by observing relative frequencies) submitted different probability estimates of H than participants who *supposed* that E occurred. Given this discrepancy on the descriptive level, we have to offer a convincing normative argument in favor of Bayesian Conditionalization.

A standard proposal, quite similar to what we have seen before, consists in **dynamic Dutch book arguments** (Teller 1973; van Fraassen 1989; Lewis 1999). In such arguments, an agent is asked how she would set her posterior $p'(H)$ if she were to learn that E. If the agent declares a value that differs from $p(\text{H}\,|\,\text{E})$, then there is a system of (conditional and unconditional) bets such that she is vulnerable to a sure loss, regardless of whether E or $\neg$E occurs.

One problem with dynamic Dutch book arguments consists in the fact that they require the agent to fix in advance which bets she is going to accept in the future if she happens to learn a certain fact about the world. In other words, dynamic Dutch book arguments are a sanity check for the preferences and commitments of an agent, instead of a proof of the irrationality of following another updating rule. Besides, dynamic Dutch books are susceptible to the same objections that have been raised against ordinary Dutch books.

Bayesian Conditionalization has also been defended from the perspective of maximizing expected epistemic utility, or minimizing (diachronic) inaccuracy (Greaves and Wallace 2006; Leitgeb and Pettigrew 2010a,b). Moreover, Bayesian Conditionalization is just a special case of a more general updating principle, namely minimizing the divergence between prior and posterior distribution. In other words, Bayesian updating is a conservative way of changing one's prior degrees of belief upon learning the constraint $p'(\mathrm{E}) = 1$ on the posterior distribution p'. Let us try to make this idea more precise. Suppose, for the sake of simplicity, that our language L contains finitely many propositional variables $A_1, A_2, A_3, \ldots, A_N$. Then there are 2^N **state descriptions** in $\mathcal{L}$, that is, maximal conjunctions $\pm \mathrm{A}_1 \wedge \pm \mathrm{A}_2 \wedge \pm \mathrm{A}_3 \wedge \ldots \wedge \pm \mathrm{A}_N$ of (negations of) pairwise distinct atomic sentences (Carnap 1950). The "$\pm$" symbol signifies that each variable A_i is assigned either the value A_i or $\neg \mathrm{A}_i$. Ω denotes the set of state descriptions and every element of $\mathcal{L}$ is logically equivalent to a disjunction of elements of Ω. The ordered pair $(\Omega, \mathcal{L})$ is a **measurable space**—that is, a set Ω of (mutually exclusive) sentences with a Boolean algebra $\mathcal{L}$ generated by their truth-functional compounds such that we can define a probability function (or probability measure, see footnote 1) on $\mathcal{L}$.

Divergence minimization is explicated with respect to the probabilities that p and p' assign to each $\omega \in \Omega$. That is, for the above partition $\Omega = \{\omega_1, \omega_2, \ldots, \omega_{2N}\}$ of logical space, we can define one of the most popular "distance measures" between probability functions, namely **Kullback–Leibler divergence** (e.g., Kullback and Leibler 1951):

$$\mathrm{D}_{\mathrm{KL}}(p', p) = \sum_{i=1}^{2^N} p'(\omega_i) \log \frac{p'(\omega_i)}{p(\omega_i)}.$$

In scientific applications, we might also encounter countably infinite partitions $\Omega = \{\omega_1, \omega_2, \omega_3, \ldots\}$, for example, when we define a probability function on an integer-valued variable (such as "the number of Earthlike planets in the universe"). In this case, we just replace the finite sum with an infinite sum. For uncountable partitions (e.g., Ω is an interval in the real numbers $\mathbb{R}$), this is not possible. Then, we usually define **density functions** $f, f' \colon \Omega \to \mathbb{R}^{\geq 0}$ that connect p and p' to the Lebesgue measure on Ω:

$$p(\mathrm{A}) = \int_A f(x)\,dx, \qquad p'(\mathrm{A}) = \int_A f'(x)\,dx.$$

Kullback–Leibler divergence is then defined as

$$D_{KL}(p', p) = \int_{\Omega} f'(x) \log \frac{f'(x)}{f(x)} \, dx.$$

With these definitions, Diaconis and Zabell (1982) have demonstrated that the following two ways of updating a given prior distribution p to a posterior distribution p' are equivalent:

(1) Bayesian Conditionalization on E, that is, $p'(\,\cdot\,) := p(\,\cdot\,|\mathrm{E})$;

(2) minimizing Kullback–Leibler divergence $D_{KL}(p', p)$ subject to the constraint that $p'(\mathrm{E}) = 1$.

Bayesian updating can thus be represented as conservative belief revision: Bayesians change their degrees of belief only in so far as newly learned constraints on those (e.g., $p'(\mathrm{E}) = 1$) force them to do so. Or in other words, they stay as close to their prior degrees of belief as these constraints allow them to do. This principle is also entrenched in non-quantitative theories of dynamic reasoning, such as the AGM-model of belief revision (Alchourrón, Gärdenfors and Makinson 1985), which operates on the binary level of belief and disbelief. Actually, Diaconis and Zabell's result holds also for more general types of divergences between prior and posterior distribution, namely all functions of the form

$$\int_{\Omega} p'(x) f\left(\frac{p(x)}{p'(x)}\right) dx,$$

where f is a convex function defined on $(0;\infty)$ with $f(1) = 0$ (Cziszár 1967, 1975). We investigate these ***f*-divergences** and their applications in more detail in Variation 4.

Hence, while it is difficult to give a fully compelling and conclusive justification for changing degrees of belief in a particular way, dynamic Dutch book arguments and the argument from distance minimization jointly make a strong case for Bayesian Conditionalization. This is especially interesting since the respective motivations are quite different: one comes from the operationalist, decision-theoretic corner, and the other from a principle of epistemic conservativity that is also used in qualitative models of belief revision.

However, Bayesian Conditionalization is somewhat restricted as a tool for learning from experience. For example, it does not describe how we should update our degrees of belief in the light of information whose propositional status is unclear, such as indicative conditionals or relative frequencies. We will address this particular challenge in Variation 4. Moreover, sometimes we do not learn that evidence E has occurred with certainty, just that it is highly likely. For instance, a look at the weather forecast may shift our degree of belief in E = "it will rain tonight" from $p(\mathrm{E}) = 1/2$ to $p'(\mathrm{E}) = 9/10$. How should our belief in other sentences, such as H = "the sun will shine tomorrow", change in the face of such evidence? We could update on the second-order proposition that the probability of E has changed, but such a move would involve great technical complications. And even then the implications for the posterior probability of H would not be clear.

To solve this challenge, Jeffrey (1965) has argued that the posterior probability $p'(\mathrm{H})$ of hypothesis H after learning E should obey the equation

$$p'(\mathrm{H}) \;=\; p'(\mathrm{E})\, p(\mathrm{H}|\mathrm{E}) \,+\, p'(\neg\mathrm{E})\, p(\mathrm{H}|\neg\mathrm{E}) \tag{JC}$$

(where "JC" stands for **Jeffrey Conditionalization**) whenever the following two equations are satisfied:

$$p(\mathrm{H}|\mathrm{E}) = p'(\mathrm{H}|\mathrm{E}), \qquad p(\mathrm{H}|\neg\mathrm{E}) = p'(\mathrm{H}|\neg\mathrm{E}). \tag{Rigidity}$$

Equation (JC) computes the new degree of belief in H as the weighted average of the conditional degrees of belief in H given E and given $\neg$E, respectively, weighted with the degrees of belief that E occurred and that it didn't, respectively. Condition (JC) follows from the Law of Total Probability together with (Rigidity). In a recent paper, Schwan and Stern (2016) argue that (Rigidity) holds whenever E screens off H from the propositional content D of the learning experience, that is, $p(\mathrm{H} \mid \mathrm{D} \wedge \mathrm{E}) = p(\mathrm{H}|\mathrm{E})$ and $p(\mathrm{H} \mid \mathrm{D} \wedge \neg\mathrm{E}) = p(\mathrm{H}|\neg\mathrm{E})$. Obviously, Jeffrey Conditionalization reduces to Bayesian Conditionalization when E is known for certain, that is, when $p'(\mathrm{E}) = 1$. The results for representing Bayesian Conditionalization as a form of distance minimization also transfer to Jeffrey Conditionalization.

A famous feature of Bayesian Conditionalization is the **"washing-out of the priors"** shown in the **merging-of-opinions theorems** by Blackwell and Dubins (1962) and Gaifman and Snir (1982) (for a discussion of their philosophical relevance, see Earman 1992). Agents with different initial degrees of belief, represented by probability functions p_1 and p_2, are gradually moving toward consensus if they both follow Bayesian Conditionalization as an updating rule. As long as they agree on the propositions which obtain measure zero (in other words, $p_1(\mathrm{X}) = 0 \Leftrightarrow p_2(\mathrm{X}) = 0$), the distance between the probability functions p_1 and p_2 will approach zero. If both agents are

Bayesian Conditionalizers, individual differences will eventually cancel out. We return to that topic in Variation 11, where we discuss the objectivity claims of Bayesian inference, and make critical glosses on the convergence theorems.

Finally, we add a third dimension to Bayesian inference. We have talked a lot about the mathematical axioms that govern the statics and dynamics of rational degrees of belief. But we have been silent on **how degrees of belief should inform rational decisions.** In many applications of Bayesian reasoning, it is emphasized that the goal of a Bayesian inference is the calculation of posterior probabilities. This is also the main result of many approaches to rational choice in economics: posterior probabilities are combined with subjective utilities in order to make an optimal choice (Savage 1972). Under certain assumptions on rational preferences, it can be shown that all relevant information for making rational decisions (in the economic, instrumental sense of rationality) is contained in an agent's posterior probability distribution. This is the practical dimension of Bayesian inference: posterior distributions form the epistemic basis for practical decisions and theory assessment, including judgments of acceptance and full ("all-or-nothing") belief (Leitgeb 2014, 2017; Easwaran 2016).

A variant of this view is the idea that posterior distributions can be used for theory assessment in terms of **verisimilitude** or **truthlikeness.** The concept of truthlikeness was introduced by Karl Popper (1959/2002), who rejected the Bayesian interpretation of probability as degree of belief. On Popper's view, all general scientific theories are false and should be assessed on the basis of the degree to which they have true contents. In other words, theories should not be judged in terms of their subjective probability, but in terms of their truthlikeness. Such a qualitative notion of truthlikeness allows us to judge, for example, whether a succession of false scientific theories $T_1, T_2, \ldots, T_n$ constitutes scientific progress in the sense of coming closer to the truth.

On Popper's account, a theory T_1 is more verisimilar than a competing theory T_2 if and only if the set of true statements of T_2 is included in the set of true statements of T_1, and the false statements of T_1 are included in the false statements of T_2. However, Tichý (1974) and Miller (1974) showed that Popper's approach entails that no false theory is closer to the truth than any other. Recent approaches in the truthlikeness paradigm take a more conciliatory stance toward Bayesian inference: theories are judged in terms of their **expected truthlikeness.** That is, one calculates the truthlikeness of a theory T with respect to each element of a partition $\mathcal{H} = \{H_1, H_2, \ldots\}$ of logical space, and weighs this value $\textsc{tl}(T \mid H_i)$ with the posterior probability of each H_i given evidence E (Maher 1993; Niiniluoto 1999, 2011):

$$\textsc{etl}(T, E) = \sum_{H_i \in \mathcal{H}} p(H_i \mid E) \times \textsc{tl}(T \mid H_i).$$

Contrary to the popular idea that Bayesian inference and search for truthlike theories are competing paradigms, it is thus possible to integrate posterior probabilities in a theory of truthlikeness of scientific theories (see also Oddie 1986, 2014; Kuipers 2000; Festa, Aliseda and Peijnenburg 2005; Cevolani and Tambolo 2013). This shows the great versatility of Bayesian inference in supporting diverse approaches to hypothesis assessment, theory acceptance and decision-making.

All in all, the core principles of subjective Bayesian inference can be aligned in three dimensions:

Static Dimension (Probabilism) The degrees of belief of a rational agent are represented by probability functions.

Dynamic Dimension Bayesian Conditionalization (or some generalization thereof) dictates how a rational agent should revise her degrees of belief.

Practical Dimension The posterior probability distribution is, together with an agent's subjective utility function, the rational basis for assessing evidence, accepting hypotheses and making decisions.

Most Bayesians accept all three of these principles (for a survey, see Weisberg 2009). For an **orthodox subjective Bayesian,** they exhaust the explicit requirements on rational degrees of belief (and decisions). Bruno de Finetti and Richard Jeffrey are often cited as representatives of this position—though see Galavotti 1989. In any case, such a theory of rationality seems to be too weak, in the sense that degrees of belief are decoupled from the aim of tracking properties of the real world:

> For the Bayesian apparatus to be relevant to scientific inference, it seems that what it needs to deliver are not mere subjective opinions but reasonable, objective, reliable degrees of belief. Thence comes the challenge: How are prior probabilities to be assigned so as to make that delivery possible? (Earman 1992, 57)

Therefore, the standard view of Bayesian rationality makes more demanding requirements on rational degrees of belief than the orthodox subjective Bayesian. This view, known as **tempered** or **calibrated subjectivism** or **(Bayesian)**

personalism, can be illustrated as follows (cf. Howson and Urbach 2006). Suppose we would like to assess a particular hypothesis H about the value of a physical parameter θ (e.g., H says that $\theta > 0$). Then a prior distribution for θ should be chosen in a "reasonable" way that is compatible with our empirical background knowledge. For instance, if θ represents the effect size of an experimental manipulation, then it can, in the absence of information to the contrary, be expected to be of a similar order of magnitude as effects in comparable experiments. In particular, it seems that our degrees of belief should track objective chances whenever we happen to know them. Such principles for coordinating credences and chances figure under names like "Principle of Direct Inference" or "Principal Principle" (e.g., Reichenbach 1949; Kyburg 1974; Levi 1977; Lewis 1980) and will be discussed in more detail in Variation 12.

All these proposals for restraining degrees of belief beyond the axioms of probability have one requirement in common: degrees of belief should mirror properties of the world around us. Jon Williamson (2007, 2010) calls this the **calibration norm** for Bayesian reasoning, added on top of the aforementioned norm of Probabilism (i.e., that degrees of belief satisfy the axioms of probability).

More demanding varieties of Bayesian inference, often subsumed under the label of **Objective Bayesianism**, go beyond the calibration norm and argue that in each situation, there is a uniquely rational, or at least a privileged, degree of belief. For instance, Williamson (2007, 2010)—building on pioneering work by Jaynes (1968, 2003) and Paris and Vencovská (1989)—recommends choosing, among all probability distributions that satisfy the calibration norm, the most equivocal, that is, the most middling, probability distribution. This **equivocation norm** is formalized by means of maximizing the **entropy** of a probability distribution, that is, by choosing the probability function which maximizes, for an appropriate partition $\Omega = \{\omega_1, \omega_2, \ldots, \omega_N\}$,

$$H(p) = -\sum_{i=1}^{N} p(\omega_i) \log p(\omega_i) \qquad \text{(Maximum Entropy—discrete case)}$$

or, in the continuous case, with probability density f,

$$H(p) = -\int_{x\in\Omega} f(x) \log f(x)\, dx \qquad \text{(Maximum Entropy—continuous case)}$$

(see also Shannon 1949). This constraint is equivalent to minimizing Kullback–Leibler divergence between the uniform distribution and the set of probability distributions that satisfy the calibration norm. Conditionalization can now be dropped: all learned propositional evidence E is taken care of by the calibration and equivocation norms taken together (i.e., choose the most equivocal probability distribution with $p(E) = 1$). As opposed to Conditionalization, the **Maximum Entropy** or **MaxEnt approach** deals easily with non-propositional evidence, such as expectation and variance of random variables, and other constraints that are hard to express in a simple propositional language. For this reason, it is particularly apt for statistical reasoning and popular with scientific practitioners. The worries are on the philosophical side: MaxEnt seems to be a revamped version of the problematic Principle of Indifference ("in the absence of evidence to the contrary, the probabilities of two mutually exclusive propositions should be equal"), which is challenged by various paradoxes (Hájek 2011). For a more detailed motivation of MaxEnt, see Williamson 2007; for criticism, see Seidenfeld 1979b, 1986.

While the MaxEnt camp accepts Probabilism (i.e., the static dimension of Bayesian inference) but rejects Conditionalization (i.e., its dynamic dimension), some Bayesian statisticians use both principles in applied work, but without commitment to the subjective interpretation of probability. Rather, they regard Bayesian inference as a convenient mathematical tool. Instead of prior distributions that represent honest subjective degrees of belief, they use **objective prior probabilities,** which are based on principles of symmetry, invariance or information minimization (Jeffreys 1961; Bernardo 1979a; Paris, Vencovská and Wilmers 1994; Vassend 2018), but do not represent any agent's degrees of belief. These prior distributions may also be "improper": they fail to sum up to 1 and violate the axioms of probability (Bernardo and Smith 1994). This does not mean that those statisticians believe Bayesian inference to be unsound; rather, they respond to the practical difficulties in eliciting meaningful subjective prior distributions when little is known about the target system. Also, they aim at screening off Bayesian inference from the charge of arbitrariness.

In practice, the link between (posterior) degree of belief and decision-making need not be strict. Many statisticians and scientists regard Bayesian inference as a foundationally sound framework for modeling rational degrees of belief, up to the point of using Bayesian models in their daily work. Nevertheless, when it comes to (high-level) theory assessment or practical decisions based on statistical evidence, they often prefer frequentist statistics, a rivalling framework for statistical inference (see Variations 9–11). The reasons are practical difficulties in coming up with meaningful prior distributions, as well as concerns about not being able to reach objective conclusions within the (subjective) Bayesian framework (e.g., Cox and Mayo 2010; US Food and Drug Administration 2010; Gelman and Shalizi 2012; Cumming 2014; Trafimow and Marks 2015).

Finally, one may decide to assess theories based on their confirmational track record, which may be derived from a Bayesian framework without being equal to a theory's posterior probability. A recent representative of such an approach is Brössel (forthcoming), who elaborates the concept of confirmation commitments (see also Hawthorne 2005).

These remarks conclude our discussion of the foundations of Bayesian inference. Incidentally, all that has been said here is compatible with extensions of Bayesian inference where agents do not entertain uniquely rational degrees of beliefs, but rather sets of degrees of belief that correspond to sets of (rational) probability functions. On this view, sharp subjective probabilities are quite unrealistic idealizations of doxastic states; a philosophical theory of uncertain reasoning should take into account the inherent imprecision of degrees of belief (e.g., Levi 1974, 1980). This is the theory of **imprecise probabilities,** which is also frequently used in scientific applications—see Bradley 2014 for a comprehensive survey, and Elga 2010 for a principled challenge.

The glossary on page 32 contains the essential concepts of Bayesian inference. We now proceed to a powerful practical tool for making Bayesian inferences: Bayesian networks. Philosophically, this does not add substantial assumptions, but since Bayesian networks will be one of our main modeling tools in the remainder of the book, it is useful to explain the principles behind them.

Causal Bayesian Networks

A (causal) Bayesian network is a directed acyclical graph (DAG), that is, a pair $G = (V, E)$ consisting of a finite set V of "vertices" or "nodes" and a finite set E of directed "edges" or "arrows" between these that does not contain cycles. Vertices stand for variables, and edges represent probabilistic dependencies between the variables (absence of an edge between two vertices thus represents independency of the corresponding variables). DAG's are a useful visual tool for describing inferential relations among variables and for supporting causal inference in science (Pearl 2000; Spirtes, Glymour and Scheines 2000).

A simple example of reasoning with a Bayesian network is given in Figure T.1. Suppose that Alice and Bob collected some mushrooms in the

Bayesian Conditionalization	The principle that learning a proposition E corresponds to replacing one's current degrees of belief p by the conditional distribution $p(\cdot\,\|\mathrm{E})$.
prior probability (of H)	The degree of belief in proposition H before a learning experience takes place, written as $p(\mathrm{H})$.
posterior probability (of H)	Degree of belief in proposition H after proposition E is learned, written as $p'_{\mathrm{E}}(\mathrm{H})$ or simply $p'(\mathrm{H})$. By Bayesian Conditionaliziation, this is equivalent to $p(\mathrm{H}\|\mathrm{E})$.
likelihood	In general, a synonym for "probability". In the context of statistical inference, "the likelihood of H on data E" has a technical meaning: it denotes the probability $p(\mathrm{E}\|\mathrm{H})$. The likelihood function for data E is the function $L_{\mathrm{E}}(\cdot) = p(\mathrm{E}\|\cdot)$.
marginal distribution	In a joint probability distribution over various variables, the marginal distribution of a variable E corresponds to the unconditional distribution of E. This distribution can be obtained by averaging over the conditional probabilities of E according to the ...
Law of Total Probability	$p(\mathrm{E}) = p(\mathrm{E}\|\mathrm{H})\,p(\mathrm{H}) + p(\mathrm{E}\|\neg\mathrm{H})\,p(\neg\mathrm{H})$
Bayes' Theorem	The identity $p(\mathrm{H}\|\mathrm{E}) = p(\mathrm{H})p(\mathrm{E}\|\mathrm{H})/\,p(\mathrm{E})$. Sometimes written in the form defined by Equation (T.2) on page 22.
Jeffrey Conditionalization	Generalization of Bayesian Conditionalization to the case of learning uncertain or noisy evidence where $p'(\mathrm{E})$ is smaller than 1.

Table T.6: A glossary of core Bayesian terms.

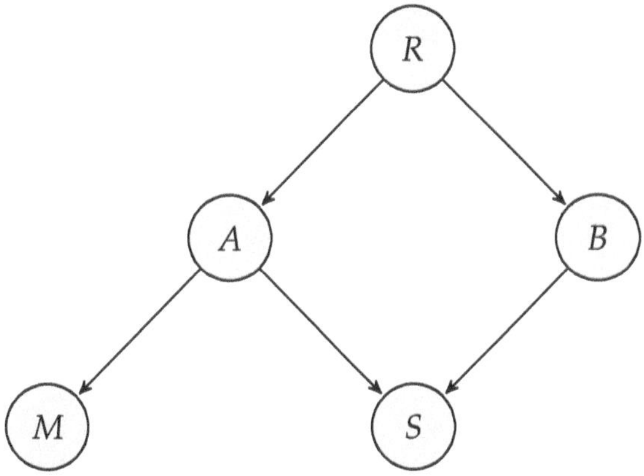

Figure T.1: The Bayesian network representation of the risotto example.

forest for cooking a risotto, but neither of them is a mushroom expert. The letter R represents the proposition that these mushrooms were slightly poisonous. The letter A represents the proposition that Alice has stomach pain after eating the risotto, and B represents the proposition that Bob has stomach pain after eating the risotto. Finally, M represents the proposition that Alice is taking a medicine to cure the stomach pain, and S represents the proposition that Alice and Bob are doing sports together on the same evening.

The edges in the graph describe dependencies between the variables, with the arrow denoting the direction of causal influence. The *descendants* of a variable are all downward variables and the *parents* of a variable are its direct causes (i.e., there is a direct arrow from the parent to the child). In Figure T.1, A, B, M and S are all descendants of R, but only A and B are parents of S. In particular, the descendant relation, but not the parent relation, is transitive. Lack of an edge between two variables indicates that they do not depend on each other directly, but only via one or several intermediate variables, if at all. Conditional on these intermediate variables, they are independent. For instance, the poisonous risotto influences Alice's decision to take a medicine only via her stomach pain. Similarly, Alice and Bob do sports together if and only if they both feel well.

Crucially, a Bayesian network comes with a probability distribution p over the variables in the DAG, and this distribution is supposed to mirror the network structure:

Markov Condition The probability distribution p is **Markovian** relative to a directed acyclical graph $G = (V, E)$ if and only if every variable is probabilistically independent of all its non-descendants in G, conditional on its parents.

For example, A and B should be probabilistically independent conditional on R, that is,

$$p(A \wedge B \,|\, R) = p(A \,|\, R)\, p(B \,|\, R), \tag{T.3}$$
$$p(A \wedge B \,|\, \neg R) = p(A \,|\, \neg R)\, p(B \,|\, \neg R), \tag{T.4}$$

Our shorthand notation for the conditional independence of A and B, given R, is $A \perp\!\!\!\perp B \,|\, R$.

By itself, the Markov condition is just an abstract mathematical property that specifies how edges between the nodes in G constrain the probability distribution p. In particular, it provides an elegant graph-theoretical representation of conditional independencies. When the Bayesian network is interpreted causally, with arrows denoting paths of causal influence, the Markov Condition is transformed into the philosophically more substantive **Causal Markov Condition:** a phenomenon is independent of its non-effects, given its direct causes. For example, if Alice and Bob have eaten a poisonous mushroom risotto, learning about Alice's stomach pain tells us nothing about whether Bob has stomach pain. In other words, eating the poisonous mushroom risotto is a *common cause* of the stomach pain of both Alice and Bob. For graphs that satisfy the (Causal) Markov Condition, we can easily read off conditional independencies and pathways of causal influence. A highly useful graphical criterion to identify which nodes are (conditionally) independent from others is **d-separation** (Pearl 2000).[5] Moreover, with the help of the Markov Condition, we can calculate joint and marginal probabilities in a straightforward way. For instance, in the above example, we can use Equations (T.3) and (T.4) to write $p(A, B, R)$—a convenient shorthand notation for the probability of the conjunction of the three propositions, $p(A \wedge B \wedge R)$—as

$$p(A, B, R) = p(A, B \,|\, R)\, p(R) = p(A \,|\, R)\, p(B \,|\, R)\, p(R), \tag{T.5}$$

and analogously for all other conjunctions of $\pm$A (that is, either A or $\neg$A), $\pm$B and $\pm$R. Summing over the different values of the variables $A = \pm$A,

[5] The Markov Condition does not state which arrows one may safely delete from a fully connected graph on the basis of probabilistic independence. For making this choice, various principles have been proposed, such as Faithfulness, Minimality and Frugality. For a recent discussion, see Forster et al. 2018.

$B = \pm\mathrm{B}$ and $R = \pm\mathrm{R}$, the **marginal probability** of proposition S can now be written as

$$\begin{aligned} p(\mathrm{S}) &= \sum_{A,B,R} p(A,B,R,\mathrm{S}) \\ &= \sum_{A,B,R} p(\mathrm{S}\,|\,A,B,R)\,p(A,B,R) \\ &= \sum_{A,B,R} p(\mathrm{S}\,|\,A,B)\,p(A\,|\,R)\,p(B\,|\,R)\,p(R). \end{aligned}$$

The derivation uses the Law of Total Probability in the first two steps and Equation (T.5) in the third.

The above equations suggest that a joint or marginal probability can always be reduced to a combination of probabilities conditional on parent variables and probabilities of root variables, that is, variables without parents. Indeed, in general, it will always be the case that for a graph G that satisfies the Markov Condition with variables $\{A_1, \ldots, A_n\}$,

$$p(\mathrm{A}_1, \ldots, \mathrm{A}_n) = \prod_{i=1}^{n} p(\mathrm{A}_i\,|\,\mathrm{Par}(A_i)).$$

Equation (T.5) is an instance of this scheme. That is, if we reason about probabilities in a Bayesian network, it suffices to know the base rates of the root variables and the conditional probability of any variable given its parents. Often, these values are much easier to elicit than joint or marginal probabilities.

The scope of applications of Bayesian networks in science is huge, and it goes beyond the scope of this extremely brief introduction to list even the most important ones. In fact, our use of Bayesian networks in this book will remain on an elementary level: representing causal relations and conditional independencies between different variables, and calculating joint and marginal probabilities in an efficient way.

Variations on a Bayesian Theme

We conclude this exposition of Bayesian reasoning with an outline of the individual book chapters. We think of them as variations on the theme of Bayesian inference in science.

The first five Variations center around the confirmation of scientific theories. To a large part, scientific theories are valued to the extent that they make accurate predictions. Their **degree of confirmation** quantifies the extent to which they have been predictively successful, and confirmation is usually explicated in probabilistic terms. Measuring degrees of confirmation is therefore a classical task for Bayesian philosophy of science. But our approach is broader: we also address challenges to Bayesian Confirmation Theory, and we demonstrate how certain argument patterns in science (e.g., the No Alternatives Argument and the No Miracles Argument) can be recast as confirmatory arguments for the theory in question.

Variation 1 is devoted to a quantitative analysis of **confirmation** and **inductive inference** in Bayesian terms: evidence E confirms theory T if and only if it raises the (subjective) probability of T. We motivate and describe the transition from qualitative, first-order logic accounts of scientific confirmation to quantitative, Bayesian models. They improve on their qualitative predecessors in various ways: by giving more nuanced and practically applicable confirmation judgments, and by addressing classical challenges to inductive inference such as the paradox of the ravens, the tacking paradox and the grue paradox. Then we discuss the plurality of Bayesian confirmation measures and how the evaluation of qualitative theses about inductive inference depends on the choice of a confirmation measure. Finally, we discuss how conceptual analysis and empirical evidence can be combined to narrow down the class of adequate confirmation measures.

Variation 2 deals with the **No Alternatives Argument**. Does the failure to find alternatives to a scientific theory confirm it? Arguments of this kind are often employed in support of string theory or other theories where it is hard to gather empirical support (e.g., in paleontology). In those cases, there have been enormous, yet fruitless, efforts to find alternative theories, suggesting that the truth (or empirical adequacy) of the theory in question may be the best explanation for the failure of these efforts. By framing the argument within a Bayesian model, we can show that longstanding failure to find alternatives supports a theory—even if the strength of the argument (i.e., the degree of confirmation it provides) is highly context-sensitive.

Variation 3 develops a probabilistic assessment of the famous **No Miracles Argument** in favor of **scientific realism**. That argument contends, in a nutshell, that the truth of scientific theories is the only viable explanation of their predictive success. We frame the No Miracles Argument as a confirmatory argument: Does the success of scientific theories license an argument to their truth or empirical adequacy? To answer this question, we set up various Bayesian models, which also correspond to different ways of interpreting the No Miracles Argument. These models take into account factors that have been neglected in reconstructions of the No Miracles Argument: the stability of theories in a specific discipline, and their past success rate. Thus we get a better grip on the circumstances when the success of science supports realist inclinations, and when it doesn't.

While Variations 2 and 3 apply the standard Bayesian framework to an analysis of confirmatory arguments, the next two Variations extend the underlying formalism. Specifically, Variation 4 describes how learning **conditional evidence** (e.g., "if intervention X is made, result Y will occur") may confirm or disconfirm a scientific theory. For instance, how should the belief in a theory T change if we learn that it makes a particular prediction E (i.e., $p(\mathrm{E}|\mathrm{T}) = 1$)? And how should we change our degrees of belief upon learning the relative frequency of E given T? To solve this challenge, we use a generalization of Bayesian Conditionalization and conceptualize the rational degree of belief change as minimizing the divergence between prior and posterior distribution. Combining this approach with considerations of causal structure allows us to handle several (counter-)examples that haunt other accounts of learning conditional evidence.

Variation 5 discusses one of the major challenges to Bayesian confirmation theory: the **Problem of Old Evidence**. How do Bayesians describe the confirmatory power of the discovery that a theory T implies evidence E when E has been known for a long time? According to the standard Bayesian model of confirmation, evidence E confirms theory T if and only if learning E raises the probability of T. But this is impossible if the evidence is already known ($p(\mathrm{E}) = 1$). We object to some classical solution attempts and propose two novel Bayesian models that demonstrate how explaining old evidence raises the rational degree of belief in theory T.

The second set of Variations abandons the topic of confirmation in favor of explicating central concepts in scientific reasoning. Variation 6 develops a Bayesian analysis of **causal strength**, building on the scientific literature on causal Bayesian networks and combining it with methods from Bayesian confirmation theory. First, we defend the choice of a framework where different measures are embedded and compared. Second, we derive representation theorems for various measures of causal strength, that is, theorems that characterize a measure of causal strength as the only measure (up to ordinal equivalence) that satisfies a certain set of adequacy conditions. Third, we make an argument for a particular measure of causal strength: the difference between $p(\mathrm{E}|\mathrm{C})$ and $p(\mathrm{E}|\neg\mathrm{C})$ (for cause C and effect E). Finally, we apply that measure to a case from epidemiology, demonstrating how closely scientific and philosophical reasoning about causal strength are intertwined.

Variation 7 is devoted to the topic of **explanatory power**. Hempel (1965a) famously postulated a structural identity between prediction and explanation: explanations show why a particular phenomenon occurred by deriving it from the theory, and explanatory power is proportional to the ability of the explanans to predict the explanandum (see also Hempel and Oppenheim 1948). We explore to what extent this classical view, fallen out of fashion in modern philosophy of science, can be rescued in a Bayesian framework, where explications of explanatory power are based on statistical relevance. What is more, we show that this approach is consistent with the view that explanations are essentially causal. Then we compare several measures of explanatory power and their respective strengths and weaknesses.

Variation 8 provides a Bayesian account of **intertheoretic reduction** between theories that operate at different levels of description. For example, in a suitable mathematical description, the equations of thermodynamics (here, the phenomenological theory) reduce to the equations of statistical mechanics (the fundamental theory). Such reductive relationships between theories at the phenomenal and the fundamental level are described by the models of Nagel (1961) and Schaffner (1967). We defend these models against popular objections and show how the establishment of reductive relationships can raise the epistemic standing of the theories involved. That is, we do not only show how reduction unifies different theories, but we also demonstrate that it has a positive effect on the subjective probability of both the fundamental and the phenomenological theory.

In Variation 9, Bayesian reasoning meets null hypothesis significance testing and Popper's Critical Rationalism. In significance tests, a null (i.e., default) hypothesis is usually tested against an unspecified alternative, and a "statistically significant" result is supposed to provide evidence against "the null" (as the null hypothesis is briefly called). Often, however, no statistically significant evidence is found. Does this mean that the null hypothesis is confirmed by the results? Neither statistics textbooks nor philosophers of science answer this question; they basically abstain from judgment. This Variation therefore engages in defining the conditions when such results confirm—or as Popper said: **corroborate**—the null hypothesis. We first show why a statistical-relevance framework cannot provide a satisfactory explication of corroboration in hypothesis tests. Then we modify the mathematical framework and derive a measure of corroboration from a parsimonious set of adequacy constraints.

Variation 10 is devoted to the value of **simplicity in statistical model selection.** First, we elucidate the general rationale behind preferring simpler to more complex models and we criticize Forster and Sober's (1994) thesis that the Akaike Information Criterion (AIC) provides an optimal trade-off rate between simplicity and goodness-of-fit. Then, we analyze the rationales behind various Bayesian model selection criteria. We observe that they are often constructed on the basis of the Bayesian formalism, but without taking seriously the philosophical commitments of Bayesian inference. Recent developments in Bayesian statistics are therefore not a solid base for general philosophical theses on the value of simplicity in scientific inference.

The last two Variations reply to pressing objections against the use of subjective Bayesian inference in science. Variation 11 investigates whether subjective Bayesian inference can ever achieve a sufficient degree of **objectivity** to counter the charge of arbitrariness and maintain the epistemic authority of science. The irreducibly subjective nature of prior probabilities and their inevitable impact on (supposedly objective) measures of evidence are often cited as a reason against. However, such arguments often presuppose an unrealistically strong conception of scientific objectivity. Our strategy for countering this criticism is twofold: we combine an up-to-date conceptual analysis of scientific objectivity with arguments that Bayesian inference is, on these accounts, no less objective than its competitors (e.g., frequentist inference).

The final Variation 12 responds to the question of how we can meaningfully reason with **Bayesian models in science**. Clearly, the subjective interpretation of the probability of a hypothesis H—an agent's degree of belief that H is true—does not apply in those cases where we reason with highly idealized Bayesian models. No rational agent would wager any amount of money on the proposition that one of those models is literally true. Yet how can we make sense of a Bayesian approach in those cases? We propose to interpret such probabilities counterfactually: as the degree of belief in a specific model on the supposition that the true model is included in the set of candidate models. As we show, this suppositional interpretation explains why Bayesian inference in science is rational and practically useful. It also explains why conditional degrees of belief often track particular statistical distributions, and more generally, how chances and credences coordinate in Bayesian inference.

The book concludes with a short recapitulation of the original theme: we count the successes and failures of Bayesian philosophy of science, make up the balance and sketch future research projects.

Variation 1: Confirmation and Induction

Confirmation of hypotheses and theories by empirical evidence is a central element of scientific reasoning. Eddington's observations of the 1919 solar eclipse confirmed Einstein's General Theory of Relativity (GTR) and strongly contributed to the endorsement of GTR among theoretical physicists. Equally spectacularly, a huge set of observations by CERN researchers confirmed the existence of the Higgs Boson, a fundamental particle hypothesized in the 1960s. In economics, Maurice Allais and Daniel Ellsberg conducted experiments about decision-making under uncertainty that undermined the empirical basis of Rational Choice Theory. But what are the exact conditions when a piece of evidence (dis-)confirms a theory?

Philosophers of science respond to this question by proposing formal criteria for a confirmatory relationship. Most often, these are cast in logical or probabilistic terms. Apart from guiding our actual scientific reasoning, such criteria facilitate the analysis and reconstruction of canonical cases of confirmation in the history of science, and they allow for a critical evaluation of experiments and observational studies in modern science. They also connect well to statistical hypothesis testing, as we will see in Variations 9 and 11. What is more, models of confirmation try to provide rules for valid inductive reasoning from empirical evidence to general hypotheses and complex theories. Not in the sense of providing a foundational, non-circular justification of inductive reasoning—this would run into Hume's problem of induction (Hume 1739, 1748)—but in the more modest sense of setting up rules that agree with successful scientific practice and that are appealing from a theoretical point of view. A good philosophical account of confirmation should successfully reconstruct classical cases of scientific reasoning, like the ones mentioned in the previous paragraph, and at the same time explain why well-confirmed theories are more likely to be true or empirically adequate.

Among the available philosophical models of confirmation, Bayesian Confirmation Theory—the analysis of confirmation in terms of (increase in) degree of belief—is the most popular one. It is the most venerable and well-worked-out branch of Bayesian philosophy of science, and it provides an excellent case for demonstrating how Bayesian models enhance scientific reasoning. In this chapter, we show how Bayesian models explicate the notion of **degree of confirmation**, and how they reconstruct crucial aspects of inductive reasoning. In doing so, we amalgamate some of our own thoughts (e.g., Hájek and Hartmann 2010; Hartmann and Sprenger 2010; Sprenger 2016c) with standard summaries and recent results by others (e.g., Earman 1992; Fitelson 2001b; Maher 2004; Crupi 2015; Crupi, Chater and Tentori 2013). We first motivate the use of the probability calculus for modeling confirmation (Section 1.1). Then we introduce different notions of confirmation. We start with the idea of confirmation as firmness of degree of belief (Section 1.2) and contrast it with the idea of confirmation as increase in firmness (Section 1.3). On the way, we show how this distinction can alleviate longstanding puzzles of inductive inference, such as the paradox of tacking by conjunction, the grue paradox and the paradox of the ravens. We examine the question of whether there is a single best confirmation measure and conclude that purely theoretical and conceptual adequacy conditions are not sufficient to single out a unique measure (Section 1.4). A discussion of future research projects concludes the chapter (Section 1.5).

1.1 Motivating Bayesian Confirmation Theory

Probability is an extremely natural model for explicating degree of confirmation and inductive reasoning. This has a number of reasons.

First, probability is, as quipped by Cicero, a guide to life. Our decisions and actions are often based on which hypotheses are more probable than others: for example, if there is a high chance of rain, we might cancel a planned beach trip. Confirmation is a guide to probability: better-confirmed hypotheses are, *ceteris paribus*, more probable than others. It is therefore natural to integrate confirmation and probability within a single mathematical formalism.

Second, probability is the preferred tool for expressing uncertainty in science. Probability distributions are used for describing measurement error and for characterizing the "noise" in a system—the part of the data which cannot be explained by reference to natural laws. Suppose you are conducting a simple linear regression analysis with an explanatory variable X and a dependent variable Y, such as $Y_i = \alpha X_i + \varepsilon_i$. In this equation, the ε_i-term models the part of the data which cannot be explained by the assumed linear relationship between X and Y. It represents the "noise" in the data (e.g., due to measurement error or omitted explanatory variables) and is usually modelled by means of a certain probability distribution (e.g., the Gaussian distribution). Properties of that distribution can then be used at various steps of the inference. By phrasing confirmation in terms of probability, we connect a philosophical analysis of inductive inference to familiar scientific models where probabilified uncertainty already plays a dominant role.

Third, statistics, the science of analyzing and interpreting data and assessing theories on the basis of data, is formulated in terms of probability theory. Statisticians have proved powerful mathematical results on the foundations of inductive inference, such as de Finetti's famous representation theorem for subjective probability (de Finetti 1974) or the convergence results by Blackwell and Dubins (1962) and Gaifman and Snir (1982). Probabilistic accounts of confirmation can directly make use of these results, leading to a beneficial interaction between philosophical and statistical work (e.g., Howson and Urbach 2006; Good 2009). For example, advantages and drawbacks of null hypothesis significance tests (NHST) can be evaluated from the standpoint of a probabilistic theory of confirmation (Royall 1997; Sprenger 2016a,b).

These considerations explain why philosophy of science has paid so much attention to confirmation theories framed in terms of probability. Among them, Bayesian Confirmation Theory is the most prominent representative. We shall now describe it in detail.

1.2 Confirmation as Firmness

We remember from the introductory chapter that Bayesians represent subjective degrees of belief by means of a probability function. The basic idea of Bayesian Confirmation Theory is that confirmation judgments are functions of an agent's conditional and unconditional degrees of belief. What is more, Bayesians describe the degree of confirmation that a piece of evidence E confers on a hypothesis H by a precise number, $c(\mathrm{H}, \mathrm{E})$. At first sight, this may appear unpalatably subjective. Two things should be noted, though. First, agents are assumed to be rational: their degrees of belief conform to the axioms of probability, take into account relevant evidence, etc. Second, that two agents entertain different posterior degrees of belief in a hypothesis does not necessarily imply that their judgments of confirmation and evidential support differ. Part of this section is devoted to making the latter intuition explicit.

We model a hypothesis H as an instantiation of a (propositional, numerical, ...) variable H that concerns a quantity of interest, for instance, the value of an unknown parameter. As stated in the introductory chapter, we write variables in italics and their instantiations in regular roman script. Ω_H denotes the set of the values that H can take, and $\mathcal{H} \subset \mathcal{P}(\Omega_H)$ stands for a σ-algebra of subsets of Ω_H so that we obtain a measurable space $(\Omega_H, \mathcal{H})$. For instance, if H is a real-valued variable ($\Omega_H = \mathbb{R}$) with the "natural" Borel σ-algebra $\mathcal{H} = \mathcal{B}(\mathbb{R})$, then elements of $\mathcal{H}$ are particular values of H such as $\{2\}$ or $\{3\}$, but also (combinations of) intervals such as $[2;3)$ or $(-1;1] \cup [10;\infty)$. The propositions over which we have degrees of belief are elements of $\mathcal{H}$. Exactly the same construction applies to evidence E with the measurable space $(\Omega_E, \mathcal{E})$.[6]

Let p denote the joint probability distribution of the variables H and E, that is, a probability function on the σ-algebra $\mathcal{H} \times \mathcal{E}$.[7] In line with the Bayesian rationale, we explicate degree of confirmation as a function of this joint probability distribution, that is, of our degrees of belief in truth-functional combinations of various values of E and H. This may be seen as a defining and very basic adequacy criterion on Bayesian confirmation measures: the exclusive dependence on degrees of belief. We omit reference to background knowledge or assumptions and just assume that they are given by the context, or shared among all rational agents.

Now we become more specific: we define degree of confirmation as a function of how observing E affects the probability of H, that is, our degree of belief that H is true. This leads us to a general adequacy constraint on confirmation measures:

Prior–Posterior Dependence There is a real-valued, continuous function $f\colon [0;1]^2 \to \mathbb{R}$ such that for any hypothesis $\mathrm{H} \in \mathcal{H}$ and piece of evidence

[6]Alternatively, we can conceptualize H and E as proper **random variables,** that is, as functions from an abstract probability space $(\Omega, \mathcal{A}, p)$ to the measurable spaces $(\Omega_H, \mathcal{H})$ and $(\Omega_E, \mathcal{E})$. This approach is closer to what mathematicians do in stochastic analysis and statistics. For the purpose of our book, though, both formulations are fine and we give the less complex and more intuitive one in the main text.

[7]In subsequent chapters, we will often embed H and E into a causal Bayesian network (G, p) where the variables in G (i.e., also H and E) satisfy the Causal Markov Condition with respect to probability distribution p. See the introductory chapter for more details. The graphical Bayesian network representation facilitates thinking about irrelevant evidence and other concepts that we discuss in this chapter.

$\mathrm{E} \in \mathcal{E}$ with probability distribution $p\colon \mathcal{H} \times \mathcal{E} \to [0;1]$, the degree $c(\mathrm{H},\mathrm{E})$ of confirmation that E confers on H can be represented as

$$c(\mathrm{H},\mathrm{E}) = f\big(p(\mathrm{H}\,|\,\mathrm{E}),\, p(\mathrm{H})\big),$$

where f is non-decreasing in the first argument and non-increasing in the second argument.

In fact, most confirmation measures that have been constructed in the Bayesian framework satisfy Prior–Posterior Dependence (see also Section 1.4).

Prior–Posterior Dependence familiarizes us with an explicative method which is paramount in this Variation and will return in later ones: to specify **adequacy conditions** on the explicandum, that is, the concept that we want to explicate (here: degree of confirmation). Such adequacy conditions express an intuitive property that a formal explication should satisfy. The idea behind this approach is that various adequacy conditions taken together may suffice to single out a precise explicatum (here: a Bayesian confirmation measure). **Representation theorems** characterize the set of measures (or possibly the unique measure) that satisfy such a set of constraints. This approach allows for a sharp demarcation and mathematically rigorous characterization of the explicandum, and at the same time for a philosophically informed discussion of the explicatum, by means of defending and criticizing the properties which are encapsulated in the adequacy conditions. Moreover, if a plausible set of adequacy conditions supports more than one measure, we can compare the relative plausibility of these conditions and use this comparison to evaluate the measures which they characterize.

A straightforward consequence of Prior–Posterior Dependence is the following principle (see, e. g., Maher 2004; Crupi 2015):

Final Probability Incrementality For any hypothesis $\mathrm{H} \in \mathcal{H}$ and possible observations $\mathrm{E}, \mathrm{E}' \in \mathcal{E}$ with probability distribution $p\colon \mathcal{H} \times \mathcal{E} \to [0;1]$,

$$c(\mathrm{H},\mathrm{E}) > c(\mathrm{H},\mathrm{E}') \quad \text{if and only if} \quad p(\mathrm{H}\,|\,\mathrm{E}) > p(\mathrm{H}\,|\,\mathrm{E}').$$

According to this principle, E confirms H more than E′ does if and only if it raises the probability of H to a higher level.

It is easy to show that Final Probability Incrementality implies that $c(\mathrm{H},\mathrm{E}) = c(\mathrm{H},\mathrm{E}')$ if and only if $p(\mathrm{H}\,|\,\mathrm{E}) = p(\mathrm{H}\,|\,\mathrm{E}')$. Given the basic intuition that degree of confirmation should covary with boost in degree of belief, satisfactory Bayesian explications of degree of confirmation should arguably satisfy this condition.

We now proceed to the next adequacy condition for confirmation measures. One of their central functions is to bridge the gap between qualitative and quantitative reasoning about confirmation, that is, between the judgment "E (dis-)confirms H" and the specific number that a confirmation measure assigns to the pair (H, E). Scientists and laypersons often reason in these qualitative categories; we need a way to connect these judgments to the precise numbers which confirmation measures yield.

Qualitative–Quantitative Bridge Principle There is a real number $t \in \mathbb{R}$ such that for any hypothesis $H \in \mathcal{H}$ and any piece of evidence $E \in \mathcal{E}$ with probability distribution $p: \mathcal{H} \times \mathcal{E} \to [0;1]$:

- E confirms/supports H if and only if $c(H, E) > t$;
- E undermines/disconfirms H if and only if $c(H, E) < t$;
- E is confirmationally neutral/irrelevant to H if and only if $c(H, E) = t$.

In other words, a measure of degree of confirmation should guide our qualitative confirmation in the sense that there is a numerical threshold for telling positive confirmation from disconfirmation (Carnap 1950, 463). From now on, as a matter of convenience, we often drop quantification over H and E and the associated probability function $p: \mathcal{H} \times \mathcal{E} \to [0;1]$, following Crupi 2015.

There are two main roads for adding more conditions, which will ultimately lead us to two different explications of confirmation (Carnap 1950, preface to the 1962 edition): confirmation as firmness of belief and confirmation as *increase* in firmness of belief, which modern writers also call "evidential support" (e.g., Fitelson 2001b).

Rank	Team	Points	Team	Points
	after 36 out of 38 rounds		after 37 out of 38 rounds	
1	Roma	78	Inter	79
2	Inter	76	Roma	78
3	Juventus	74	Juventus	74

Table 1.1: A motivating example for the principle of Local Equivalence. Top of the *Serie A* after 36 and 37 out of 38 rounds, respectively.

We begin with **confirmation as firmness,** or "absolute" confirmation. Consider the football standings from Table 1.1. Three teams in the Italian *Serie A*, AS Roma, FC Internazionale ("Inter") and Juventus Turin, are competing for the *scudetto*, the national soccer championship. A win gives three points, a draw one point, a loss none. The penultimate match day is derby day. Inter beats Milan in the *Derby della Madonnina*, Juventus loses to FC Torino in the *Derby della Mole* and Roma loses to Lazio in the *Derby della Capitale*. Call this conjunction of propositions E. Let H = "Inter will win the championship" and H′ = "Roma will be the runner-up". Given evidence E, the hypotheses H and H′ are logically equivalent: we can derive them from each other using E as an additional premise. (Juventus can't surpass Inter or Roma anymore.) When H and H′ are indistinguishable in this sense, E seems to confirm them to an equal degree. This intuition is expressed in the following adequacy condition:

Local Equivalence If H and H′ are logically equivalent given E (that is, $E \wedge H \vDash H'$ and $E \wedge H' \vDash H$), then $c(H, E) = c(H', E)$.

Local Equivalence allows for a neat representation theorem: all confirmation measures that satisfy Prior–Posterior Dependence and Local Equivalence agree in their confirmation rankings (i. e., which hypotheses are more and which less confirmed) with the posterior probability $p(H \mid E)$:

Theorem 1.1 (Confirmation as Firmness). *All confirmation measures* $c(H, E)$ *that satisfy Prior–Posterior Dependence and Local Equivalence are ordinally equivalent to the posterior probability of* H: $c^*(H, E) = p(H \mid E)$.

A key concept in the formulation of the theorem is **ordinal equivalence.** Two confirmation measures c and c' are ordinally equivalent if and only if for all propositions H_1, H_2, E_1, E_2,

$$c(H_1, E_1) > c(H_2, E_2) \quad \text{if and only if} \quad c'(H_1, E_1) > c'(H_2, E_2).$$

In particular, ordinal equivalence entails that the two measures can be represented as monotonically increasing functions of each other: $c'(H, E) = f(c(H, E))$. So two ordinally equivalent measures may use different scales, but they produce the same confirmation rankings and share most philosophically interesting properties.

To the extent that Local Equivalence is plausible, Theorem 1.1 shows that confirmation as firmness boils down to high probability. A very similar result based on Local Equivalence, where Prior–Posterior Dependence is replaced by two other conditions, has been proved by Michael Schippers (2017). Since confirmation as firmness is increasing in $p(H \mid E)$, it follows from the Qualitative–Quantitative Bridge Principle that E confirms H (in the qualitative sense) if and only if $p(\mathrm{H}|\mathrm{E}) \geq t$ for some $t \in [0;1]$. In other words, we call a hypothesis H confirmed when $p(\mathrm{H}|\mathrm{E})$ exceeds a certain, possibly context-relative, threshold. This corresponds to Carnap's concept of probability$_1$ or "degree of confirmation" in his system of inductive logic (Carnap 1950).

The Bayesian explication of confirmation as firmness dispels some problems that have plagued qualitative accounts of confirmation. One of the most popular ones is **hypothetico-deductive (H-D) confirmation:** hypotheses are confirmed if they make a deductive prediction and this prediction comes true. In other words, an observed piece of evidence E H-D-confirms H if and only if we can deduce E from H (possibly relative to a body of auxiliary assumptions which connect theory and observation). The idea is actually quite old and already made explicit in writings by William Whewell:

> our hypotheses ought to *foretel* [sic!] phenomena which have not yet been observed ... the truth and accuracy of these predictions were a proof that the hypothesis was valuable and, at least to a great extent, true. (Whewell 1847, 62–63)

The prominent place that deductive reasoning from theory to observation takes in this approach aligns well with Popper's falsificationist view of scientific method as consisting of bold conjectures and subsequent refutations (e.g., Popper 1959/2002). Popper's view has, in turn, been hugely influential in the scientific community, which still has a weakness for H-D reasoning (Gelman and Shalizi 2013).

However, H-D confirmation runs into the **paradox of tacking by conjunction:** If E confirms H (because $\mathrm{H} \models \mathrm{E}$), then E confirms also $\mathrm{H} \wedge \mathrm{X}$, for an arbitrary hypothesis X, even if X stems from a completely different domain of science and is irrelevant for E. This is clearly too permissive, since confirmation is allowed to spread in an uncontrolled way. Phenomena from the behavioral sciences could suddenly acquire relevance for particle physics, and vice versa. The tacking by conjunction paradox—and its close cousin where irrelevant *disjunctions* are tacked to the evidence E—is therefore regarded as a major blow for the hypothetico-deductive approach to confirmation, notwithstanding recent solution attempts (Schurz 1991; Gemes 1993, 1998; Sprenger 2011, 2013a).

The Bayesian account of confirmation as firmness dissolves the tacking paradox. For any irrelevant X, it will be the case that $p(\mathrm{H} \wedge \mathrm{X} \mid \mathrm{E}) \leq p(\mathrm{H}|\mathrm{E})$. Theorem 1.1 then tells us that there exists an increasing function g that maps the conditional probability of a hypothesis to its degree of confirmation.

Hence, we can infer

$$c(\mathrm{H} \wedge \mathrm{X}, \mathrm{E}) = g(p(\mathrm{H} \wedge \mathrm{X} \mid \mathrm{E})) \leq g(p(\mathrm{H} | \mathrm{E})) = c(\mathrm{H}, \mathrm{E}),$$

demonstrating that the conjunction is *confirmed to a lower degree* than the original hypothesis H, especially so for an unlikely, far-fetched proposition X. Confirmation as firmness gives the intuitively correct response to the tacking paradox. It does not deny that $\mathrm{H} \wedge \mathrm{X}$ is confirmed as well—after all, E is still relevant for H—but the paradox is mitigated by decreasing the amount of confirmation.

On the other hand, confirmation as firmness does not always agree with the use of that concept in scientific reasoning. To be sure, relative to the totality of observed evidence, we would call a theory well-confirmed if and only if it is sufficiently probable, conditional on the evidence. This is captured in Carnap's

Requirement of Total Evidence Inductive inference and the assessment of degree of confirmation at time *t* should be based on the totality of evidence available at *t*. (Carnap 1947, 1950, our paraphrase)

But often, scientists are interested in whether a particular experiment or observation supports a hypothesis—independent of the overall epistemic standing of the hypothesis. Even if the hypothesis is, all things considered, very unlikely, we may want to ask whether the outcome of an experiment provides a good reason to believe it. For instance, when Einstein invented the General Theory of Relativity (GTR), most scientists adopted a quite skeptical attitude because of its counterintuitive nature. However, when the predictive and explanatory successes of GTR became evident, such as explaining the Mercury perihelion shift and the bending of starlight by the sun, their attitudes shifted substantially (Earman 1992). The account of confirmation as firmness fails to capture this intuition and to give an adequate reconstruction of such cases from the history of science. It also fails to align with modern scientific practice, where the confirmatory strength of the evidence is evaluated on the basis of whether the results are statistically significant in a particular experiment.

All this suggests that the Requirement of Total Evidence should be relaxed. It is not suited for describing cases where we want to judge the evidential relevance of a specific body of data for a hypothesis. See also Variation 5, where we discuss the Problem of Old Evidence. But according to confirmation as firmness, evidence E could confirm hypothesis H even if it *lowers* the probability of H, as long as $p(\mathrm{H}|\mathrm{E})$ is still large enough. Few people would call such an experiment a confirmation of H.

In a now classical debate in philosophy of science, Karl R. Popper (1954, 1959/2002) raised these points against Carnap: degree of confirmation cannot be (posterior) probability. As a reaction, Carnap distinguished two concepts of confirmation in the second edition (1962) of *Logical Foundations of Probability*: confirmation as firmness and **confirmation as increase in firmness.** The second concept shall be discussed in the rest of this Variation.

1.3 Confirmation as Increase in Firmness and the Paradoxes of Confirmation

The preceding paragraphs have motivated the search for Bayesian explications of confirmation that go beyond probability conditional on the evidence. A natural candidate is probability-raising, which corresponds to confirmation as increase in firmness. Qualitatively, we can describe this concept as follows:

Confirmation as Increase in Firmness For two propositions H and E,

1. evidence E **confirms/supports** hypothesis H if and only if E raises the probability of H: $p(\mathrm{H}|\mathrm{E}) > p(\mathrm{H})$;
2. evidence E **disconfirms/undermines**hypothesis H if and only if E lowers the probability of H: $p(\mathrm{H}|\mathrm{E}) < p(\mathrm{H})$;
3. evidence E is **neutral** with respect to H if and only if E leaves the probability of H unchanged: $p(\mathrm{H}|\mathrm{E}) = p(\mathrm{H})$.

In other words, E confirms H if and only if E raises our degree of belief in H. Such explications of confirmation are also called **statistical-relevance accounts of confirmation** because the neutral point is determined by the statistical independence of H and E. They measure the **evidential support** that H receives from E. The increase-in-firmness explication of confirmation receives empirical support from findings by Tentori et al. (2007): ordinary people use the concept of confirmation in a way which can be dissociated from posterior probability and is strongly correlated with measures of evidential support. In the remaining Variations, we will standardly use increase in firmness, or evidential support, when modeling the confirmation of scientific hypotheses and theories.

Confirmation as increase in firmness has interesting relations to qualitative accounts of confirmation and longstanding paradoxes of confirmation.

For instance, hypothetico-deductive confirmation emerges as a special case: if H entails E and $p(\mathrm{E}) < 1$, then $p(\mathrm{E}\,|\,\mathrm{H}) = 1$, and by Bayes' Theorem, $p(\mathrm{H}\,|\,\mathrm{E}) > p(\mathrm{H})$. We will also show that confirmation as increase in firmness can address the tacking paradox. But first, we will demonstrate how confirmation as increase in firmness handles the longstanding **paradox of the ravens.**

Let $\mathrm{H} = \forall x\colon (Rx \rightarrow Bx)$ stand for the hypothesis that all ravens are black, formulated in first-order predicate logic (see, e.g., Williamson 2017, for an extension of Bayesian reasoning from propositional to predicate logic). H is logically equivalent to the hypothesis $\mathrm{H}' = \forall x\colon (\neg Bx \rightarrow \neg Rx)$ that no non-black object is a raven. It is highly intuitive that logically equivalent hypotheses are confirmed or disconfirmed to the same degree; nothing in a formal theory of confirmation should depend on the particular formulation of the hypothesis. Hence, anything that confirms H also confirms H′ and vice versa (Nicod 1925/61). It is also intuitive that universal conditionals such as "all ravens are black" are confirmed by their instances, that is, black ravens. However, as Hempel (1945a,b) observed, the conjunction of both principles leads to paradoxical results: A black raven is an instance of H and confirms the raven hypothesis. A white shoe is an instance of H′ and confirms the hypothesis that non-black objects are not ravens. But because of the aforementioned equivalence condition, the white shoe also confirms the hypothesis that all ravens are black! This result is known as the paradox of the ravens, or alternatively, Hempel's paradox. Formally, there is a tension between the following three statements, which cannot be jointly true (see also Maher 1999; Fitelson 2006; Sprenger 2010a):

Nicod's Condition (Confirmation by Instances) Universal conditionals of the form "$\forall x\colon (Fx \rightarrow Gx)$" are confirmed by their instances, that is, propositions such as $Fa \wedge Ga$.

Equivalence Condition Logically equivalent hypotheses are confirmed equally by given evidence.

Ravens Intuition Observation reports pertaining to a white shoe or other non-black non-ravens do *not* confirm the hypothesis that all ravens are black.

The Bayesian account of confirmation as firmness allows us to spot what is wrong with confirmation by instances, and thereby resolves the paradox. While that intuition is certainly valid for *some* background assumptions, not all instances of universal conditionals raise their probability. I. J. Good

	World 1 (W_1)	World 2 (W_2)
Black ravens	100	1,000
Non-black ravens	0	1
Other objects	1,000,000	1,000,000

Table 1.2: I. J. Good's (1967) counterexample to the Nicod Condition: a universal conditional is *disconfirmed* by one of its instances.

(1967) constructed a simple counterexample: Assume that there are only two possible worlds, W_1 and W_2, whose properties are described by Table 1.2.

In this scenario, the raven hypothesis $H = \forall x \colon (Rx \rightarrow Bx)$ is true whenever W_1 is the case, and false whenever W_2 is the case. Moreover, the observation of a black raven is evidence that W_2 is the case and therefore evidence that not all ravens are black:

$$p(\mathrm{Ra} \wedge \mathrm{Ba} \mid W_1) = \frac{100}{1{,}000{,}100} < \frac{1{,}000}{1{,}001{,}001} = p(\mathrm{Ra} \wedge \mathrm{Ba} \mid W_2).$$

By an application of Bayes' Theorem, we infer $p(W_1 \mid \mathrm{Ra} \wedge \mathrm{Ba}) < p(W_1)$ and $p(H \mid \mathrm{Ra} \wedge \mathrm{Ba}) < p(H)$. There are cases where the observation of a black raven *disconfirms* the raven hypothesis. The raven paradox may thus be resolved by rejecting one of its assumptions, namely Nicod's Condition that universal conditionals are always confirmed by their instances. The explication of confirmation as increase in firmness has helped us to correct our pre-theoretic intuitions regarding the theory–evidence relation.

While Nicod's Condition is not generally valid on the Bayesian account, there are certainly some contexts where a Bayesian analysis should classify the observation of concrete instances (e. g., black ravens) as confirming evidence. Then, the paradox persists. Hempel's own response was to bite the bullet and to reject the Ravens Intuition: white shoes, too, provide support to the raven hypothesis because they rule out potential counterexamples. In other words, we may end up rejecting *two* of the three assumptions: Nicod's Condition and the Ravens Intuition.

The raven paradox is twofold, however. Apart from the qualitative question of whether or not the observation of a white shoe confirms the raven hypothesis, there is also the **comparative version of the paradox:** shouldn't the observation of a black raven confirm the raven hypothesis to a higher degree than the observation of a white shoe? Fitelson and Hawthorne (2011) conduct a Bayesian analysis and show in their Theorem 2 (op. cit.) that this is indeed the case if one makes plausible assumptions on the base rate of ravens and black objects in the real world. If their assumptions are satisfied, $\text{Ra} \wedge \text{Ba}$ raises the probability of the raven hypothesis H to a higher level than $\neg\text{Ba} \wedge \neg\text{Ra}$ does. By Final Probability Incrementality, $\text{Ra} \wedge \text{Ba}$ then confirms H to a higher degree than $\neg\text{Ba} \wedge \neg\text{Ra}$ does. This shows, ultimately, why we consider a black raven to be more important evidence for the raven hypothesis than a white shoe.

The Bayesian analysis also addresses another notorious paradox, Nelson Goodman's (1955) **new riddle of induction.** There is considerable discussion about what this paradox shows and how it should be understood (e.g., Jackson 1975; Okasha 2007; Fitelson 2008b). We will adopt a traditional confirmation-theoretic reading, where Goodman's riddle attacks a plausible scheme for inductive inference, deriving the paradoxical conclusion that one and the same evidence confirms two hypotheses with incompatible predictions.

Consider the following innocuous case:

$$\begin{array}{l} \text{Observation at } t=t_1\text{: emerald } e_1 \text{ is green.} \\ \text{Observation at } t=t_2\text{: emerald } e_2 \text{ is green.} \\ \qquad\qquad\qquad\qquad \vdots \\ \text{Observation at } t=t_n\text{: emerald } e_n \text{ is green.} \\ \hline \text{General hypothesis: All emeralds are green.} \end{array}$$

This seems to be a perfect example of a valid inference by enumerative induction (i.e., the "straight rule" of induction). We have no reason to assume that we are in a situation where instances do not confirm a hypothesis, such as Good's raven example. So the observation of the emeralds seems to confirm the general hypothesis that all emeralds are green.

Now define the predicate "grue", which applies to an object (1) if it is green and has been observed up to time $t_{\text{now}} = t_n$, or (2) it is blue and is observed after t_{now}. This is just a description of the extension of the predicate "grue"; no object is supposed to change its color. The following inductive inference satisfies the same logical scheme as the previous one:

$$\begin{array}{l} \text{Observation at } t=t_1\text{: emerald } e_1 \text{ is grue.} \\ \text{Observation at } t=t_2\text{: emerald } e_2 \text{ is grue.} \\ \qquad\qquad\qquad\qquad \vdots \\ \text{Observation at } t=t_n\text{: emerald } e_n \text{ is grue.} \\ \hline \text{General hypothesis: All emeralds are grue.} \end{array}$$

In spite of the gerrymandered nature of the "grue" predicate, the inference is sound: it satisfies the scheme of enumerative induction, and the premises are undoubtedly true. But then, it is paradoxical that two valid inductive inferences support flatly opposite conclusions.

Goodman describes the paradox as follows:

> Thus according to our definition, the prediction that all emeralds subsequently examined will be green and the prediction that all will be grue are alike confirmed by evidence statements describing the same observations. But if an emerald subsequently examined is grue, it is blue and hence not green. Thus although we are well aware which of the two incompatible predictions is genuinely confirmed, they are equally well confirmed according to our present definition. (Goodman 1955)

How do we escape from this dilemma? Again, there is a qualitative question (is the "grue" hypothesis confirmed at all?) and a comparative question (are both hypotheses confirmed equally?; Fitelson 2008b). Goodman gives affirmative answers to both questions.

One may propose that in virtue of its gerrymandered nature, the predicate "grue" should not enter inductive inferences. Goodman notes, however, that it is perfectly possible to redefine the standard predicates "green" and "blue" in terms of "grue" and its conjugate predicate "bleen" (i.e., blue if observed prior to t_{now}, else green). Hence, any preference for the "natural" predicates and the "natural" inductive inference seems to be arbitrary, or at least conditional on the choice of a specific language.

Goodman's own solution proposal consists in restricting the confirmation relation to generalizable, "projectible" predicates, which have a successful prediction history. This distinction drives a wedge between "green" and "grue". But as Bayesians, we do not have to follow Goodman's radical proposal. From the point of view of confirmation as firmness, which is a function of posterior probability, one can just deny that the "green" and the "grue" hypothesis are equally plausible. The "grue" hypothesis may fail to meet the probability threshold for confirmation, while the "green" hypothesis does. And from the point of view of confirmation as *increase* in firmness, we can bite the bullet: both hypotheses (the "green" and the "grue" hypothesis) may count as confirmed. We need to abandon the idea that evidence cannot confirm incompatible hypotheses. For example, Einstein's work on the photoelectric effect raised our degree of belief in the hypothesis that electromagnetic radiation can be divided into a finite number of quanta, and thereby also our degree of belief in different versions of quantum theory—e.g., those that were compatible with relativity theory and those that weren't. For the Bayesian, Goodman's paradox is not as threatening as for the confirmation theorists in the 1940s and 1950s, who heavily drew on the concept of instantial relevance (e. g., Hempel 1945a,b).

Moreover, the Bayesian need not agree with the comparative aspect of Goodman's new riddle of induction, namely that both hypotheses, and both predictions for the color of the next observed emerald, are supported *equally* by the observed evidence. This is in general only true if both hypotheses are equally probable beforehand. However, that is rarely the case, and the "green" hypothesis is much more plausible than the "grue" hypothesis. By means of their influence on the posterior distribution, prior probabilities keep influencing our predictions when the observations cannot distinguish between two hypotheses. Because of the different priors, the degree of confirmation (as firmness and as increase in firmness) can be higher for the "green" than for the "grue" hypothesis. A different solution, which rejects the Requirement of Total Evidence, is offered by Fitelson (2008b).

Note that the choice of priors cannot be based on Bayesian reasoning itself; they have to come from theoretical principles and past track record, in short: scientific judgment. Bayesian Confirmation Theory explains how to amalgamate prior degrees of belief with observed evidence, but it does not tell you which prior degrees of belief are reasonable. In this sense, Goodman shows a general problem for formal reasoning about confirmation and evidence: there is no viable complete theory of inductive support (see also Norton forthcoming).

The three showcases above—the tacking paradox, the paradox of the ravens and Goodman's new riddle of induction—make clear that Bayesian Confirmation Theory can successfully address longstanding puzzles in inductive reasoning. However, there is one question we have evaded so far, and now we shall turn to it: How can we quantify evidential support, that is, confirmation as increase in firmness?

1.4 The Plurality of Bayesian Confirmation Measures

For experimentally working scientists, quantifying the strength of the observed evidence with respect to the tested hypothesis is an essential element of their daily work. Practical and methodological aspects of reporting experimental findings put aside, such confirmation measures are also crucial for resolving properly philosophical problems, such as answering the notorious Duhem–Quine problem (Duhem 1914; Quine 1951). Suppose an experiment

Difference Measure	$d(\mathrm{H},\mathrm{E}) = p(\mathrm{H}\,\vert\,\mathrm{E}) - p(\mathrm{H})$
Log-Ratio Measure	$r(\mathrm{H},\mathrm{E}) = \log \frac{p(\mathrm{H}\vert\mathrm{E})}{p(\mathrm{H})}$
Log-Likelihood Measure	$l(\mathrm{H},\mathrm{E}) = \log \frac{p(\mathrm{E}\vert\mathrm{H})}{p(\mathrm{E}\vert\neg\mathrm{H})}$
Kemeny–Oppenheim Measure	$k(\mathrm{H},\mathrm{E}) = \frac{p(\mathrm{E}\vert\mathrm{H})-p(\mathrm{E}\vert\neg\mathrm{H})}{p(\mathrm{E}\vert\mathrm{H})+p(\mathrm{E}\vert\neg\mathrm{H})}$
Generalized Entailment Measure	$z(\mathrm{H},\mathrm{E}) = \begin{cases} \frac{p(\mathrm{H}\vert\mathrm{E})-p(\mathrm{H})}{1-p(\mathrm{H})}, & \text{if } p(\mathrm{H}\,\vert\,\mathrm{E}) \geq p(\mathrm{H}); \\ \frac{p(\mathrm{H}\vert\mathrm{E})-p(\mathrm{H})}{p(\mathrm{H})}, & \text{if } p(\mathrm{H}\,\vert\,\mathrm{E}) < p(\mathrm{H}) \end{cases}$
Christensen–Joyce Measure	$s(\mathrm{H},\mathrm{E}) = p(\mathrm{H}\,\vert\,\mathrm{E}) - p(\mathrm{H}\,\vert\,\neg\mathrm{E})$
Carnap's Relevance Measure	$c'(\mathrm{H},\mathrm{E}) = p(\mathrm{E})\big(p(\mathrm{H}\,\vert\,\mathrm{E}) - p(\mathrm{H})\big)$
Rips Measure	$r'(\mathrm{H},\mathrm{E}) = 1 - \frac{p(\neg\mathrm{H}\vert\mathrm{E})}{p(\neg\mathrm{H})}$

Table 1.3: A list of popular measures of evidential support, that is, confirmation as increase in firmness.

does not yield the predicted result. Which of the involved principal and auxiliary hypotheses should we then reject? Their degree of (dis-)confirmation can be used to evaluate their standing and to indicate which of them we should discard. For this reason, the search for a proper confirmation measure is more than a sterile technical exercise: it is of vital importance for distributing praise and blame between different hypotheses that bear on an observation. Since such assessments may vary with the used confirmation measure, characterizing their mathematical properties and comparing them on a normative basis is an important philosophical research program (e.g., Fitelson 1999, 2001a,b; Eells and Fitelson 2000, 2002; Crupi 2015).

Table 1.3 gives a survey of measures that are frequently discussed in the literature. We have normalized them such that for each measure $c(\mathrm{H},\mathrm{E})$, confirmation amounts to $c(\mathrm{H},\mathrm{E}) > 0$, neutrality to $c(\mathrm{H},\mathrm{E}) = 0$ and disconfirmation to $c(\mathrm{H},\mathrm{E}) < 0$, in alignment with the Qualitative–Quantitative Bridge Principle. This allows for a better comparison of their properties.

Evidently, these measures all have quite distinct properties. Only the log-likelihood measure and the Kemeny–Oppenheim measure are ordinally equivalent. It thus makes sense to apply the methodology that we used for confirmation as firmness, and to characterize them in terms of representation theorems where, as before, Prior–Posterior Dependence and Final Probability Incrementality will serve as minimal reasonable constraints on any measure of evidential support. Notably, these conditions already rule out two of the measures in the list, namely Carnap's relevance measure $c'(H,E) = p(E)(p(H|E) - p(H))$ and the Christensen–Joyce measure $s(H,E) = p(H|E) - p(H|\neg E)$ (Christensen 1999). Carnap's relevance measure is also problematic because it satisfies the following condition:

Symmetrical Confirmation For all propositions E and H, $c(H,E) = c(E,H)$.

In other words, E confirms H as much as H confirms E. Many intuitive confirmation judgments violate this equality. For example, knowing that a specific card in the deck is the ace of spades confirms the hypothesis that this card is a spade much stronger than the other way round. The same problem affects the (log-)ratio measure $r(H,E)$ (Eells and Fitelson 2002).

We will not provide representation results for all confirmation measures in the literature, but we present, *pars pro toto*, four adequacy conditions and corresponding representation theorems for popular explications of confirmation as increase in firmness. The first condition is the

Law of Likelihood

$$c(H,E) > c(H',E) \quad \text{if and only if} \quad p(E|H) > p(E|H').$$

This condition has a long history of discussion in philosophy and statistics. The idea is that E favors H over H′ if and only if the likelihood of H on E is greater than the likelihood of H′ on E (Hacking 1965). In other words, E is more expected under H than under H′. Law of Likelihood is also at the basis of the **likelihoodist theory of confirmation,** expressed as a condition on evidential favoring: E favors H over a competing hypothesis H′ if and only if $p(E|H)$ exceeds $p(E|H')$. Likelihoodists do not make reference to prior probabilities, replace confirmation by the contrastive relation of evidential favoring and refuse to evaluate how much E confirms H simpliciter (Edwards 1972; Royall 1997; Sober 2008).

The second condition demands that conditioning on E′ does not affect the confirmation relation between H and E′, as long as E′ is sufficiently independent from E and H:

Modularity If H screens off E from E′ (that is, $p(E \mid \pm H \wedge E') = p(E|\pm H)$), then $c(H,E) = c_{|E'}(H,E)$, where $c_{|E'}$ denotes confirmation relative to the probability distribution $p(\cdot\,|E')$ that emerges by conditionalizing on E′.

That is, if E' does not affect the likelihoods that H and $\neg$H have on E, then conditioning on E—now supposedly irrelevant evidence—does not alter the degree of confirmation (Heckerman 1988; Crupi, Chater and Tentori 2013). The intuition behind Modularity is that adding probabilistically irrelevant information should neither raise nor lower the degree of confirmation.

A third condition concerns the question of how the confirmation of hypothesis H by evidence E relates to the confirmation of the disjunction $H \vee H'$ by the same evidence. The idea is that the logical weakening of a hypothesis contributes to the confirmation of the compound if and only if the added disjunct is confirmed by the evidence.

Disjunction of Alternative Hypotheses Assume that H and H' are inconsistent with each other. Then,

$$c(H \vee H', E) > c(H, E) \quad \text{if and only if} \quad p(H' | E) > p(H').$$

Analogous conditions hold for the cases $c(H \vee H', E) = c(H, E)$ and $c(H \vee H', E) < c(H, E)$.

Finally, the fourth condition is inspired by the analogy between deductive and inductive logic: confirmation is viewed as a generalization of logical entailment to uncertain reasoning (Crupi, Tentori and González 2007; Crupi and Tentori 2013). Degree of confirmation should therefore display the symmetry that contraposition expresses for logical entailment: if E confirms H, then $\neg$H confirms $\neg$E, and to the same degree. Similarly, disconfirmation is treated as a generalization of logical inconsistency and modeled as a symmetrical relation.

Contraposition/Commutativity If E confirms H then $c(H, E) = c(\neg E, \neg H)$; and if E disconfirms H then $c(H, E) = c(E, H)$.

Combined with Prior–Posterior Dependence, each of these four principles gives rise to a representation theorem that singles out a particular measure (for very similar theorems, see Heckerman 1988; Crupi, Chater and Tentori 2013; Crupi 2015).

Theorem 1.2 (Confirmation as Increase in Firmness).

1. *All confirmation measures that satisfy Prior–Posterior Dependence and Law of Likelihood are ordinally equivalent to*

$$r(H, E) = \log \frac{p(H | E)}{p(H)}.$$

2. *All confirmation measures that satisfy Prior–Posterior Dependence and Modularity are ordinally equivalent to*

$$l(\mathrm{H},\mathrm{E}) = \log \frac{p(\mathrm{E}\,|\,\mathrm{H})}{p(\mathrm{E}\,|\,\neg\mathrm{H})} \quad \textit{and to} \quad k(\mathrm{H},\mathrm{E}) = \frac{p(\mathrm{E}\,|\,\mathrm{H}) - p(\mathrm{E}\,|\,\neg\mathrm{H})}{p(\mathrm{E}\,|\,\mathrm{H}) + p(\mathrm{E}\,|\,\neg\mathrm{H})}.$$

3. *All confirmation measures that satisfy Prior–Posterior Dependence and Disjunction of Alternative Hypotheses are ordinally equivalent to*

$$d(\mathrm{H},\mathrm{E}) = p(\mathrm{H}\,|\,\mathrm{E}) - p(\mathrm{H}).$$

4. *All confirmation measures that satisfy Prior–Posterior Dependence and Contraposition/Commutativity are ordinally equivalent to*

$$z(\mathrm{H},\mathrm{E}) = \begin{cases} \frac{p(\mathrm{H}|\mathrm{E})-p(\mathrm{H})}{1-p(\mathrm{H})}, & \textit{if } p(\mathrm{H}\,|\,\mathrm{E}) \geq p(\mathrm{H}); \\ \frac{p(\mathrm{H}|\mathrm{E})-p(\mathrm{H})}{p(\mathrm{H})}, & \textit{if } p(\mathrm{H}\,|\,\mathrm{E}) < p(\mathrm{H}). \end{cases}$$

It should also be noted that the **Bayes factor,** a popular measure of evidential support in Bayesian statistics (Kass and Raftery 1995; Goodman 1999b), falls under the scope of the theorem. For mutually exclusive hypotheses H_0 and H_1 and evidence E, the Bayes factor in favor of H_0 is defined as

$$B_{01}(\mathrm{E}) := \frac{p(\mathrm{H}_0\,|\,\mathrm{E})}{p(\mathrm{H}_1\,|\,\mathrm{E})} \cdot \frac{p(\mathrm{H}_1)}{p(\mathrm{H}_0)} = \frac{p(\mathrm{E}\,|\,\mathrm{H}_0)}{p(\mathrm{E}\,|\,\mathrm{H}_1)}. \tag{1.1}$$

It is not difficult to see that this quantity is ordinally equivalent to the log-likelihood measure l and the Kemeny–Oppenheim measure k (Kemeny and Oppenheim 1952) when H_0 and H_1 exhaust the space of hypotheses: just substitute H and ¬H for H_0 and H_1.

To underline that the differences between the various confirmation measures have substantial philosophical ramifications, let us go back to the tacking by conjunction paradox. If we analyze this problem in terms of the ratio measure r and assume that H entails E, then we obtain that for any hypothesis H′,

$$\begin{aligned} r(\mathrm{H}\wedge\mathrm{H}', \mathrm{E}) &= p(\mathrm{H}\wedge\mathrm{H}'\,|\,\mathrm{E})/p(\mathrm{H}\wedge\mathrm{H}') \\ &= p(\mathrm{E}\,|\,\mathrm{H}\wedge\mathrm{H}')/p(\mathrm{E}) = 1/p(\mathrm{E}) = p(\mathrm{E}\,|\,\mathrm{H})/p(\mathrm{E}) = r(\mathrm{H},\mathrm{E}). \end{aligned}$$

In other words, the conjunction $\mathrm{H}\wedge\mathrm{H}'$ is supported to the same degree as the original hypothesis H. Since H′ could be literally any hypothesis unrelated to the evidence, such as "the star Sirius is a giant light bulb", this judgment of evidential support is unacceptable. Moreover, deductive entailment between theory and evidence is not even required for the problem to arise: whenever the likelihoods of H and $H \wedge H'$ on E are the same (i.e., $p(E \mid H \wedge H') = p(E \mid H)$), E confirms H to the same degree as it confirms $H \wedge H'$ (Hawthorne and Fitelson 2004, revised Theorem 2).

The other measures fare better in this respect: whenever $p(E \mid H \wedge H') = p(E \mid H)$, all other measures in Theorem 1.2 reach the conclusion that $c(H \wedge H', E) < c(H, E)$ (ibid.). In this way, we can see how Bayesian Confirmation Theory improves on H-D confirmation and other qualitative accounts of confirmation: the existence of the paradox is acknowledged, but at the same time, it is demonstrated how it can be mitigated.

The previous investigations leave us with a choice between several measures of confirmation as increase in firmness, such as d, l and z. Which one should we prefer? **Confirmational monists** claim that there is a definite answer to this question: there is a single confirmation measure which outperforms its competitors, at least in ordinary circumstances. **Confirmational pluralists** object that we cannot resolve the question on purely theoretical grounds. Which measure performs best will depend on the context or the goals of inquiry (Vassend 2018). Indeed, the adequacy conditions in the representation theorems have quite divergent motivations (cf. Huber 2005b), and a straightforward comparison is unlikely to lead to conclusive results. In particular, it has been shown that no confirmation measure satisfies the following two conditions: (i) degree of confirmation is maximal if E implies H; (ii) the a priori informativity (cashed out in terms of predictive content and improbability) of a hypothesis contributes to degree of confirmation (Brössel 2013, 389–390: see also Variation 9). Both conditions are intuitively plausible: Condition (i) captures the idea that degree of confirmation generalizes logical entailment, Condition (ii) rewards hypotheses with informative predictions. But we have to choose, and our choice will depend on what we value in scientific reasoning.

The idea that there is "the one true measure of confirmation" (Milne 1996) is therefore problematic. We may abandon such a confirmational monism in favor of pluralism (Fitelson 1999, 2001b): there are different senses of degree of confirmation that correspond to different explications. For example, d is a natural explication of increase in subjective confidence, z generalizes the principles of deductive entailment to evidential support, and l and k quantify how much the evidence discriminates between H and ¬H. While $d(H, E)$ indicates how confidence in a hypothesis grew over time, $l(H, E)$ assesses the probative value of a hypothesis in a single experiment.

Ultimately, the choice between the measures may also depend on empirical findings. Crupi, Tentori and González (2007) and Tentori et al. (2007) compare different confirmation measures in an experiment where white and black balls are drawn from an urn and the participants must assess the confirmation of different hypotheses about the composition of balls in the urn. Their results favor the z-measure, followed by the l-measure, whereas the difference measure d is at the bottom of the list. In a similar vein, a recent experiment by Colombo, Postma and Sprenger (2016) has shown that judgments of confirmation are enhanced by the prior plausibility of a hypothesis if the probabilistic relevance relations are held constant. This phenomenon, consistent with the findings of Crupi, Tentori and González (2007), is also called the *Matthew effect* (Festa 2012; Roche 2014; Festa and Cevolani 2017): "For unto every one that hath shall be given, and he shall have abundance: but from him that hath not shall be taken even that which he hath" (Matthew 25:29). For Bayesian confirmation measures, this means that measures which do not assign a *ceteris paribus* bonus to logically stronger and more informative hypotheses are probably more in line with our empirical confirmation judgments. If there is hope for confirmational monism, it might come from empirical research on confirmation judgments, showing that participants share the motivation behind a specific measure.

1.5 Discussion

This Variation has introduced Bayesian Confirmation Theory and its relation to inductive reasoning. We have examined two principal ways of explicating the concept of degrees of confirmation, namely confirmation as firmness and confirmation as increase in firmness. This has led us to several important representation results and provided us with responses to pertinent paradoxes of inductive reasoning, such as the tacking paradox, the paradox of the ravens and Goodman's new riddle of induction. Because of all these successes, Bayesian Confirmation Theory is the most popular approach to reasoning about confirmation and evidence these days.

That said, it is also a research paradigm that connects well to various scientific disciplines: Bayesian explications of confirmation can be applied in statistical inference, because two popular confirmation measures (log-likelihood and the Kemeny–Oppenheim measure) are ordinally equivalent to the Bayes factor. Bayesian reasoning has sparked interest among psychologists and is used to explain cognitive fallacies, the processing of linguistic structures, and so on (e.g., Oaksford and Chater 2000; Doya et al. 2007; Douven 2016). There is a large number of interdisciplinary papers on probabilistic reasoning, where both cognitive scientists and philosophers have been involved (e.g., Tentori, Crupi and Osherson 2007; Crupi, Fitelson and Tentori 2008; Zhao et al. 2012). But on the theoretical side, too, there is ample room for future research. Questions that are just about to be explored include an analysis of confirmation measures in information-theoretic terms (Crupi and Tentori 2014) and the use of confirmation measures for analyzing the diagnostic value of medical tests (Crupi, Tentori and Lombardi 2009). Especially the latter question, which deals with designing medical tests that lead to a high amount of (dis-)confirmation upon revealing the results, strikes us as an exciting combination of Bayesian philosophy of science with clinical practice.

The concept of confirmation also has numerous relations to other central topics of scientific reasoning. For example, Variations 6 and 7 expose substantial links between degree of confirmation, causal effect and explanatory power, following up on Carl G. Hempel's (1965) postulate of a structural identity between explanation and prediction. In Variation 8, we show how establishing intertheoretic relations between different theories (e.g., Nagelian reduction) may confirm a theory and raise our confidence in it.

Moreover, scientific argumentation can often be cast in confirmatory terms. This brings us to an outlook on the next chapters. In Variation 2, we show how the failure to find satisfactory alternatives may confirm a theory, even if there is no positive empirical evidence in its favor. Furthermore, Bayesian Confirmation Theory allows for a critical analysis of the famous No Miracles Argument (NMA). This argument claims that the astonishing success of science in recent centuries indeed confirms the hypothesis that our best scientific theories genuinely refer and constitute knowledge of the world. More on this is said in Variation 3. Variations 4 and 5 demonstrate how theories are confirmed by learning conditional information, and by learning that they predict "old evidence"—previously known phenomena for which there was no satisfactory explanation. Finally, the Bayesian literature on confirmation can be connected to work in epistemology that searches for probabilistic conditions for rational acceptance and the all-or-nothing, dichotomous concept of belief (Leitgeb 2017; Fitelson 2018).

Appendix: Proofs of the Theorems

Proof of Theorem 1.1: It is easy to verify that $c^*(H|E) = p(H|E)$ satisfies Prior–Posterior Dependence and Local Equivalence. Therefore we only show the opposite direction: any measure that satisfies both assumptions must be ordinally equivalent to c^*. We know by assumption that any measure that satisfies Prior–Posterior Dependence is of the form

$$c(H,E) = f(p(H|E), p(H)).$$

Suppose now that there are x, y, $y' \in [0;1]$ with $y \neq y'$ such that $f(x,y) \neq f(x,y')$. In that case, we can choose propositions E, H_1 and H_2 and a probability distribution p such that $x = p(H_1|E)$, $y = p(H_1)$ and $y' = p(H_2)$ as well as $H_1 \wedge E \models H_2$ and $H_2 \wedge E \models H_1$. Then, $p(H_1|E) = p(H_1 \wedge H_2 | E) = p(H_2|E)$ and

$$\begin{aligned} c(H_1,E) &= f(p(H_1|E), p(H_1)) \\ &= f(p(H_1 \wedge H_2 | E), p(H_1)) = f(p(H_2|E), p(H_1)). \end{aligned}$$

By Local Equivalence, we also know

$$c(H_1,E) = c(H_2,E) = f(p(H_2|E), p(H_2)).$$

Taking both systems of equations together leads to the equality

$$f(p(H_2|E), p(H_1)) = f(p(H_2|E), p(H_2)).$$

But this contradicts our assumption $f(p(E_2|C), p(E_1)) \neq f(p(E_2|C), p(E_2))$. So f cannot depend on its second argument. Hence, all Bayesian confirmation measures that satisfy Prior–Posterior Dependence and Local Equivalence must be ordinally equivalent to (that is, increasing functions of) $c^*(H|E) = p(H|E)$. □

Proof of Theorem 1.2: The theorem consists of four statements, which will be proved one by one. In general, we will omit the proofs that the measures satisfy the conditions that characterize them (this is left to the reader as an exercise). Instead, we only show that they are jointly sufficient to characterize a confirmation measure up to ordinal equivalence.

1. Difference Measure $d(H,E) = p(H|E) - p(H)$: Suppose that for mutually exclusive H, $H' \in \mathcal{H}$, $p(H'|E) = p(H')$. Then we infer with the help of Prior–Posterior Dependence:

$$c(H,E) = f(p(H|E), p(H)),$$

$$\begin{aligned} c(\mathrm{H}\vee\mathrm{H}', \mathrm{E}) &= f\big(p(\mathrm{H}\vee\mathrm{H}' \,|\, \mathrm{E}),\, p(\mathrm{H}\vee\mathrm{H}')\big) \\ &= f\big(p(\mathrm{H}|\mathrm{E}) + p(\mathrm{H}'|\mathrm{E}),\, p(\mathrm{H}) + p(\mathrm{H}')\big) \\ &= f\big(p(\mathrm{H}|\mathrm{E}) + p(\mathrm{H}'|\mathrm{E}),\, p(\mathrm{H}) + p(\mathrm{H}')\big). \end{aligned}$$

Applying Disjunction of Alternative Hypotheses implies $c(\mathrm{H}\vee\mathrm{H}', \mathrm{E}) = \eta(\mathrm{H}, \mathrm{E})$ and leads to the equality

$$f(p(\mathrm{H}|\mathrm{E}), p(\mathrm{H})) = f\big(p(\mathrm{H}|\mathrm{E}) + p(\mathrm{H}'|\mathrm{E}),\, p(\mathrm{H}) + p(\mathrm{H}')\big).$$

Since we have made no assumptions about the values of these conditional probabilities, f satisfies the formula $f(x, x') = f(x+y, x'+y)$ in full generality. It is then easy to see (e.g., by looking at the indifference curves of f) that there must be a function $g\colon [0;1]^2 \to \mathbb{R}$ such that $f(x, x') = g(x - x')$. Hence,

$$c(\mathrm{H}, \mathrm{E}) = f(p(\mathrm{H}|\mathrm{E}), p(\mathrm{H})) = g(p(\mathrm{H}|\mathrm{E}) - p(\mathrm{H})),$$

showing the representation result. Here and in the consecutive proofs, ordinal equivalence follows from the requirement that $c(\mathrm{H}, \mathrm{E})$ be non-decreasing in $p(\mathrm{H}|\mathrm{E})$ and non-increasing in $p(\mathrm{H})$.

2. **Ratio Measure** $r(\mathrm{H}, \mathrm{E}) = p(\mathrm{H}|\mathrm{E})/p(\mathrm{H})$: Assume that $p(\mathrm{H}), p(\mathrm{H}') \neq 0$. By Prior–Posterior Dependence, there must be a continuous function $g\colon [0;1]^2 \to \mathbb{R}$ such that

$$c(\mathrm{H}, \mathrm{E}) = g(p(\mathrm{H}|\mathrm{E})/p(\mathrm{H}), p(\mathrm{H})) = g(p(\mathrm{E}|\mathrm{H})/p(\mathrm{E}), p(\mathrm{H}))$$

(namely, $g(x, y) = f(xy, y)$). Assume now that $p(\mathrm{E}|\mathrm{H}) = p(\mathrm{E}|\mathrm{H}')$. This implies, by Law of Likelihood, $c(\mathrm{H}, \mathrm{E}) = c(\mathrm{H}', \mathrm{E})$, and we can derive

$$g(p(\mathrm{E}|\mathrm{H})/p(\mathrm{E}), p(\mathrm{H})) = c(\mathrm{H}, \mathrm{E}) = c(\mathrm{H}', \mathrm{E}) = g(p(\mathrm{E}|\mathrm{H}')/p(\mathrm{E}), p(\mathrm{H}')).$$

Since the first argument of g is the same on both sides of the equation, and we have not made any assumptions on $p(\mathrm{H})$ and $p(\mathrm{H}')$ apart from the fact that they should not be zero, we can infer that g depends on its first argument only. (The extension to $p(\mathrm{H}) = 0$ follows from the continuity of g.) Thus there is a function $g'\colon [0;1] \to \mathbb{R}$ such that $c(\mathrm{H}, \mathrm{E}) = g'(p(\mathrm{E}|\mathrm{H})/p(\mathrm{E}))$, or equivalently, $c(\mathrm{H}, \mathrm{E}) = g'(p(\mathrm{H}|\mathrm{E})/p(\mathrm{H}))$.

3. **Likelihood Ratio Measure** $l(\mathrm{H}, \mathrm{E}) = p(\mathrm{E}|\mathrm{H})/p(\mathrm{E}|\neg\mathrm{H})$: Suppose the conditions of Modularity are satisfied:

$$p(\mathrm{E} \,|\, \mathrm{H}\wedge\mathrm{E}') = p(\mathrm{E}|\mathrm{H}), \qquad p(\mathrm{E} \,|\, \neg\mathrm{H}\wedge\mathrm{E}') = p(\mathrm{E}|\neg\mathrm{H}).$$

By Modularity (first and fifth line), Prior–Posterior Dependence (second line) and an application of Bayes' Theorem (third line), we can then derive

$$\begin{aligned} c(\mathrm{H},\mathrm{E}) &= c_{|\mathrm{E}'}(\mathrm{H},\mathrm{E}) \\ &= f(p(\mathrm{H}\,|\,\mathrm{E}\wedge\mathrm{E}'),\, p(\mathrm{H}|\mathrm{E}')) \\ &= f\left(\frac{p(\mathrm{E}\,|\,\mathrm{H}\wedge\mathrm{E}')\,p(\mathrm{H}|\mathrm{E}')}{p(\mathrm{E}|\mathrm{E}')},\, p(\mathrm{H}|\mathrm{E}')\right) \\ &= f\left(\left(1+\frac{p(\mathrm{E}\,|\,\neg\mathrm{H}\wedge\mathrm{E}')}{p(\mathrm{E}\,|\,\mathrm{H}\wedge\mathrm{E}')}\cdot\frac{p(\neg\mathrm{H}|\mathrm{E}')}{p(\mathrm{H}|\mathrm{E}')}\right)^{-1},\, p(\mathrm{H}|\mathrm{E}')\right) \\ &= f\left(\left(1+\frac{p(\mathrm{E}|\neg\mathrm{H})}{p(\mathrm{E}|\mathrm{H})}\cdot\frac{1-p(\mathrm{H}|\mathrm{E}')}{p(\mathrm{H}|\mathrm{E}')}\right)^{-1},\, p(\mathrm{H}|\mathrm{E}')\right). \end{aligned}$$

We see that f can be represented as a function of two quantities: the likelihood ratio $p(\mathrm{E}|\mathrm{H})/p(\mathrm{E}|\neg\mathrm{H})$ and $p(\mathrm{H}|\mathrm{E}')$. If f were allowed to vary as a function of the latter, Prior–Posterior Dependence for $c(\mathrm{H},\mathrm{E})$ would be violated since $p(\mathrm{H}|\mathrm{E}')$ cannot be expressed as a function of $p(\mathrm{H}|\mathrm{E})$ and $p(\mathrm{H})$. However, the former expression is a function of $p(\mathrm{H}|\mathrm{E})$ and $p(\mathrm{H})$:

$$\frac{p(\mathrm{E}|\mathrm{H})}{p(\mathrm{E}|\neg\mathrm{H})} = \frac{p(\mathrm{H}|\mathrm{E})}{1-p(\mathrm{H}|\mathrm{E})}\cdot\frac{1-p(\mathrm{H})}{p(\mathrm{H})}.$$

Thus there must be a function $g\colon [0;1]\to\mathbb{R}$ such that

$$c(\mathrm{H},\mathrm{E}) = g\left(\frac{p(\mathrm{E}|\mathrm{H})}{p(\mathrm{E}|\neg\mathrm{H})}\right).$$

4. **Generalized Entailment Measure** $z(\mathrm{H},\mathrm{E}) = (p(\mathrm{H}|\mathrm{E})-p(\mathrm{H}))/\,(1-p(\mathrm{H}))$ **(for positive confirmation):** Suppose that E confirms H. Using Bayes' Theorem, we begin by writing $p(\neg\mathrm{E}|\neg\mathrm{H})$ in a slightly different way, namely as

$$\begin{aligned} p(\neg\mathrm{E}|\neg\mathrm{H}) &= \frac{p(\neg\mathrm{H}|\neg\mathrm{E})\,p(\neg\mathrm{E})}{p(\neg\mathrm{H})} \\ &= \frac{p(\neg\mathrm{H}) - p(\neg\mathrm{H}|\mathrm{E})\,p(\mathrm{E})}{1-p(\mathrm{H})} \\ &= \frac{1-p(\mathrm{H}) - (1-p(\mathrm{H}|\mathrm{E}))\,p(\mathrm{E})}{1-p(\mathrm{H})} \\ &= 1 + p(\mathrm{E})\frac{p(\mathrm{H}|\mathrm{E})-1}{1-p(\mathrm{H})} \\ &= 1 + p(\mathrm{E})\frac{p(\mathrm{H}|\mathrm{E})-p(\mathrm{H})}{1-p(\mathrm{H})} - p(\mathrm{E})\frac{p(\neg\mathrm{H})}{1-p(\mathrm{H})} \end{aligned}$$

$$= 1 - p(\mathrm{E}) + p(\mathrm{E})\frac{p(\mathrm{H}|\mathrm{E}) - p(\mathrm{H})}{1 - p(\mathrm{H})}.$$

By Contraposition/Commutativity and Prior–Posterior Dependence, we know that

$$c(\mathrm{H},\mathrm{E}) = c(\neg\mathrm{E},\neg\mathrm{H}) = f(p(\neg\mathrm{E}|\neg\mathrm{H}), p(\neg\mathrm{E})).$$

Using the above equation, we can then infer

$$c(\mathrm{H},\mathrm{E}) = f\left(1 - p(\mathrm{E}) + p(\mathrm{E})\frac{p(\mathrm{H}|\mathrm{E}) - p(\mathrm{H})}{1 - p(\mathrm{H})},\ 1 - p(\mathrm{E})\right).$$

We now use the same argument as in the previous proof. $c(\mathrm{H},\mathrm{E})$ can be represented as a function of $(p(\mathrm{H}|\mathrm{E}) - p(\mathrm{H}))/(1 - p(\mathrm{H}))$ and $p(\mathrm{E})$. At the same time, it must be a function of $p(\mathrm{H}|\mathrm{E})$ and $p(\mathrm{H})$ only, and degenerate cases aside, an arbitrary pair of $p(\mathrm{H}|\mathrm{E})$ and $p(\mathrm{H})$ allows for different values of $p(\mathrm{E})$. So it is a function of $(p(\mathrm{H}|\mathrm{E}) - p(\mathrm{H}))/(1 - p(\mathrm{H}))$ only.

The case of disconfirmation runs analogously: one needs to notice that

$$c(\mathrm{H},\mathrm{E}) = c(\mathrm{E},\mathrm{H}) = f(p(\mathrm{E}|\mathrm{H}), p(\mathrm{E})) = f\big(p(\mathrm{E})\,p(\mathrm{H}|\mathrm{E})/p(\mathrm{H}),\ p(\mathrm{E})\big)$$

and that $(p(\mathrm{H}|\mathrm{E}) - p(\mathrm{H}))/p(\mathrm{H}) = p(\mathrm{H}|\mathrm{E})/p(\mathrm{H}) - 1$ is ordinally equivalent to the ratio $p(\mathrm{H}|\mathrm{E})/p(\mathrm{H})$. □

Variation 2: The No Alternatives Argument

Probability-raising is the cornerstone of Bayesian confirmation theory. For example, the observation of a black raven raises our degree of belief in the hypothesis that all ravens are black, and thereby confirms it. Similarly, certain clicks in a particle detector make us more confident in the existence of the top quark. However, probability-raising requires empirical evidence, which may be unavailable over long periods of time. In contemporary high-energy physics, the empirical signatures of string theory or a Grand Unified Theory exceed the reach of present-day experimental technology by several orders of magnitude. The same diagnosis holds for scientific fields such as palaeontology or anthropology, where scientists rely on the scarce and haphazard empirical evidence they happen to find in the ground.

Nonetheless, scientists often have confidence in their theories when empirical evidence is largely or entirely absent. In such cases, they base their trust on what we call **non-empirical evidence:** evidence for T that neither falls into the (broadly construed) intended domain of T, nor is logically or probabilistically related to T. Such evidence can, for example, consist in observations about the research process leading up to the construction of T, or the standing of T in the research community. Thus, the term "non-empirical" is not supposed to connote a rationalist or idealist concept of theory confirmation; more often, it will relate to sociological or historical facts about the development and standing of T.

From an empiricist point of view, it is tempting to discard arguments from non-empirical evidence as metaphysical speculation. After all, it is unclear how such evidence relates to the empirical content and actual predictions of T. Therefore, arguments from non-empirical evidence neither have objective scientific weight nor do they contribute to confirming T.

We challenge this claim by exploring the following case: scientists develop considerable trust in a theory T because they did not find sensible alternatives to T, despite considerable efforts. We call this argument the **No Alternatives Argument (NAA)** and set up a Bayesian model to prove its validity. The argument does not intend to prove that there are no alternatives to T; rather, its name stems from one of its premises: the lack of viable alternatives to T. Also, we do not claim that scientific theories should be confirmed from the armchair. Nothing can replace rigorous experimental or observational testing. Still, there is considerable philosophical interest in extending the range of relevant evidence beyond the immediate or indirect predictions of T. The Bayesian framework enables us to demonstrate the confirmatory relevance of such observations.

A pragmatic variant of NAA is often used in politics and is well-known under the acronym **TINA: "There is no alternative."** The idea behind TINA is that failure to develop viable alternatives justifies a particular policy—allegedly the only one with a chance of avoiding disaster. The former British prime minister Margaret Thatcher was famous for defending her politics of privatization and economic liberalization by means of this slogan. More recently, the German chancellor, Angela Merkel, used TINA to rally support for financial rescue packages in the European sovereign debt crisis: "If the Euro falls, Europe will fall, too" (May 19, 2010, German Bundestag). In both cases, the proposed policy was not described as desirable in the first place, but as without alternatives and therefore imperative. Thus, an investigation of the NAA does not only shed light on patterns of scientific reasoning and the possibility of non-empirical theory confirmation, but it also elucidates the validity of argument patterns in a political debate.

The setup of this Variation is very simple: Section 2.1 introduces a formal model of the NAA and makes plausible assumptions for the epistemic consequences of non-empirical evidence. Section 2.2 presents our main results. Section 2.3 discusses their significance and explores an application to Inference to the Best Explanation. Proofs are given in the Appendix. For more details on the NAA and related modes of reasoning, see Dawid, Hartmann and Sprenger 2015.

2.1 Modeling the No Alternatives Argument

We would like to investigate whether observing a lack of alternatives to T confirms the empirical adequacy of T, in the sense of raising the probability of T (i.e., increase in firmness). Here, we consider two theories identical if they make the same predictions. Moreover, we would like to sidestep debates about scientific realism (see Variation 3) and focus on the empirical adequacy of T rather than on its truth. This means that we focus on the observable implications of T:

> a theory is empirically adequate exactly if what it says about the observable things and events in the world is true—exactly if it "saves the phenomena." (van Fraassen 1980, 12)

Skepticism about the NAA is not unfounded. After all, a lack of alternatives is neither deductively nor probabilistically implied by T. It does not even fall into the intended domain of T. Does this observation then qualify as (non-empirical) evidence in an *argument from ignorance*, such as: if T were not empirically adequate, then we would have disproved it before (Walton 1995; Hahn and Oaksford 2007; Sober 2009)? Also, there may always be unconceived alternatives to T which explain the available evidence as well as T, or even better (Stanford 2006). How can we confirm T in such a situation?

The most plausible way to solve this problem is to deploy a two-step process. First, we define a statement X that predicts the failure to find alternatives to T. Then, we show that X supports the empirical adequacy of T. Via X, confirmation is then transmitted from the failure to find alternatives to the empirical adequacy of T.

In the case of NAA, our non-empirical evidence consists in the fact that scientists have not found any alternatives to a specific solution of a research problem, despite looking for them with considerable energy and for a long time. Obviously, the number of satisfactory, yet-to-be-discovered alternatives to T matters here. The smaller this number is, the more likely it becomes that scientists have not yet found one of them. And the more alternatives exist, the less likely it is that our actual theory is empirically adequate. We argue that such a statement about the available number of alternatives links failure to find alternatives to the empirical adequacy of T.

In other words, the observation that scientists have not yet found an alternative to T indicates that there are not too many alternatives to T, and thus figures as an argument for T. A lower number of possible scientific theories that accounts for a certain set of empirical data increases the degree of belief that our actual theory is empirically adequate—see also the *argument from no choice* (Dawid 2006, 2009).

Inferences about the number of alternatives to a theory T naturally depend on what counts as a genuine alternative. This is, in turn, sensitive to the specific scientific context. Apart from some general constraints given below, we leave the problem of individuating alternative theories to the relevant scientific communities. They usually have the best grip on what should count as a distinct theory. Moreover, for the No Alternatives Argument we only require the premise that the number of alternatives to T possibly be finite. In other words, we must not be certain a priori that there are infinitely many alternatives.

In order to motivate this assumption, we assume that different theories provide different solutions to a given research problem. That is, theories which only differ in a detail, such as the precise value of a parameter, do not count as different theories. For example, the simple Higgs model in particle physics is treated as *one* theory, although the mass of the Higgs particle may take different numerical values. Similarly, theories which do not differ except for theoretically meaningless variables or predictions (e.g., about the existence of a Twin Earth in a remote corner of the universe) do not count as distinct theories. Generally, if we could create new theories by modifying the value of a parameter or adding a dummy prediction, then coming up with new theories would be an easy and not very creative task. It is much harder to invent a novel mechanism, or to tell a new story of why a certain phenomenon occurred.

Indeed, scientists often formulate no alternatives arguments at the level of general conceptual principles while allowing for a large spectrum of specific realizations of those principles. For example, since the 1980s particle physicists strongly supported an argument from the lack of alternatives with respect to the Higgs mechanism. That is, they believed that no alternatives to a gauge theory that was spontaneously broken by a Higgs sector of scalar fields could account for the available empirical data. Based on such an NAA, physicists strongly believed that the Higgs sector would be observed in the recent Large Hadron Collider experiments, but they did not have particular trust in any of the specific models. In this case, the NAA was clearly operating at the level of physical principles rather than specific models.

Following this line of reasoning, we reconstruct the NAA in Bayesian terms. A crucial concept in our model is the unknown number of possible alternative theories. We introduce a random variable A that takes values in the natural numbers and expresses the number of feasible alternatives to T. More precisely, $\mathrm{A}_k := \{A = k\}$ expresses the proposition that there are k adequate and distinct alternatives which satisfy a set of theoretical constraints $\mathcal{C}$, are consistent with the existing data $\mathcal{D}$, and give distinguishable predictions for the outcomes of some set $\mathcal{E}$ of future experiments. We will then show that failure to find an alternative to T raises our degree of belief that A takes a low value and thus confirms the empirical adequacy of T (see also Dawid, Hartmann and Sprenger 2015, 215–216).

To do so, we introduce two propositional variables: H and F_A. As before, H takes the values

H: Theory T is empirically adequate;

¬H: Theory T is not empirically adequate,

and F_A takes the values

F_A: The scientific community has found an alternative to T that fulfills $\mathcal{C}$, explains $\mathcal{D}$ and predicts the outcomes of $\mathcal{E}$;

$\neg\mathrm{F}_A$: The scientific community has not yet found an alternative to T that fulfills $\mathcal{C}$, explains $\mathcal{D}$ and predicts the outcomes of $\mathcal{E}$.

Figure 2.1: The Bayesian network representation of the two-propositions model of the No Alternatives Argument. Adapted under the terms of the Creative Commons Attribution 4.0 International (CC BY 4.0) license from "The No Alternatives Argument" by Richard Dawid, Stephan Hartmann and Jan Sprenger, *British Journal for the Philosophy of Science*, Volume 66, No. 1, pp. 213–234. © 2014, doi:10.1093/bjps/axt045.

We would now like to explore under which conditions $\neg\mathrm{F}_A$ confirms H in the Bayesian sense, that is, when

$$p(\mathrm{H}|\neg\mathrm{F}_A) > p(\mathrm{H}).$$

Here p is a subjective probability distribution over the product algebra of the variables involved, like in the previous Variation.

Figure 2.1 suggests a direct influence of H on F_A. But since a direct influence is blocked by the non-empirical nature of F_A, we introduce a third variable A which mediates the connection between H and F_A. The variable A has values in the natural numbers, and A_k corresponds to the proposition that there are exactly k hypotheses that fulfill $\mathcal{C}$, explain $\mathcal{D}$ and predict the outcomes of $\mathcal{E}$.

We should also note that the value of F_A—that scientists find/do not find an alternative to T—does not depend only on the number of available alternatives, but also on the difficulty of the problem, the cleverness of the scientists, or the available computational, experimental and mathematical resources. Call the variable that captures these complementary factors D, and let it take values in the natural numbers, with $\mathrm{D}_j := \{D{=}j\}$ and $d_j := p(\mathrm{D}_j)$. The higher the value of D, the more difficult the problem. For the purpose of our argument, it is not necessary to assign a precise operational meaning to the various levels of D—see Condition **A3** later on. It is clear that D has no direct influence on A and H (or vice versa), but that it matters for F_A and that this influence has to be represented in our Bayesian network.

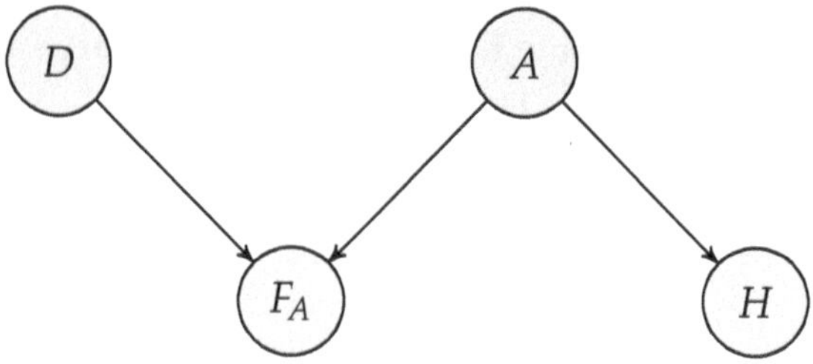

Figure 2.2: The Bayesian network representation of the four-propositions model of the No Alternatives Argument. Adapted under the terms of the Creative Commons Attribution 4.0 International (CC BY 4.0) license from "The No Alternatives Argument" by Richard Dawid, Stephan Hartmann and Jan Sprenger, *British Journal for the Philosophy of Science*, Volume 66, No. 1, pp. 213–234. © 2014, doi:10.1093/bjps/axt045.

We now list five plausible assumptions that we need for showing the validity of the No Alternatives Argument.

A1 The variable H is conditionally independent of F_A given A:

$$H \perp\!\!\!\perp F_A \,|\, A.$$

Hence, learning that the scientific community has or has not found an alternative to T does not alter our belief in the empirical adequacy of T if we already know the value of A (e. g., that there are exactly k viable alternatives).

A2 The variable D is (unconditionally) independent of A:

$$D \perp\!\!\!\perp A.$$

The variable D represents the aggregate of those context-sensitive factors that affect whether scientists find an alternative to T, apart from the number of suitable alternatives. So D and A are independent of each other by construction. However, they become dependent when the value of F_A is known.

These are our most important assumptions, and we consider them to be eminently sensible. Figure 2.2 represents the independence assumptions graphically as a Bayesian network. To complete it, we have to specify the prior distribution over D and A and the conditional distributions over F_A and T, given the values of their parents. This is done in the following three assumptions.

A3 The conditional probabilities

$$f_{kj} := p(\neg \mathrm{F}_A \,|\, \mathrm{A}_k, \mathrm{D}_j)$$

are non-increasing in k for all $j \in \mathbb{N}$ and non-decreasing in j for all $k \in \mathbb{N}$.

The (weak) monotonicity in the first argument reflects the intuition that for fixed difficulty of a problem, a higher number of available alternatives increases the chance of finding one of them. In other words, the more reasonable alternatives to T are around, the more likely it is that scientists will find one. The (weak) monotonicity in the second argument reflects the intuition that increasing difficulty of a problem does not increase the likelihood of finding an alternative to T, provided that the number of alternatives to T is fixed.

A4 The conditional probabilities

$$t_k := p(\mathrm{H} | \mathrm{A}_k)$$

are non-increasing in k.

This assumption reflects the intuition that an increase in the number of alternative theories does not make it more likely that the current theory is empirically adequate.

A5 With the definition

$$a_k := p(\mathrm{A}_k),$$

there is at least one pair (i, k) with $i < k$ for which (i) $a_i,\ a_k > 0$, (ii) $f_{ij} > f_{kj}$ for some $j \in \mathbb{N}$, and (iii) $t_i > t_k$.

This assumption demands that the probability distribution of A is not concentrated on a single number (i). It slightly strengthens **A3** and **A4**, by demanding that the f_{ij} (ii) and t_i (iii) not be constant in i: for some pair of numbers, an increase in the number of alternatives raises the probability of finding an alternative to T and lowers the probability that T is empirically adequate.

2.2 Results

The previous section has set up a formal model of the NAA in a Bayesian network (see Figure 2.2) and made five assumptions on how the variables in that network hang together (see **A1–A5**). With these assumptions in hand, we can now show the following main result:

Theorem 2.1 (Validity of the NAA). *Let H, F_A and D be propositional variables, and let A be an integer-valued variable. If the probability distribution p over these variables satisfies assumptions* **A1** *to* **A5**, *then* $\neg F_A$ *confirms* H, *that is,* $p(H|\neg F_A) > p(H)$.

We have therefore shown that $\neg F_A$ confirms the empirical adequacy of T under rather weak and plausible assumptions.

In line with the introduction of A in Section 2.1, we have assumed that A only takes values in the natural numbers. This might be seen as evading the skeptical argument that there may be infinitely many (theoretically adequate, empirically successful, ...) alternatives to T. Therefore we now extend the theorem by explicitly allowing for the possibility $A_\infty := \{A = \infty\}$, and we modify our assumptions accordingly. In particular, we observe that **A5** entails $p(A_\infty) < 1$, we define $f_{\infty j} := p(\neg F_A \mid A_\infty, D_j)$ and $t_\infty := p(H|A_\infty)$ and we demand that

$$f_{ij} \geq f_{\infty j} \quad \text{for all } i, j \in \mathbb{N}, \tag{2.1}$$

$$f_{\infty i} \leq f_{\infty j} \quad \text{for all } i, j \in \mathbb{N} \text{ with } i < j, \tag{2.2}$$

$$t_i \geq t_\infty \quad \text{for all } i \in \mathbb{N}. \tag{2.3}$$

These requirements naturally extend assumptions **A3** and **A4** to the case of infinitely many alternatives. Then we obtain the following generalization of the NAA:

Theorem 2.2 (Validity of the NAA, infinitely many alternatives). *Let H, F_A and D be propositional variables, and let A take values in $\mathbb{N} \cup \{\infty\}$. If the probability distribution p over these variables satisfies assumptions* **A1** *to* **A5** *together with their extensions* (2.1)–(2.3), *then* $\neg F_A$ *confirms* H, *that is,* $p(H|\neg F_A) > p(H)$.

In other words, even if we concede to the skeptic that there may be infinitely many alternatives to T, she must still acknowledge the validity of the No Alternatives Argument as long as her degrees of belief satisfy $p(A_\infty) < 1$. This is, to our mind, a quite substantial and surprising result. For a long time, philosophy of science has focused on logical and probabilistic relations between theory and evidence and neglected other forms of theory confirmation. However, the above theorem demonstrates that non-empirical evidence (in our specific sense of the word) can raise our confidence in the empirical adequacy of a theory.

Note that only a dogmatic skeptic who insists on $p(\mathrm{A}_\infty) = 1$ can deny the validity of NAA. Theorem 2.2 convinces anyone who is not committed a priori to the existence of infinitely many genuine alternatives to T. (Recall that theories do not count as distinct when they are just different realizations of the same mechanism or principle.) Convincing such a fair and non-committal skeptic is, to our mind, much more important than convincing a dogmatic who just denies our premises by insisting on $p(\mathrm{A}_\infty) = 1$.

2.3 Discussion

We have seen that the NAA can be used in support of a proposed theory under quite general circumstances. Non-empirical evidence (in our technical sense of the word) can raise the rational degree of belief that we assign to the empirical adequacy of a scientific theory. This is a result of foundational philosophical interest. The question remains, however, whether the degree of support is significant enough to be practically meaningful: does the NAA raise our degrees of belief substantially or only marginally? To facilitate matters, we conduct this analysis for the finite case (Theorem 2.1); the infinite case is analogous.

The Bayesian network representation of NAA in Figure 2.2 suggests that the NAA cannot easily obtain confirmatory significance without supportive reasoning. According to Figure 2.2, $\neg \mathrm{F}_A$ may confirm an instance of D—limitations on the scientists' abilities to solve difficult problems—as well as an instance of A, such as limitations on the number of possible theories. It is then easy to see that for all $l \in \mathbb{N}$,

$$p(\mathrm{D}_l | \neg \mathrm{F}_A) = \frac{p(\mathrm{D}_l, \neg \mathrm{F}_A)}{p(\neg \mathrm{F}_A)} = \frac{d_l \cdot \sum_k a_k f_{kl}}{\sum_{j,k} d_j a_k f_{kj}},$$

which may be greater than $p(\mathrm{D}_l)$ for plausible assignments of parameter values. To become confident of a theory by the NAA, one has to amend the *qualitative* claim shown above with a *comparative* claim, namely that $\neg \mathrm{F}_A$ confirms T more than the claim $\{D > K\}$ ("the problem is just very difficult") for some threshold K. But such a statement is sensitive to the specific parameter assignments as well as to the chosen confirmation measure—and therefore hard to prove in general. Applied to the political context (the TINA variant of NAA), this result means that the failure to find viable alternatives to a particular policy does indeed confirm that the chosen policy may be the best one, in the sense of confirmation as increase in firmness. But without additional assumptions, it would be invalid to conclude that the probability has increased substantially, let alone that we should now be *confident* (e.g., with a degree of belief greater than 1/2) that the chosen policy is the best one.

The NAA also opens up interesting philosophical perspectives. First, **Inference to the Best Explanation (IBE)** (Lipton 2004; Douven 2017) can, to a certain extent, be explicated in terms of NAA. We conjecture that some Inferences to the Best Explanation in science are actually NAA's in disguise: they take the failure of attempts to find an alternative as a reason to infer the truth or empirical adequacy of the only available hypothesis that explains a phenomenon $\mathcal{E}$. The relevant variables then read as follows:

H: The hypothesis T is empirically adequate.

A_k: There are k distinct alternatives to T that explain $\mathcal{E}$.

F_A: The scientific community has found an alternative to T that explains $\mathcal{E}$.

$\neg F_A$: The scientific community has not yet found an alternative to T that explains $\mathcal{E}$.

It is not difficult to motivate analogues of **A1–A5** for this interpretation of our propositional variables, and to derive that $\neg F_A$ confirms H. Does this prove the validity of IBE? Only partially so. The NAA demonstrates the relevance of explanatory reasoning for theory confirmation, but only in the sense of confirmation as increase in firmness, that is, probability-raising. It does not establish that the probability of T is higher than a certain threshold. In general, the absence of alternative explanations is not a sufficient reason to conclude that T must be true, or empirically adequate. Still, the relationship between NAA and IBE is a fertile field for future research.

Second, the reasoning scheme of the NAA is similar to eliminative induction in the style of Francis Bacon, or more recently, Arthur Conan Doyle's character Sherlock Holmes: "When you have eliminated the impossible, whatever remains, however improbable, must be the truth." Could we use the NAA as a case study for creating deeper links between Bayesian and eliminative inference (Earman 1992; Forber 2011)?

Third, our theoretical analysis should be complemented by more case studies in science: from string theory as a classical application of the NAA

(Dawid 2006, 2009), and also from other disciplines where empirical evidence is scarce, such as palaeontology, archeology or anthropology.

Fourth and last, the relationship between NAA and the TINA argument in public policy should be investigated more closely. Can the endorsement of a policy really be defended with a type of NAA? Or does "failure to find viable alternatives" mean something very different in the political context, invalidating the application of the NAA in that domain? Similarly, can the NAA be used for validating argument patterns such as "no reason for is a reason against" (Eva and Hartmann 2018b)?

In the next Variation, we will address the issue of scientific realism, which looms behind the NAA and related reasoning schemes. In particular, we will show how the model underlying NAA, with an explicit probability distribution over the number of alternatives to T, can be used to develop a sophisticated Bayesian version of the No Miracles Argument (NMA).

Appendix: Proofs of the Theorems

Proof of Theorem 2.1: $\neg F_A$ confirms H if and only if $p(H|\neg F_A) - p(H) > 0$, that is, if and only if

$$\Delta := p(H, \neg F_A) - p(H)\, p(\neg F_A) > 0.$$

We now apply the theory of Bayesian Networks to the structure depicted in Figure 2.2, using assumptions **A1** ($H \perp\!\!\!\perp F_A \,|\, A$) and **A2** ($D \perp\!\!\!\perp A$):

$$\begin{aligned}
p(\neg F_A) &= \sum_{i=0}^{\infty}\sum_{j=0}^{\infty} p(\neg F_A \,|\, A_i, D_j)\, p(A_i, D_j) = \sum_{i=0}^{\infty}\sum_{j=0}^{\infty} d_j a_i f_{ij}, \\
p(H) &= \sum_{k=0}^{\infty} p(H|A_k)\, p(A_k) = \sum_{k=0}^{\infty} t_k a_k, \\
p(H, \neg F_A) &= \sum_{i=0}^{\infty} p(\neg F_A, H \,|\, A_i)\, p(A_i) = \sum_{i=0}^{\infty} a_i\, p(\neg F_A|A_i)\, p(H|A_i) \\
&= \sum_{i=0}^{\infty} a_i t_i \sum_{j=0}^{\infty} p(\neg F_A \,|\, A_i, D_j)\, p(D_j|A_i) = \sum_{i=0}^{\infty}\sum_{j=0}^{\infty} d_j a_i t_i f_{ij}.
\end{aligned}$$

Hence we obtain, using $\sum_{k\in\mathbb{N}} a_k = 1$,

$$\begin{aligned}
\Delta &= \sum_{i=0}^{\infty}\sum_{j=0}^{\infty} d_j a_i t_i f_{ij} - \left(\sum_{i=0}^{\infty}\sum_{j=0}^{\infty} d_j a_i f_{ij}\right)\sum_{k=0}^{\infty} a_k t_k \\
&= \left(\sum_{i=0}^{\infty}\sum_{j=0}^{\infty} d_j a_i t_i f_{ij}\right)\sum_{k=0}^{\infty} a_k - \left(\sum_{i=0}^{\infty}\sum_{j=0}^{\infty} d_j a_i f_{ij}\right)\sum_{k=0}^{\infty} t_k a_k \\
&= \sum_{i=0}^{\infty}\sum_{j=0}^{\infty}\sum_{k=0}^{\infty}\left(d_j a_i a_k t_i f_{ij} - d_j a_i a_k t_k f_{ij}\right) \\
&= \sum_{j=0}^{\infty} d_j \sum_{i=0}^{\infty}\sum_{\substack{k=0\\k\neq i}}^{\infty} a_i a_k f_{ij}(t_i - t_k) \\
&= \sum_{j=0}^{\infty} d_j \sum_{i=0}^{\infty}\sum_{k>i}\left(a_i a_k f_{ij}(t_i - t_k) + a_k a_i f_{kj}(t_k - t_i)\right) \\
&= \sum_{j=0}^{\infty} d_j \sum_{i=0}^{\infty}\sum_{k>i} a_i a_k\left(f_{ij}(t_i - t_k) + f_{kj}(t_k - t_i)\right) \\
&= \sum_{j=0}^{\infty} d_j \sum_{i=0}^{\infty}\sum_{k>i} a_i a_k (t_i - t_k)(f_{ij} - f_{kj}) \\
&> 0,
\end{aligned}$$

because of **A3–A5** taken together: **A3** entails that the differences $f_{ij}-f_{kj}$ are non-negative, **A4** does the same for the t_i-t_k, and **A5** entails that these differences are strictly positive for at least one pair (i,k). Hence the entire double sum is strictly positive. □

Proof of Theorem 2.2: We perform essentially the same calculations as in the proof of Theorem 2.1 and additionally include the possibility $\mathrm{A}_\infty := \{A=\infty\}$.[8] Defining $f_{\infty j} := p(\neg \mathrm{F}_A \,|\, \mathrm{D}_j, \mathrm{A}_\infty)$ leads us to the equalities

$$\begin{aligned} p(\neg \mathrm{F}_A) &= \sum_{i=0}^{\infty}\sum_{j=0}^{\infty} d_j a_i f_{ij} + \sum_{j=0}^{\infty} d_j a_\infty f_{\infty j}, \\ p(\mathrm{H}) &= \sum_{k=0}^{\infty} t_k a_k + t_\infty a_\infty, \\ p(\neg \mathrm{F}_A, \mathrm{H}) &= \sum_{i=0}^{\infty}\sum_{j=0}^{\infty} d_j a_i t_i f_{ij} + \sum_{j=0}^{\infty} d_j t_\infty a_\infty f_{\infty j}, \end{aligned}$$

from which it follows, using $\lim_{K\to\infty}\sum_{k=1}^{K} a_k = 1-a_\infty$, that

$$\begin{aligned} p(\neg \mathrm{F}_A)p(\mathrm{H}) &= \sum_{i=0}^{\infty}\sum_{j=0}^{\infty}\sum_{k=0}^{\infty} d_j t_k a_i a_k f_{ij} + \sum_{i=0}^{\infty}\sum_{j=0}^{\infty} d_j t_\infty a_i a_\infty f_{ij} + \\ &\quad \sum_{j=0}^{\infty}\sum_{k=0}^{\infty} d_j t_k a_k a_\infty f_{\infty j} + \sum_{j=0}^{\infty} d_j t_\infty a_\infty^2 f_{\infty j} \end{aligned}$$

and

$$p(\neg \mathrm{F}_A, \mathrm{H}) = \frac{1}{1-a_\infty}\sum_{i=0}^{\infty}\sum_{j=0}^{\infty}\sum_{k=0}^{\infty} d_j t_i a_i a_k f_{ij} + \sum_{j=0}^{\infty} d_j t_\infty a_\infty f_{\infty j}.$$

With the definition

$$\Delta := \sum_{i=0}^{\infty}\sum_{j=0}^{\infty}\sum_{k=0}^{\infty} d_j t_i a_i a_k f_{ij} - \sum_{i=0}^{\infty}\sum_{j=0}^{\infty}\sum_{k=0}^{\infty} d_j t_k a_i a_k f_{ij}$$

we observe that $\Delta > 0$, as shown above in the proof of Theorem 2.1 (the parameter values satisfy the relevant conditions **A3–A5**). Noting that **A5** requires $a_\infty < 1$, it follows that

$$p(\neg \mathrm{F}_A, \mathrm{H}) - p(\mathrm{H})p(\neg \mathrm{F}_A)$$

[8] The notation suggests that ∞ is already included in the summation index, but the infinity sign on top of the Σ is just the shortcut for the limit of the sequence of all natural numbers. Thus the case $A=\infty$ has to be treated separately.

$$
\begin{aligned}
&= \Delta + \frac{a_\infty}{1-a_\infty} \sum_{i=0}^{\infty} \sum_{j=0}^{\infty} \sum_{k=0}^{\infty} d_j t_i a_i a_k f_{ij} + \sum_{j=0}^{\infty} d_j t_\infty a_\infty (1-a_\infty) f_{\infty j} - \\
&\quad \sum_{i=0}^{\infty} \sum_{j=0}^{\infty} d_j t_\infty a_i a_\infty f_{ij} - \sum_{i=0}^{\infty} \sum_{j=0}^{\infty} d_j t_i a_i a_\infty f_{\infty j} \\
&= \Delta + \frac{a_\infty}{1-a_\infty} \sum_{i=0}^{\infty} \sum_{j=0}^{\infty} \sum_{k=0}^{\infty} d_j t_i a_i a_k f_{ij} + \frac{a_\infty}{1-a_\infty} \sum_{i=0}^{\infty} \sum_{j=0}^{\infty} \sum_{k=0}^{\infty} d_j t_\infty a_i a_k f_{\infty j} - \\
&\quad \frac{a_\infty}{1-a_\infty} \sum_{i=0}^{\infty} \sum_{j=0}^{\infty} \sum_{k=0}^{\infty} d_j t_\infty a_i a_k f_{ij} - \frac{a_\infty}{1-a_\infty} \sum_{i=0}^{\infty} \sum_{j=0}^{\infty} \sum_{k=0}^{\infty} d_j t_i a_i a_k f_{\infty j} \\
&= \Delta + \frac{a_\infty}{1-a_\infty} \sum_{j=0}^{\infty} d_j \sum_{i=0}^{\infty} \sum_{k>i} a_i a_k \Big(t_i f_{ij} + t_\infty f_{\infty j} - t_\infty f_{ij} - t_i f_{\infty j} \Big) \\
&= \Delta + \frac{a_\infty}{1-a_\infty} \sum_{j=0}^{\infty} d_j \sum_{i=0}^{\infty} \sum_{k>i} a_i a_k (t_i - t_\infty)(f_{ij} - f_{\infty j}) \\
&> 0,
\end{aligned}
$$

since the extensions of **A3** and **A4** imply $f_{ij} \geq f_{\infty j}$ and $t_i \geq t_\infty$ (Equations (2.1) and (2.3)), independent of the values of i and j. □

Variation 3: Scientific Realism and the No Miracles Argument

The debate between scientific realists and anti-realists is one of the classics of philosophy of science, comparable to a soccer match between England and Germany or Brazil and Argentina. Realism is no monolithic block, but comes in different varieties, such as metaphysical, semantic and epistemological realism (for a survey, see Chakravartty 2017). Modest varieties of scientific realism posit the existence of a mind-independent world (metaphysical realism) and the reference of theoretical terms to mind-independent entities (semantic realism). The most ambitious variety of scientific realism, however, concerns the epistemic status of our best scientific theories: Are we justified to believe in their literal truth? Do they constitute knowledge of the external world? Realists respond positively to these questions (e.g., Boyd 1983; Psillos 1999, 2009), anti-realists urge for caution.

In this Variation, we demonstrate how Bayesian models clarify and sharpen the debate between epistemological realists and anti-realists, synthesizing the arguments we presented in Sprenger 2016d and Dawid and Hartmann 2018. In Section 3.1, we expound the No Miracles Argument (NMA) and present the objection that the realist commits the base rate fallacy (Howson 2000; Magnus and Callender 2004). The following sections respond to this worry. In Section 3.2, we argue that the observed stability of scientific theories supports a claim to scientific realism. We also demonstrate that the strength of the NMA crucially depends on the context of a specific scientific discipline. In Section 3.3, we investigate the frequency-based No Miracles Argument: we show how shifting from individual theories to a *series* of theories in a scientific discipline may alleviate the base rate fallacy. Both arguments are refinements of the NMA within a Bayesian model. Note that none of our arguments makes a sufficient case for scientific realism:

the scope of the Bayesian NMA does not extend beyond claims to empirical adequacy. However, the NMA is necessary for parrying vital threats to the realist view, such as Laudan's argument from Pessimistic Meta-Induction. The scope of our Bayesian models, and directions for future research, are discussed in Section 3.4. All in all, we do not argue that Bayesian philosophy of science should be aligned with a realist stance. Rather, we show how Bayesian methods can be used to clearly articulate the realist argument, to investigate its validity and to determine its scope.

3.1 The Bayesian No Miracles Argument

Perhaps the best-known argument in favor of scientific realism is the **No Miracles Argument (NMA).** It contends that the truth of our best scientific theories is the only hypothesis that does not make the astonishing predictive, retrodictive and explanatory successes of science a mystery (Putnam 1975). If our best scientific theories do not correctly describe the world, why are they so often successful? The truth of our best theories is an excellent, and perhaps the only, explanation of their success. Therefore we should accept the realist hypothesis: Our best scientific theories are true and constitute knowledge of the world.

For the sake of convenience, we summarize the empirical merits of a scientific theory under the label of predictive success. Even so, it is not entirely clear whether the NMA is an empirical or a super-empirical argument. As an argument from the successes of our best scientific theories to their truth, it involves two major steps: the step from past and present predictive success to justified belief in empirical adequacy, and the step from justified belief in empirical adequacy to justified belief in truth (see Figure 3.1). The first one is an empirical inference, while the second goes far beyond our current experimental knowledge: empirical evidence cannot distinguish between different theoretical structures that yield the same observable consequences.

Much philosophical discussion has been devoted to the second step of the NMA (e.g., Psillos 1999; Lipton 2004; Stanford 2006), which seems in greater need of a philosophical defense. After all, the realist has to address the problem of underdetermination of theory by evidence. But also the first step of the NMA is far from trivial, and strengthening it against criticism is vital for the scientific realist. For instance, Laudan (1981) has argued that there have always been lots of temporarily successful scientific theories which were later found to be non-referring and empirically inadequate (e.g., the phlogiston

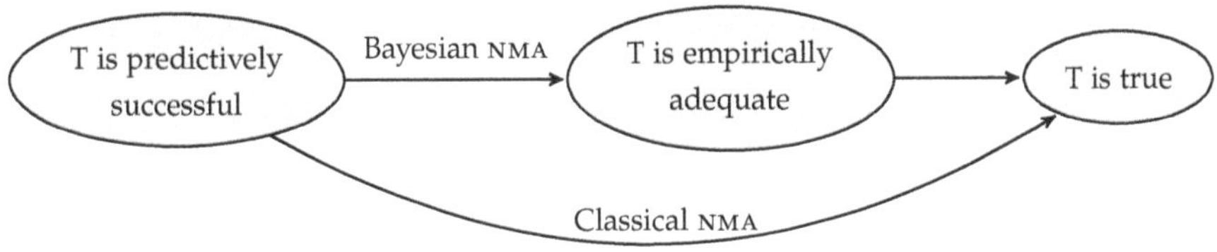

Figure 3.1: The structure of the NMA as a two-step argument from the predictive success of T to its truth. We conceptualize the NMA as an argument for the first inference in this figure, that is, for an inference from T's predictive success to its empirical adequacy (i.e., the observable parts of T are true).

theory in early modern chemistry, or the aether theory in nineteenth-century physics). According to Laudan, these observations undermine the inference from predictive success to empirical adequacy. Thus, even if the gap between empirical adequacy and truth could be bridged successfully, the NMA as a whole may still in danger.

Arguments against the inference from predictive success to empirical adequacy threaten not only full-blown scientific realism but also structural realism (Worrall 1989) and some varieties of anti-realism. One of them is Bas van Fraassen's **Constructive Empiricism** (van Fraassen 1980; Monton and Mohler 2017). Proponents of this view deny that we have reasons to believe that our best scientific theories are literally true, but they affirm that we are justified to believe in their observable consequences. Evidently, criticism and defense of the inference to empirical adequacy also affects such positions.

Hence, the first step of the NMA does not draw a sharp divide between realists and anti-realists. Rather, the debate takes place between those who commit themselves to a scientific theory on the basis of its predictive success, and those who don't. This Variation explores whether such epistemic commitments are justified. For convenience, we stick to the traditional terminology and refer to the first group as "realists" and to the second group as "anti-realists".

We begin with a Bayesian analysis of a simple NMA. Assume that theory T is predictively and explanatorily successful in a certain scientific domain. Since we only investigate arguments for the empirical adequacy of T, we introduce a propositional variable H, which represents whether T is empirically adequate. See Figure 3.2 for a simple Bayesian network representation of the dependence between H and the propositional variable S that represents whether T is predictively successful.

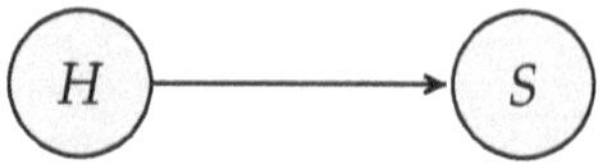

Figure 3.2: The Bayesian network representation of the impact of H (i.e., T is/is not empirically adequate) on the predictive success of T, denoted by S.

Expressed as a Bayesian argument, the simple NMA then runs as follows: S is much more probable if T is empirically adequate than if it is not. This can be expressed by the following two assumptions:

F1 $s := p(\mathrm{S}|\mathrm{H})$ is large.

F2 $s' := p(\mathrm{S}|\neg\mathrm{H}) < k \ll 1$.

From Bayes' Theorem, we can then infer

$$p(\mathrm{H}|\mathrm{S}) > p(\mathrm{H}).$$

In other words, S confirms H: our degree of belief in the empirical adequacy of T is increased if T is successful. As before, p is a probability distribution on the sentences of a propositional language, with elements S and H.

Anti-realists object to the above argument that the inequality $p(\mathrm{H}|\mathrm{S}) > p(\mathrm{H})$ falls short of establishing the first step of the NMA. We are primarily interested in whether H is sufficiently probable given S, not in whether S raises our degree of belief in H. After all, the increase in probability could be negligibly small. The result $p(\mathrm{H}|\mathrm{S}) > p(\mathrm{H})$ does not establish that $p(\mathrm{H}|\mathrm{S}) > K$ for a critical threshold K, e.g., $K = 1/2$. We already know this distinction between posterior probability and incremental confirmation from Variation 1, under the name of confirmation as firmness vs. confirmation as increase in firmness.

More specifically, it has been argued that the NMA commits the **base rate fallacy** (Howson 2000; Magnus and Callender 2004). This is a type of inference that frequently misleads people in probabilistic and statistical reasoning (e.g., Kahneman, Slovic and Tversky 1982; Goodman 1999a). On page 22 in the introductory chapter, we explained it in detail, but we restate the basic idea: taken by itself, evidential support is no reliable guide to posterior probability. If the prior probability of the hypothesis is sufficiently low, posterior probability will be low even when evidential support is high. However, people tend to discard prior (im-)plausibility in favor of apparently sweeping evidence (e.g., eyewitness testimonials, test results, and so on).

This objection to the NMA can be explicated using Bayes' Theorem. Our quantity of interest is the posterior probability $p(\mathrm{H}|\mathrm{S})$, our confidence in H given S. This quantity can be written as

$$p(\mathrm{H}|\mathrm{S}) = \frac{p(\mathrm{H})\,p(\mathrm{S}|\mathrm{H})}{p(\mathrm{S})} = \left(1 + \frac{1-p(\mathrm{H})}{p(\mathrm{H})} \cdot \frac{p(\mathrm{S}|\neg\mathrm{H})}{p(\mathrm{S}|\mathrm{H})}\right)^{-1}, \qquad (3.1)$$

which shows that $p(\mathrm{H}|\mathrm{S})$ is not only increasing in $p(\mathrm{S}|\mathrm{H})$ and decreasing in $p(\mathrm{S}|\neg\mathrm{H})$: its value crucially depends on the base rate $p(\mathrm{H})$ of H (i.e., the prior plausibility of H).

Anti-realists claim that NMA is built on a base rate fallacy: from the high value of $p(\mathrm{H}|\mathrm{S})$ ("the empirical adequacy of T explains its success") and the low value of $p(\mathrm{S}|\neg\mathrm{H})$ ("success of T would be a miracle if T were not empirically adequate"), the NMA infers justified belief in H ("T is empirically adequate"). The Bayesian model demonstrates, however, that we need additional assumptions about $p(\mathrm{H})$ to warrant this inference. In the absence of such assumptions, the NMA does not entitle us to accept T as empirically adequate.

What do these considerations show? First of all, the NMA does not refute the anti-realist standpoint when reconstructed as a Bayesian inference to the posterior probability of H. Any weight of evidence in favor of H can be counterbalanced by a sufficiently skeptical prior, that is, a sufficiently low value assigned to $p(\mathrm{H})$. To convince the anti-realist, the realist needs to provide good reasons why $p(\mathrm{H})$ should not be arbitrarily close to zero. Since such reasons will typically presuppose realist inclinations (e.g., "our best scientific theories are generally reliable"), the NMA is threatened by circularity. This is a substantial problem for realists who claim that the NMA is an intersubjectively compelling argument in favor of scientific realism. Howson (2013, 211) concludes that due to the dependence on unconstrained prior degrees of belief, the NMA is, "as a supposedly objective argument, ... dead in the water" (see also Howson 2000, ch. 3; Lipton 2004, 196–198; and Chakravartty 2017).

Even more, given our knowledge of the history of science, having a low prior degree of belief in H may actually be a rational attitude, and be more rational than a high prior degree of belief. Take, for example, Larry Laudan's aforementioned argument from **Pessimistic Meta-Induction (PMI):** "I believe that for every highly successful theory in the past of science which we now believe to be a genuinely referring theory, one could find half a dozen successful theories which we now regard as substantially non-referring" (Laudan 1981, 35). Why should our currently best theory $\mathrm{T}_n = \mathrm{T}$

not suffer the same fate as its predecessors $T_1, \ldots, T_{n-1}$, which proved to be empirically inadequate although they were once the best scientific theories? What justifies our optimism with respect to our currently best theories?

Laudan's PMI affects the values of $p(S|\neg H)$ and $p(H)$ as follows: On the one hand, history teaches us that there have often been false theories that explained the data well (and were superseded later). In other words, empirically non-adequate theories can be highly successful, undermining the argument that $p(S|\neg H)$ needs to be low. On the other hand, PMI advises a low degree of belief that T is empirically adequate, since comparable predecessor theories turned out to be false in the past.

To substantiate these anti-realist concerns, we conduct a numerical analysis of the Bayesian NMA. For the sake of simplicity and being generous to the realist (i.e., our imagined opponent), let us sharpen **F1** to $s = p(S|H) = 1$: if theory T is empirically adequate, then it is also successful. Furthermore, define $s' := p(S|\neg H)$ and let $h := p(H)$ be the prior probability of H. We now ask the question: which range of values for of s' and h leads to a posterior probability $p(H|S)$ of H greater than 1/2? That is, when would it be more plausible to believe that T is empirically adequate than to deny it? Satisfying this condition is arguably a minimal requirement for claiming that the success of T warrants justified belief in its empirical adequacy.

By using Bayes' Theorem, we can easily calculate when the inequality $p(H|S) > 1/2$ is satisfied. Making use of Equation (3.1) allows us to rewrite this inequality as

$$\frac{1}{2} < \left(1 + s'\frac{1-h}{h}\right)^{-1},$$

which simplifies to

$$s' < \frac{h}{1-h}. \tag{3.2}$$

See Figure 3.3 for a graphical illustration.

However, Inequality (3.2) is not easy to satisfy. As mentioned above, false theories and models often make accurate predictions and perform well on other cognitive values (for an overview, see Frigg and Hartmann 2012). Classical examples that are still used today involve Newtonian mechanics, the Lotka–Volterra model from population biology (e.g., Weisberg 2007) and Rational Choice Theory. Hence, the value of $s' = p(S|\neg H)$ should not be too low. If we choose, for example, $s' = 1/4$, then we would require $p(H) \in [1/3; 1]$ to satisfy Inequality (3.2) and to make the NMA work. In other words, the NMA only works for theories which are already likely to be empirically adequate. If we were mildly skeptical about H and adopted the prior $p(H) = 0.05$, we

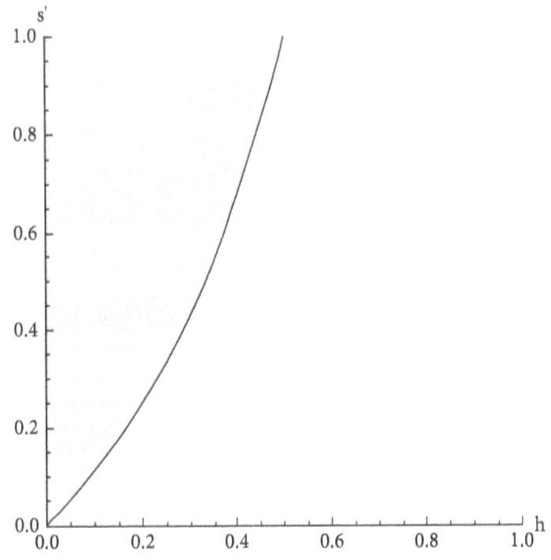

Figure 3.3: The scope of the No Miracles Argument in a naïve Bayesian model, represented graphically. $p(\mathrm{H}|\mathrm{S}) > 1/2$ is the case in the white area below the line. Reproduced under the terms of the Creative Commons Attribution 4.0 International (CC BY 4.0) license from "The Probabilistic No Miracles Argument" by Jan Sprenger, *European Journal for Philosophy of Science*, Volume 6, No. 2, pp. 173–189. © 2015, doi: 10.1007/s13194-015-0122-0.

would require s' to lie in the range $[0; 0.053]$. This would amount to making the assumption that only the empirical adequacy of a scientific theory can explain its predictive success. But this is essentially a realist premise, which the anti-realist could refuse to accept. She could point to the existence of unconceived alternatives (Stanford 2006, ch. 6), the explanatory successes of false theories, and so on. Summing up, our simple Bayesian model of the NMA exposes a dilemma for the realist: to the extent that the NMA is valid, its premises presuppose realist inclinations; to the extent that the NMA builds on premises that are neutral between the realist and the anti-realist, it fails to be valid.

Are things thus hopeless for the realist who wants to convince the anti-realist that the NMA is a good argument? Does "all realistic hope of resuscitating the [no miracles] argument" fail, as Howson (2013, 211) writes? Not necessarily. The Bayesian NMA only took into account the predictive and explanatory success of T. Now we also consider the *stability* of scientific theories as evidence for scientific realism. This move is related to the No Alternatives Argument (NAA, see Variation 2), and in fact, our Bayesian model will be inspired by NAA-type reasoning.

3.2 Extending the No Miracles Argument to Stable Scientific Theories

Recently, Ludwig Fahrbach (2009, 2011) has argued that the stability of major scientific theories in the second half of the twentieth century provides a strong argument against PMI and in favor of scientific realism. In this section, we show how observing **theoretical stability in a scientific discipline** can give a boost to the Bayesian NMA.

Fahrbach's argument is mainly based on scientometric data. He observes an exponential growth of scientific activity in the twentieth century, with a doubling of scientific output every twenty years (Meadows 1974). He also notes that at least 80 % of all scientific work ever done has been conducted after the year 1950 and he observes that our best scientific theories (e.g., the periodic table of elements, optical and acoustic theories, the theory of evolution, etc.) were stable during that period of time. That is, they were neither replaced by competing theories nor did they undergo major conceptual change. Indeed, Laudan's examples in favor of PMI such as the caloric theory of heat, the ether theory in physics or the humoral theory in medicine, stem from the early periods of modern science.

For giving a fair assessment of PMI, we have to take into account the amount of scientific work done in a particular period. Due to the exponential growth of the scientific enterprise, this implies that the period 1800–1820 should, for example, receive much less weight than the period 1950–1970 or 1970–1990. According to Fahrbach, PMI fails because for the items on Laudan's list, "all corresponding theory changes occurred during the time of the first 5 % of all scientific work ever done by scientists" (Fahrbach 2011, 149). If PMI were valid as a general pattern for scientific theory change, we should have observed more turnovers or scientific revolutions in recent decades. However, although the theories of modern science often encounter difficulties, they do, by and large, remain stable. Revolutionary changes do not, or only very rarely, happen. For example, quantum mechanics and relativity theory have been part of modern physics since more than a hundred years. The real scientific challenge does not consist in replacing them, but in integrating them into the unified framework of the Standard Model or a Grand Unified Theory. According to Fahrbach, pessimistic meta-induction stands refuted—or at least, it is no more rational than optimistic meta-induction.

Certainly Fahrbach's model is very simplified. First of all, there is no distinction between fields with a strong theoretical basis (e.g., particle physics) and disciplines with more contested theoretical foundations (e.g., parts of behavioral science). Moreover, many new disciplines emerged in the last sixty to seventy years, and this may also contribute to the effect that Fahrbach observes.

Second, the observed stability of major scientific theories may have sociological as well as epistemic reasons. In the second half of the twentieth century, science has become very professionalized and industrialized. As a consequence, the number of published papers need not be a reliable indicator of the amount of theoretical work done in a discipline—many papers might just apply an existing theory to practical problems (e.g., in engineering). Moreover, feasibility and cost–benefit ratio of a research project are major criteria for evaluating and funding research projects, and this favours "convergent", normal-science research over projects that try to question the received view. Modern science may just be more conservative than most of its history.

Third, also some refuted theories in the history of science, such as Newtonian mechanics, have been stable over a long period of time and enjoyed extraordinary theoretical and predictive successes. Theoretical stability is no guarantee for empirical adequacy. In light of these observations, one may also conclude that the history of science is too complex to tell us as simple a lesson as Fahrbach draws.

However, we are not primarily interested in the descriptive accuracy of Fahrbach's picture of twentieth-century science. Rather than engaging in an actual defense of the realist view, we would like to show a *possibility result*, and to explore whether observations of long-term theoretical stability can buttress the NMA. To this end, we refine our Bayesian model from the previous section.

As before, the propositional variable H expresses the empirical adequacy of theory T, and S denotes the predictive, retrodictive and explanatory success of T. The integer-valued variable A expresses the number of satisfactory alternatives to T, and A_j is our shorthand for the proposition $A = j$. Like in the previous chapter on the NAA, we demand that genuine alternatives satisfy a set of (context-dependent) theoretical constraints $\mathcal{C}$, be consistent with the currently available data $\mathcal{D}$ and give distinguishable predictions for the outcome of some set $\mathcal{E}$ of future experiments. In line with our focus on empirical adequacy rather than truth, we do not distinguish between empirically equivalent theories with different theoretical structures. Finally, major theory change in the domain of T is denoted by C, and absence of change and theoretical stability by $\neg$C. "Theory change" is understood in a broad sense, including scenarios where rivaling theories emerge and end up co-existing with T.

The dependency between these four propositional variables—A, C, H and S—is given by the Bayesian network in Figure 3.4. Proposition S, the success of theory T, depends only on the empirical adequacy of T, that is, on H. The probability of H depends on the number of distinct alternatives that are also consistent with the current data, etc. Finally, the probability of C, that is, of observing substantial theory change, depends on S and A: the predictive success of T and the number of available alternatives. To rule out preservation of a theory by a series of degenerative accommodating moves, the variable C should be evaluated over a longer period (e.g., thirty to fifty years).

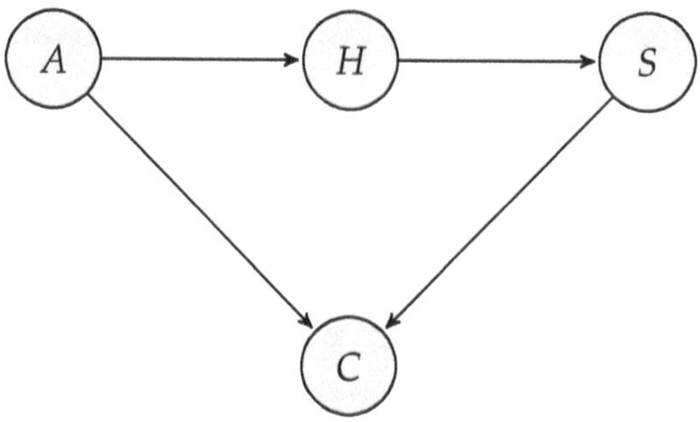

Figure 3.4: The Bayesian network representation of the relation between the variables A (the number of alternatives to T), H (empirical adequacy of theory T), S (predictive success of T) and C (major theory change).

We abbreviate the probability distribution of these variables as follows:

- Denote by $a_j := p(\mathrm{A}_j)$ the probability that there are exactly j alternatives to T that satisfy the theoretical constraints $\mathcal{C}$, are consistent with current data $\mathcal{D}$ and give definite predictions for future experiments $\mathcal{E}$, etc.

- Denote by $h_j := p(\mathrm{H}|\mathrm{A}_j)$ the probability that T is empirically adequate if there are exactly j alternatives to T.

- As before, denote by $s := p(\mathrm{S}|\mathrm{H})$ and $s' := p(\mathrm{S}|\neg\mathrm{H})$ the probability that T is successful if it is (not) empirically adequate.

- Denote by $c_j := p(\neg\mathrm{C}\,|\,\mathrm{A}_j, \mathrm{S})$ the probability that no substantial theory change occurs if T is successful and there are exactly j alternatives to T.

Suppose that we now observe ¬C (no substantial theory change has occurred in the last few decades) and S (theory T is successful). The Bayesian network structure allows for a simple calculation of the posterior probability of H:

Proposition 3.1. *The posterior probability of* H *given* ¬C *and* S *is given by*

$$p(\mathrm{H}\,|\,\neg\mathrm{C},\mathrm{S}) = \frac{\sum_{j=0}^{\infty} a_j c_j s h_j}{\sum_{j=0}^{\infty} a_j c_j \left(s h_j + s'(1-h_j)\right)}. \tag{3.3}$$

In order to study how stable predictive success affects our confidence in the empirical adequacy of T, we make some assumptions on the variables in Equation (3.3).

B1 The variables A, C, H and S satisfy the (conditional) independencies in the Bayesian network structure of Figure 3.4.

B2 If T is empirically adequate then it will be successful in the long run: $p(\mathrm{S}\,|\,\mathrm{H}) = 1$.

B3 The empirical adequacy of T is no more or less probable than the empirical adequacy of an alternative which satisfies the same set of theoretical and empirical constraints: $h_j := p(\mathrm{H}\,|\,\mathrm{A}_j) = 1/(j+1)$. In other words, there is no "actualist bias" in favor of T.

B4 The more satisfactory alternatives exist, the less likely is an extended period of theoretical stability. In other words, $c_j := p(\neg\mathrm{C}\,|\,\mathrm{A}_j)$ is a decreasing function of j. For convenience, we choose $c_j = 1/(j+1)$. (This particular assignment will be relaxed later on.)

B5 Assume that T is our currently best theory and we happen to find a satisfactory alternative T′. Then, the probability of finding another alternative T″ is the same as the probability of finding T′ in the first place. Formally:

$$p\Big(A>j\,\Big|\,\bigvee_{k=j}^{\infty}\mathrm{A}_k\Big) = p\Big(A>j+1\,\Big|\,\bigvee_{k=j+1}^{\infty}\mathrm{A}_k\Big) \quad \text{for all } j\geq 0. \tag{3.4}$$

In other words, finding an alternative does not, in itself, raise or lower the probability of finding another alternative.

Note that **B1–B5** are equally plausible for the realist and the anti-realist. In other words, no realist bias has been incorporated into the assumptions. We can now show the following proposition (all proofs are in the Appendix):

Proposition 3.2. **B5** *implies that* $a_j = a_0 \cdot (1-a_0)^j$.

Together with this proposition, **B1–B5** allow us to rewrite Equation (3.3) as follows:

$$p(\mathrm{H}\,|\,\neg\mathrm{C},\mathrm{S}) = \frac{\sum_{j=0}^{\infty}(1-a_0)^j \frac{1}{(j+1)^2}}{\sum_{j=0}^{\infty}(1-a_0)^j \frac{1-s'j}{(j+1)^2}}. \qquad (3.5)$$

With the help of this formula, we can now rehearse the NMA once more and determine its scope, that is, those parameter values where $p(\mathrm{H}\,|\,\neg\mathrm{C},\mathrm{S}) > 1/2$. The two relevant parameters are a_0, the prior probability that there are no satisfactory alternatives to T, and s', the probability that T is successful although not empirically adequate. Since an analytical solution of Equation (3.5) is not feasible, we conduct a numerical analysis. Results are plotted in Figure 3.5.

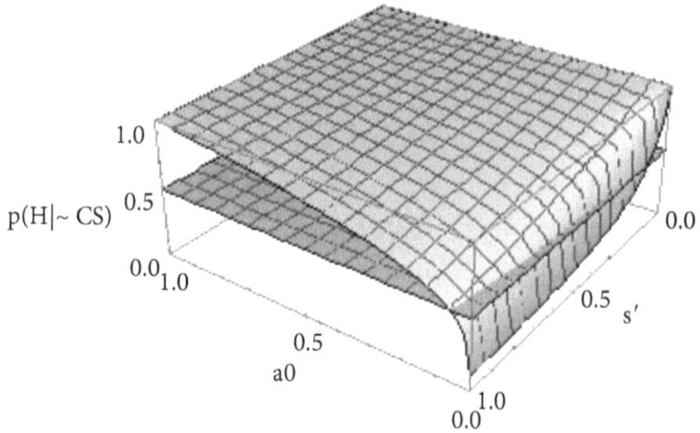

Figure 3.5: The scope of the No Miracles Argument in the revised formulation of the Bayesian model. The posterior probability $p(\mathrm{H}\,|\,\neg\mathrm{C},\mathrm{S})$ of H is plotted as a function of (1) the prior probability that T is empirically adequate (a_0); (2) the probability that T is successful if T is false ($s' = p(\mathrm{S}|\neg\mathrm{H})$). The hyperplane $z = 1/2$ is inserted in order to show for which parameter values $p(\mathrm{H}\,|\,\neg\mathrm{C},\mathrm{S})$ is greater than 1/2. Reproduced under the terms of the Creative Commons Attribution 4.0 International (CC BY 4.0) license from "The Probabilistic No Miracles Argument" by Jan Sprenger, *European Journal for Philosophy of Science*, Volume 6, No. 2, pp. 173–189. Copyright © 2015, doi: 10.1007/s13194-015-0122-0.

These results are very different from the ones in the previous section. With the hyperplane $z = 0.5$ dividing the graph into a region where T may be accepted and a region where this is not the case, we see that the scope of the NMA has increased substantially compared to Figure 3.3. For instance,

$a_0 = p(\mathrm{H}) > 0.1$ suffices for a posterior probability greater than 1/2, almost regardless of the value of s'. This is a striking difference to the previous analysis, where more optimistic values had to be assumed in order to validate the NMA.

So far, the analysis has been conducted in terms of confirmation as firmness, that is, the posterior probability of H. We now complement it by an analysis in terms of confirmation as increase in firmness. That is, we calculate the **evidential support that** $\neg\mathrm{C} \wedge \mathrm{S}$ **confers on** H. We use the log-likelihood measure $l'(\mathrm{H}, \mathrm{E}) = \log_2(p(\mathrm{E}|\mathrm{H})/p(\mathrm{E}|\neg\mathrm{H}))$, which has a good reputation in confirmation theory and a firm standing in scientific practice (see Variation 1 or Royall 1997; Good 2009). Also, it is a confirmation measure that is relatively insensitive to prior probabilities and that describes the discriminative power of the evidence with respect to the realist and the anti-realist hypothesis. The calculations are in the Appendix.

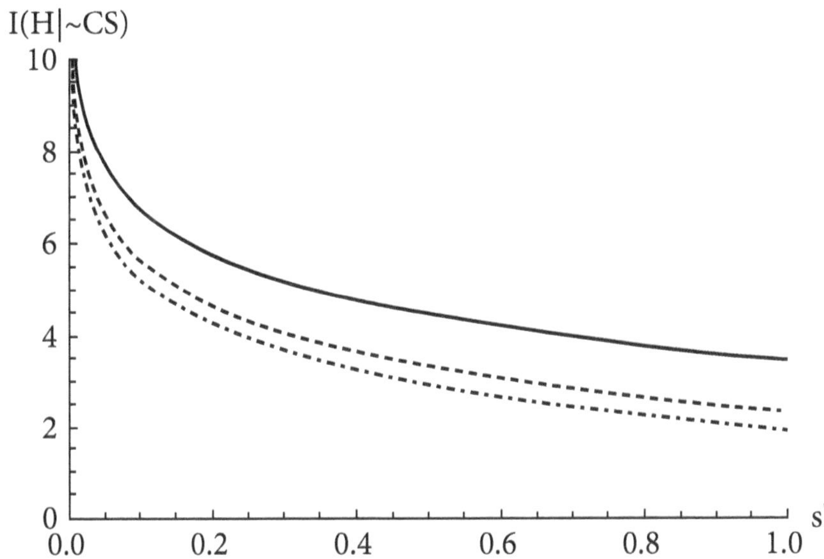

Figure 3.6: The degree of confirmation $l'(\mathrm{H}, \neg\mathrm{C}, \mathrm{S}) = \log_2(p(\neg\mathrm{C}, \mathrm{S}|\mathrm{H})/p(\neg\mathrm{C}, \mathrm{S}|\neg\mathrm{H}))$, that C and S confer on H for three different values of a_0. Full line: $a_0 = 0.01$. Dashed line: $a_0 = 0.05$. Dot-dashed line: $a_0 = 0.1$. Reproduced under the terms of the Creative Commons Attribution 4.0 International (CC BY 4.0) license from "The Probabilistic No Miracles Argument" by Jan Sprenger, *European Journal for Philosophy of Science*, Volume 6, No. 2, pp. 173–189. Copyright © 2015, doi: 10.1007/s13194-015-0122-0.

In Figure 3.6, we have plotted the degree of confirmation as a function of the value of s', for three different values of a_0, namely 0.01, 0.05 and 0.1. As visible from the graph, the (logarithmic) degree of confirmation is substantial for all three cases, even for large values of s'. In particular, it is robust vis-à-vis the values of a_0 and s' and able to withstand the anti-realist argument that plagued the original version of the NMA. Note that if s' is small, as will often be the case in practice, the logarithmic (!) degree of confirmation comes close to 10, which corresponds to a likelihood ratio of more than $1,000$! And even if an anti-realist insists that $s' \approx .2$—not a very plausible assumption—the likelihood ratio hovers in the range between fifteen and thirty. This finding accounts for the realist intuition that the stability of scientific theories over time, together with their predictive success, is strong evidence for their empirical adequacy.

Finally, we relax our assumptions **B1–B5**. Qualitatively, our results do not change if we replace **B2** with the more cautious formulation $p(\mathrm{S}|\mathrm{H}) = 1-\varepsilon$. More interesting is a robustness analysis regarding **B4**. Arguably, the function $c_j := p(\neg\mathrm{C}\,|\,\mathrm{A}_j, \mathrm{S}) = 1/(j+1)$ suggests that scientists are quite ready to give up on their currently best theory in favor of a good alternative. But as many have philosophers and historians of science have argued (e.g., Kuhn 1977b), scientists may be more conservative and continue to work in the standard framework, even if good alternatives exist. Therefore we also analyze a different choice of the c_j, namely $c_j := e^{-1/2(x/\alpha)^2}$, where c_j falls more gently in j. This choice can then be plugged into Equation (3.3), yielding values of $p(\mathrm{H}\,|\,\neg\mathrm{C}, \mathrm{S})$ that are different from the ones in Equation (3.5).

The corresponding graph of $p(\mathrm{H}\,|\,\neg\mathrm{C}, \mathrm{S})$, as a function of a_0 and s', is presented in Figure 3.7. We have set $\alpha = 4$, corresponding to a high degree of reluctance to reject the currently best theory. Yet the results match those from Figure 3.5: the scope of the NMA is much larger than in the simple version of the Bayesian NMA. Hence our findings are more robust with regard to different choices of c_j.

All in all, our model shows that a Bayesian NMA need not be doomed. Its validity depends crucially on the disciplinary context it operates in: What are our expectations regarding the invention of satisfactory alternatives to T? Has the discipline been in a long period of theoretical stability? And so on. Of course our model makes simplifying assumptions, but unlike the assumptions in the original model, they do not carry a realist bias. This allows for a more nuanced and context-sensitive assessment of the realist arguments. The first step of the NMA, from predictive success to empirical adequacy, is valid when theories are stable and few potential

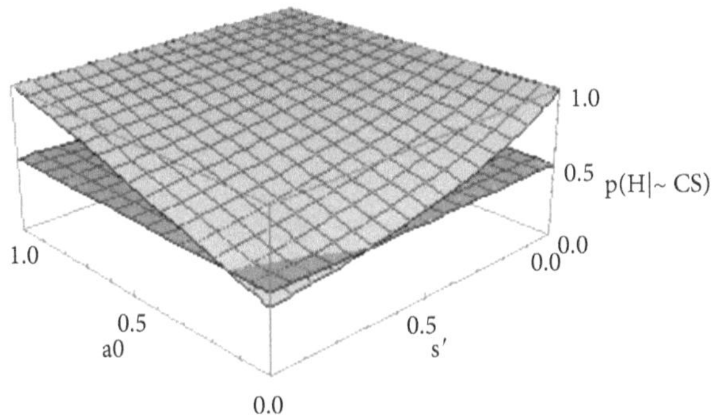

Figure 3.7: The scope of the No Miracles Argument in the revised formulation of the Bayesian model, with $c_j := e^{-1/2(x/\alpha)^2}$. The posterior probability $p(\mathrm{H} \mid \neg\mathrm{C}, \mathrm{S})$ of H is plotted as a function of a_0 and s', like in Figure 3.5, and contrasted with the hyperplane $z = 1/2$. Reproduced under the terms of the Creative Commons Attribution 4.0 International (CC BY 4.0) license from "The Probabilistic No Miracles Argument" by Jan Sprenger, *European Journal for Philosophy of Science*, Volume 6, No. 2, pp. 173–189. Copyright © 2015, doi: 10.1007/s13194-015-0122-0.

alternative explanations of observed phenomena can be conceived. Anti-realist objections are supported by case studies where scientific theories have been volatile or one of our assumptions **B1–B5** is implausible. To our mind, this context-sensitivity is not a vice but a virtue. It explains why realists and anti-realists often talk past each other, and it sketches a fruitful research program for future case studies. In particular, more research is needed into which areas of science have been theoretically stable, and whether the stability indicated by Fahrbach (2009, 2011) is genuine or based on a superficial continuity that hides substantial shifts in the meaning of theoretical terms.

3.3 The Frequency-Based No Miracles Argument

In the above analysis, the empirical adequacy of a particular theory T explained why T is predictively successful. We shall call an NMA that tries to derive the empirical adequacy of theory T from its predictive success an **individual-theory-based No Miracles Argument.** The previous two sections were devoted to analyzing such arguments. However, there is another way of conceptualizing the NMA. Following this second understanding, what is to be explained by the realist conjecture is not the empirical adequacy of a particular theory (e.g., the Standard Model of modern particle physics), but the *tendency* of theories in mature science to be empirically adequate. In this version, the NMA primarily relies on observed characteristics of science as a whole, or of a specific segment of science. Theories in that segment, such as theories that are part of a mature research field, are expected to have a high rate of being empirically adequate. We will call an NMA based on the frequency of predictive success a **frequency-based No Miracles Argument.** This is actually the type of argument that Hilary Putnam put forward in his famous characterization of the NMA:

> The positive argument for scientific realism is that it is the only philosophy that does not make the success of science a miracle. That terms in a mature science *typically* refer (this formulation is due to Richard Boyd), that the theories accepted in a mature science are *typically* approximately true, that the same terms can refer to the same even when they occur in different theories—these statements are viewed [by the scientific realist] not as necessary truths but as part of the only scientific explanation of the *success of science* and hence as part of any adequate description of science and its relations to its objects. (Putnam 1975, 73, our emphasis)

Note that Putnam speaks of the success of *science* rather than of the success of an individual scientific theory. He clearly understands the success of science as a general and observable phenomenon. There is a high success rate of scientific theories based on our observations of the history of (mature) science. Putnam infers that mature scientific theories are typically approximately true, or at least empirically adequate.

Another early main exponent of the NMA, Richard Boyd (1980, 1983, 1984), is committed to the frequency-based NMA as well. Boyd emphasizes that only what he calls the "predictive reliability of well-confirmed scientific theories" and the "reliability of scientific methodology in identifying predictively reliable theories" provides the basis for the NMA.

It is important to note that the frequency-based NMA is not adequately captured by Howson's reconstruction. Howson takes into account the prior probability of a theory T, but he does not consider the success frequency of the predecessors of T. An accurate Bayesian reconstruction of the frequency-based NMA must include updating under the observation of scientific successes and failures in the entire research field. This will be done now.

Consider a scientific discipline or research field. We count all n_E theories in the field that have been empirically tested, and determine the number n_S of theories that were predictively successful. We can thus state the following observation O:

O: n_S out of n_E theories in the research field are predictively successful.

Let us assume that we are confronted with a new and so far empirically untested theory T in that research field. We want to extract the probability $p(S|O)$ for the predictive success S of T given observation O. In order not to beg the question by assuming predictive success a priori, we assume a prior probability $p(S) = \varepsilon$, where ε can be an arbitrarily small number.

We then assume that each new theory that appears in the research field can be treated as a random pick with respect to predictive success. That is, we assume that there is a certain overall rate of predictively successful theories in the research field, and in the absence of further knowledge, the success chances of a new theory should be estimated according to our best estimate r of that success rate:

$$r = p(S|O).$$

The value of r will be based on observation O. The most straightforward assessment of r is to use the long-run information about the frequency of success in a discipline and to identify r with n_S/n_E.

Moreover, we make two assumptions similar to the individual-theory-based NMA:

F1′ $p(S|H,O)$ is quite large.

F2′ $p(S|\neg H,O) < k \ll 1$.

Note that realists assume that the empirical adequacy of T is the dominating element in explaining the theory's predictive success. If that is so, then S is roughly conditionally independent of O given H (i.e., given that theory T is empirically adequate) and we have $p(S|H,O) \approx p(S|H)$ and $p(S|\neg H,O) \approx p(S|\neg H)$. The conditions **F1′** and **F2′** then roughly correspond to **F1** and **F2**.

We now come to the crucial point of our analysis: *Accounting for observation* O *blocks the base rate fallacy*. The base rate fallacy in the individual-theory-based NMA consisted in disregarding the possibility of arbitrarily small priors $p(H)$. In the frequency-based NMA, however, the crucial probability is $p(H|O)$ rather than $p(H)$.

Proposition 3.3. *If Conditions* **F1′** *and* **F2′** *are satisfied, then the following inequality holds:*

$$p(\mathrm{H}\,|\,\mathrm{O}) > r - k. \tag{3.6}$$

The frequency-based NMA takes it as a premise, as an observed fact about (parts of) mature science, that $n_\mathrm{S}/n_\mathrm{E}$ is fairly large. Hence, r is fairly large and we can infer from Equation (3.6) that $p(\mathrm{H}\,|\,\mathrm{O})$ is also substantially greater than zero. Thus, the base rate fallacy is avoided. Note that the first and crucial inference in the frequency-based NMA is made before accounting for the predictive success of T itself: it relates $p(\mathrm{S}\,|\,\mathrm{O})$ to $p(\mathrm{H}\,|\,\mathrm{O})$ by the Law of Total Probability.

However, the final strength of NMA is expressed by the value $p(\mathrm{H}\,|\,\mathrm{S},\mathrm{O})$. In other words, the realist has to show that

$$p(\mathrm{H}\,|\,\mathrm{S},\mathrm{O}) > K, \tag{3.7}$$

where K is, as before, some reasonably high probability value. $K = 1/2$ may be viewed as a plausible condition for taking the NMA seriously. How does a condition on $p(\mathrm{H}\,|\,\mathrm{S},\mathrm{O})$ translate into a condition on $p(\mathrm{S}\,|\,\mathrm{O})$ and therefore on the observed success frequency r? First we observe the following result:

Proposition 3.4.

$$p(\mathrm{H}\,|\,\mathrm{S},\mathrm{O}) = \frac{p(\mathrm{S}\,|\,\mathrm{H},\mathrm{O})}{p(\mathrm{S}\,|\,\mathrm{O})} \cdot \left(\frac{p(\mathrm{S}\,|\,\mathrm{O}) - p(\mathrm{S}\,|\,\neg\mathrm{H},\mathrm{O})}{p(\mathrm{S}\,|\,\mathrm{H},\mathrm{O}) - p(\mathrm{S}\,|\,\neg\mathrm{H},\mathrm{O})} \right).$$

Then we observe that $p(\mathrm{H}\,|\,\mathrm{S},\mathrm{O})$ is decreasing in $p(\mathrm{S}\,|\,\mathrm{H},\mathrm{O})$ if $p(\mathrm{S}\,|\,\mathrm{O})$ and $p(\mathrm{S}\,|\,\neg\mathrm{H},\mathrm{O})$ are held fixed. It thus makes sense to focus on the case where it is most difficult for the realist to make the frequency-based NMA work, namely the case $p(\mathrm{S}\,|\,\mathrm{H},\mathrm{O}) = 1$. Then, the following theorem describes a sufficient condition for $p(\mathrm{H}\,|\,\mathrm{S},\mathrm{O})$ to exceed the threshold K:

Theorem 3.1 (Frequency-Based NMA). *If Conditions* **F1′** *and* **F2′** *are satisfied, if* $p(\mathrm{S}\,|\,\mathrm{H},\mathrm{O}) = 1$, *and if the following inequality holds:*

$$p(\mathrm{S}\,|\,\mathrm{O}) > \frac{p(\mathrm{S}\,|\,\neg\mathrm{H},\mathrm{O})}{1 - K + K \cdot p(\mathrm{S}\,|\,\neg\mathrm{H},\mathrm{O})},$$

then Inequality (3.7) is satisfied as well: $p(\mathrm{H}\,|\,\mathrm{S},\mathrm{O}) > \mathrm{K}$.

For $K = 1/2$, and using the base rate estimate $p(\mathrm{S}\,|\,\mathrm{O}) = n_\mathrm{S}/n_\mathrm{E}$, this is the case if and only if

$$n_\mathrm{S}/n_\mathrm{E} > 2\frac{p(\mathrm{S}\,|\,\neg\mathrm{H},\mathrm{O})}{1 + p(\mathrm{S}\,|\,\neg\mathrm{H},\mathrm{O})}. \tag{3.8}$$

In particular, $2p(\mathrm{S} \mid \neg\mathrm{H}, \mathrm{O}) < n_{\mathrm{S}}/n_{\mathrm{E}}$ is sufficient for satisfying Equations (3.8) and (3.7). Thus we don't need an impressively high rate of predictive success to buttress an argument in favor of scientific realism: as long as (1) predictive success is relatively unlikely if T is not empirically adequate and (2) past theories in the discipline did well on average, we should accept T. A defender of the NMA can avoid the base rate fallacy by taking a global perspective on the success of science. This perspective is also faithful to the intentions of those philosophers who put the NMA forward in the first place—namely Hilary Putnam and Richard Boyd.

All this does not imply that the NMA is valid. A supporter of the frequency-based NMA must specify on which grounds she takes a high frequency of predictive success in science to be borne out by the data. And she must undertake the difficult task of justifying assumptions **F1′** and (especially) **F2′**. But we have shown, contra Howson, that the NMA still has a fighting chance.

3.4 Discussion

This Variation has investigated the scope and limits of the No Miracles Argument (NMA) when formalized as a Bayesian argument aiming at the empirical adequacy of a particular theory T. For philosophers like ourselves, who are not committed to a particular position in the debate between realists and anti-realists, the Bayesian reconstruction of the NMA offers the chance to understand the argumentative mechanics behind the realist intuition, to explain and to guide the strategies that realists and anti-realists pursue, and to critically evaluate the merits of realist and anti-realist standpoints.

In the simple Bayesian model of the NMA, we have confirmed the diagnosis that it does not hold water as an objective argument (Howson 2000, 2013): too much depends on the choice of the prior probability $p(\mathrm{H})$, assuming what is supposed to be shown. We have supported this diagnosis by a detailed analysis of the Bayesian mechanics of NMA.

Then, we showed two ways out of the dilemma: one for the NMA based on individual theories and one for the frequency-based NMA. First, we have investigated how the stability of a scientific theory T influences the probability that T is empirically adequate. We have shown that observations of theoretical stability over an extended period of time can greatly increase the range of prior probabilities for which the NMA leads to acceptance of T. This finding mitigates Howson's pessimistic conclusion and gives new hope to the realist.

In a second model, we have taken the base rate of predictively successful theories into account. This mirrors the view of those scientific realists (Boyd, Putnam) who conceive of the NMA as a global argument based on the high frequency of successful theories in science. In both cases, we have supplemented the classical NMA reasoning with novel and distinct kinds of evidence that can be embedded into the Bayesian framework. Using our models, the realist thesis (or at least the part leading up to empirical adequacy) can be defended with much weaker assumptions than in the simple version of the NMA.

It is noteworthy that all formalizations of the NMA stress the scientific track record in the particular discipline to which T belongs. Instead of reading NMA as a "wholesale argument" for scientific realism that is valid across the board, we should understand it as a "retail argument" (Magnus and Callender 2004), that is, as an argument that may be strong for some scientific theories and weak for others. While context-sensitive assumptions are required in our arguments, their relative weakness leaves open the possibility of a coherent, non-circular realist position in philosophy of science. It also makes the realist argument more sensitive to scientific practice, which is, ultimately, something that all formal reconstructions of scientific reasoning should aim at.

This observation brings us to projects for future research. Obviously the formal models presented in this Variation would greatly benefit from empirical input. First, one has to examine whether theories have really been quite stable during the second half of the twentieth century: this is a crucial premise of the Bayesian individual-theory-based NMA. For tackling this research question, a combination of case studies about particular research programs and scientometric analysis (e.g., along the lines of Herfeld and Doehne forthcoming) strikes us as a promising approach.

Second, we have seen that the Bayesian versions of the NMA are highly sensitive to the probability of predictive success if T is not empirically adequate. The evaluation of this quantity is itself a point of contention between realists and anti-realists; after all, anti-realists often stress the explanatory success of false models, whereas realists are usually committed to the thesis that only true models yield stable predictive success. But how often do such cases occur in practice?

Third, there is a certain analogy between the argument from theoretical stability and the No Alternatives Argument (NAA)—the claim that the continuous failure to find satisfactory alternatives to a theory provides evidence for it. In the previous Variation, we have shown that under plausible assumptions, this observation raises the probability that theory T is empirically adequate. The NAA can also be seen as a variation of the NMA: the empirical adequacy of T is the only explanation for why scientists have not yet found an alternative. It would be exciting to investigate the parallel between the NAA and the NMA in greater detail, and to proceed to a general analysis of argument patterns that take non-empirical evidence (in our technical sense of the word) as their premise.

Fourth, it would be good to explore whether the Bayesian NMA can be developed into an argument for the full realist position, that is, the view that T is true (and not only empirically adequate). At present, we do not see an obvious way of doing so, but we invite realists to take our formalism and to extend it to a full-fledged defense of the realist view.

Finally, one could replace empirical adequacy and truth as goals of scientific theorizing by the notion of **verisimilitude** or **truthlikeness** (Popper 1959/2002; Niiniluoto 1999; Zamora Bonilla 2000; Cevolani and Tambolo 2013). According to this approach, which combines realist and anti-realist intuitions, we do not have to assume that there is one true or empirically adequate theory among our best scientific theories. It suffices to be able to rank them by means of their distance to the true theory, and their predictive track record gives us a way to estimate which theory may have the smallest distance to the truth. It is therefore highly interesting to investigate whether we can use our Bayesian models for investigating whether, and under which circumstances, predictively successful theories are more often truthlike.

Together with the preceding Variation, this Variation has shown how Bayesian reasoning can model and vindicate non-empirical evidence, and how Bayesian models can contribute to a fair assessment of the debate between realists and anti-realists. We now move on to modeling theory confirmation by yet another type of evidence: evidence that is presented in the form of indicative conditionals (Variation 4), and "old evidence" which is explained by a new theory (Variation 5).

Appendix: Proofs of the Theorems

Proof of Proposition 3.1:

$$\begin{aligned}
p(\neg C, S, H) &= \sum_A p(A)\, p(\neg C \,|\, A, S)\, p(S|H)\, p(H|A) \\
&= \sum_{j=0}^{\infty} a_j c_j s h_j, \\
p(\neg C, S) &= \sum_{A,H} p(A)\, p(\neg C \,|\, A, S)\, p(S|H)\, p(H|A) \\
&= \sum_A p(A)\, p(\neg C \,|\, A, S)\, p(S|H)\, p(H|A) \; + \\
&\qquad \sum_A p(A)\, p(\neg C \,|\, A, S)\, p(S|\neg H) p(\neg H|A) \\
&= \sum_{j=0}^{\infty} a_j c_j \big(s h_j + s'(1-h_j)\big).
\end{aligned}$$

With the help of Bayes' Theorem, these equations allows us to calculate the posterior probability of H conditional on C and S:

$$p(H|\neg C, S) = \frac{p(\neg C, S, H)}{p(\neg C, S)} = \frac{\sum_{j=0}^{\infty} a_j c_j s h_j}{\sum_{j=0}^{\infty} a_j c_j \big(s h_j + s'(1-h_j)\big)}. \qquad \square$$

Proof of Proposition 3.2: Assumption **B4** is equivalent to the following claim:

$$p\big(A_j \,\big|\, \textstyle\bigvee_{k=j}^{\infty} A_k\big) = p\big(A_{j+1} \,\big|\, \textstyle\bigvee_{k=j+1}^{\infty} A_k\big) \qquad \text{for all } j \geq 0,$$

which entails that for all $j \geq 0$, we have

$$\frac{p(A_j)}{p\big(\bigvee_{k=j}^{\infty} A_k\big)} = \frac{p(A_{j+1})}{p\big(\bigvee_{k=j+1}^{\infty} A_k\big)}.$$

This implies in turn

$$p(A_{j+1}) = p(A_j) \frac{p\big(\bigvee_{k=j+1}^{\infty} A_k\big)}{p\big(\bigvee_{k=j}^{\infty} A_k\big)} = p(A_j) \frac{1 - \sum_{k=0}^{j} p(A_k)}{1 - \sum_{k=0}^{j-1} p(A_k)}.$$

By a simple induction proof, we can now show

$$p(A_n) = p(A_0)\left(1 - \sum_{k=0}^{n-1} p(A_k)\right). \qquad (3.9)$$

For $n=1$, Equation (3.9) follows immediately. Assuming that it holds for level n, we then obtain

$$p(\mathrm{A}_{n+1}) = p(\mathrm{A}_n)\frac{1-\sum_{k=0}^{n} p(\mathrm{A}_k)}{1-\sum_{k=0}^{n-1} p(\mathrm{A}_k)}$$
$$= p(\mathrm{A}_0)\Big(1-\sum_{k=0}^{n-1} p(\mathrm{A}_k)\Big)\frac{1-\sum_{k=0}^{n} p(\mathrm{A}_k)}{1-\sum_{k=0}^{n-1} p(\mathrm{A}_k)} = p(\mathrm{A}_0)\Big(1-\sum_{k=0}^{n} p(\mathrm{A}_k)\Big),$$

where we have used the inductive premise in the second step. Finally, we use straight induction once more to show that

$$p(\mathrm{A}_n) = p(\mathrm{A}_0)(1-p(\mathrm{A}_0))^n, \tag{3.10}$$

where the case $n=0$ is trivial and the inductive step $n \to n+1$ is proven as follows:

$$\begin{aligned} p(\mathrm{A}_{n+1}) &= p(\mathrm{A}_0)\left(1-\sum_{k=0}^{n} p(\mathrm{A}_k)\right) \\ &= p(\mathrm{A}_0)\left(1-\sum_{k=0}^{n} p(\mathrm{A}_0)(1-p(\mathrm{A}_0))^k\right) \\ &= p(\mathrm{A}_0)\left(1-p(\mathrm{A}_0)\frac{1-(1-p(\mathrm{A}_0))^{n+1}}{1-(1-p(\mathrm{A}_0))}\right) \\ &= p(\mathrm{A}_0)\left(1-\left(1-(1-p(\mathrm{A}_0))^{n+1}\right)\right) \\ &= p(\mathrm{A}_0)(1-p(\mathrm{A}_0))^{n+1}. \end{aligned}$$

In the second line, we have applied the inductive premise to $p(\mathrm{A_k})$, and in the third line, we have used the well-known formula for the geometric series:

$$\sum_{k=0}^{n} q^k = \frac{1-q^{n+1}}{1-q}.$$

This shows Equation (3.10) and completes the proof of the proposition. □

Calculation of the degree of confirmation (Figure 3.6):

$$\begin{aligned} p(\neg\mathrm{C},\mathrm{S}\,|\,\mathrm{H}) &= \frac{p(\neg\mathrm{C},\mathrm{S},\mathrm{H})}{p(\mathrm{H})} \\ &= \frac{\sum_A p(A)\,p(\neg\mathrm{C}\,|\,A,\mathrm{S})\,p(\mathrm{S}\,|\,\mathrm{H})\,p(\mathrm{H}\,|\,A)}{\sum_A p(A)\,p(\mathrm{H}\,|\,A)} \end{aligned}$$

$$= \frac{\sum_{j=0}^{\infty} a_j c_j s h_j}{\sum_{j=0}^{\infty} a_j h_j} = \frac{\sum_{j=0}^{\infty} (1-a_0)^j \frac{1}{(1+j)^2}}{\sum_{j=0}^{\infty} (1-a_0)^j \frac{1}{1+j}},$$

$$\begin{aligned} p(\neg C, S \mid \neg H) &= \frac{p(\neg C, S, \neg H)}{p(\neg H)} \\ &= \frac{\sum_A p(A)\, p(\neg C \mid A, S)\, p(S \mid \neg H)\, p(\neg H \mid A)}{\sum_A p(A)\, p(\neg H \mid A)} \\ &= \frac{\sum_{j=0}^{\infty} a_j c_j s'(1-h_j)}{\sum_{j=0}^{\infty} a_j (1-h_j)} = \frac{\sum_{j=0}^{\infty} (1-a_0)^j \frac{s' j}{(1+j)^2}}{\sum_{j=0}^{\infty} (1-a_0)^j \frac{j}{1+j}}. \qquad \square \end{aligned}$$

Proof of Proposition 3.3: We first apply the Law of Total Probability and obtain

$$\begin{aligned} p(S \mid O) &= p(S \mid H, O)\, p(H \mid O) + p(S \mid \neg H, O)\, p(\neg H, O) \\ &= p(S \mid H, O)\, p(H \mid O) + p(S \mid \neg H, O) \cdot (1 - p(H, O)) \\ &= p(H, O) \cdot \big(p(S \mid H, O) - p(S \mid \neg H, O) \big) + p(S \mid \neg H, O). \end{aligned}$$

Hence,

$$p(H \mid O) = \frac{p(S \mid O) - p(S \mid \neg H, O)}{p(S \mid H, O) - p(S \mid \neg H, O)}. \tag{3.11}$$

From Equation (3.11), we derive

$$\begin{aligned} p(H \mid O) &= \frac{p(S \mid O) - p(S \mid \neg H, O)}{p(S \mid H, O) - p(S \mid \neg H, O)} \\ &\geq \frac{p(S \mid O) - p(S \mid \neg H, O)}{1 - p(S \mid \neg H, O)} \\ &\geq p(S \mid O) - p(S \mid \neg H, O) &\quad (3.12) \\ &> p(S \mid O) - k. &\quad (3.13) \end{aligned}$$

Here Equation (3.12) follows from assumptions **F1′** and **F2′** and Equation (3.13) follows from assumption **F2′**. This leads directly to the desired result:

$$p(H \mid O) > p(S \mid O) - k = R - k. \qquad \square$$

Proof of Proposition 3.4: From Bayes' Theorem and Equation (3.11), we infer

$$\begin{aligned} p(H \mid S, O) &= \frac{p(H \mid O)\, p(S \mid H, O)}{p(S \mid O)} \\ &= \frac{p(S \mid H, O)}{p(S \mid O)} \cdot \frac{p(S \mid O) - p(S \mid \neg H, O)}{p(S \mid H, O) - p(S \mid \neg H, O)}. \end{aligned} \tag{3.14}$$

□

Proof of Theorem 3.1: We have assumed that $p(\mathrm{S} \mid \mathrm{H}, \mathrm{O}) = 1$. Inserting this equality into Equation (3.14) gives us

$$p(\mathrm{H} \mid \mathrm{S}, \mathrm{O}) = \frac{1}{p(\mathrm{S} \mid \mathrm{O})} \cdot \frac{p(\mathrm{S} \mid \mathrm{O}) - p(\mathrm{S} \mid \neg\mathrm{H}, \mathrm{O})}{1 - p(\mathrm{S} \mid \neg\mathrm{H}, \mathrm{O})}.$$

Hence we can write the condition $p(\mathrm{H} \mid \mathrm{S}, \mathrm{O}) > K$ as

$$\frac{1}{p(\mathrm{S} \mid \mathrm{O})} \cdot \frac{p(\mathrm{S}, \mathrm{O}) - p(\mathrm{S} \mid \neg\mathrm{H}, \mathrm{O})}{1 - p(\mathrm{S} \mid \neg\mathrm{H}, \mathrm{O})} > K.$$

Rewriting this inequality a couple of times, we obtain

$$\begin{aligned}
\frac{1}{p(\mathrm{S} \mid \mathrm{O})} \cdot \Big(p(\mathrm{S} \mid \mathrm{O}) - p(\mathrm{S} \mid \neg\mathrm{H}, \mathrm{O})\Big) &> K(1 - p(\mathrm{S} \mid \neg\mathrm{H}, \mathrm{O})), \\
1 - \frac{p(\mathrm{S} \mid \neg\mathrm{H}, \mathrm{O})}{p(\mathrm{S} \mid \mathrm{O})} &> K(1 - p(\mathrm{S} \mid \neg\mathrm{H}, \mathrm{O})), \\
1 - K(1 - p(\mathrm{S} \mid \neg\mathrm{H}, \mathrm{O})) &> \frac{p(\mathrm{S} \mid \neg\mathrm{H}, \mathrm{O})}{p(\mathrm{S} \mid \mathrm{O})}, \\
p(\mathrm{S} \mid \mathrm{O}) \cdot \Big(1 - K(1 - p(\mathrm{S} \mid \neg\mathrm{H}, \mathrm{O}))\Big) &> p(\mathrm{S} \mid \neg\mathrm{H}, \mathrm{O}), \\
p(\mathrm{S} \mid \mathrm{O}) &> \frac{p(\mathrm{S} \mid \neg\mathrm{H}, \mathrm{O})}{1 - K + K \cdot p(\mathrm{S} \mid \neg\mathrm{H}, \mathrm{O})}.
\end{aligned}$$

This was exactly one of the assumptions of the theorem. Thus we can infer that, indeed, $p(\mathrm{H} \mid \mathrm{S}, \mathrm{O}) > K$. □

Variation 4: Learning Conditional Evidence

Indicative conditionals of the form "if A, then C" (shorthand notation: $A \rightarrow C$) constitute a substantial part of scientific evidence. Often, we describe the results of experimental interventions in this way: If a piece of sugar is immersed into a glass of water, it will dissolve. If plants are not watered, they will die. If the framing of a decision problem is changed, many agents will reverse their preferences.

Similarly, the results of scientific studies are often expressed in terms of **relative frequencies.** For example, 20 % of all patients treated with a particular drug report a particular side effect, such as severe headache. Or 50 % of the participants in a simulation of the Ultimatum Game in economics decide to reject a 70/30 offer. This information can be expressed as conditional probabilities: for example, the probability of headache (E) for patients in the treatment group (H) is $p(E|H) = .2$.

Learning indicative conditionals is closely related to learning relative frequencies: the proposition "if H, then E" is commonly held to imply that $p(E|H) = 1$. In the literature on conditionals, there is broad consensus that anybody who accepts "if H, then E" should also accept that the conditional probability of E, given H, is 1, or a value very close to it (e.g., Adams 1965; Edgington 1995; Bennett 2003). This assumption is compatible with agnosticism about the propositional status of conditionals. It is also weaker than Stalnaker's Thesis, which identifies the probability of a conditional with its conditional probability (Stalnaker 1968, 1970, 1975)—a claim that is vulnerable to triviality arguments in the style of Lewis 1976.

Moreover, universal generalizations in science such as "All planets move in elliptical orbits" or "All whales are mammals" can be expressed either as indicative conditionals or as relative frequencies. Their logical form is $\forall x$: $(Wx \rightarrow Mx)$. This can be rephrased either as "If something is a whale, then it is a mammal" or as $p(\text{Mammal}|\text{Whale}) = 1$. Hence an account of learning indicative conditionals and relative frequencies has also direct implications on how we should change our belief upon learning universal generalizations in science.

All these types of evidence can be described as **conditional evidence:** information that is relative to supposing a certain proposition or reference class. This Variation describes how we should revise our beliefs upon learning such conditional evidence. Within statistics, a large amount of literature has been devoted to the question of learning relative frequencies, but without conclusive results. Some proposals, such as R. A. Fisher's (1935) famous "fiducial argument", try to infer directly from an observed relative frequency to the probability of the hypothesis in question (see also Fisher 1935/74; Seidenfeld 1979a). Other authors try to reconcile Bayesian and frequentist inference by integrating the learning of relative frequencies into an objective Bayesian framework (e.g., Kyburg and Teng 2001; Williamson 2013). Finally, some approaches are motivated by the question of how we should reason with indicative conditionals. We will now expound two classical proposals along the latter lines.

First and most straightforwardly, one might identify the natural language indicative conditional $\mathrm{H} \rightarrow \mathrm{E}$ with the material conditional $\mathrm{H} \supset \mathrm{E}$, which is equivalent to $\neg\mathrm{H} \vee \mathrm{E}$. Learning the relative frequency $p(\mathrm{E}\,|\,\mathrm{H}) = 1$ can then be conceptualized as conditionalizing on $\neg(\mathrm{H} \wedge \neg\mathrm{E}) = \neg\mathrm{H} \vee \mathrm{E}$. This proposition is a natural way of conceptualizing both the indicative conditional and the associated relative frequency.

Popper and Miller (1983) have challenged this proposal with the following argument: Consider two propositions H and E and a probability distribution p with $0 < p(\mathrm{H}) < 1$ and $p(\mathrm{E}\,|\,\mathrm{H}) < 1$. We now learn the conditional evidence "if H, then E" (alternatively, $p(\mathrm{E}\,|\,\mathrm{H}) = 1$), which implies that we use Bayesian Conditionalization on the proposition $\neg\mathrm{H} \vee \mathrm{E}$ in order to obtain a posterior probability distribution $p'(\mathrm{H}) := p(\mathrm{H}\,|\,\neg\mathrm{H} \vee \mathrm{E})$. Then, it is always the case that $p'(\mathrm{H}) < p(\mathrm{H})$. The proof is elementary; so we reproduce it here. Bayes' Theorem implies that

$$p(\mathrm{H}\,|\,\neg\mathrm{H} \vee \mathrm{E}) \;=\; p(\neg\mathrm{H} \vee \mathrm{E}\,|\,\mathrm{H}) \cdot \frac{p(\mathrm{H})}{p(\neg\mathrm{H} \vee \mathrm{E})},$$

and hence it is sufficient to show that $p(\neg\mathrm{H} \vee \mathrm{E}\,|\,\mathrm{H}) < p(\neg\mathrm{H} \vee \mathrm{E})$:

$$\begin{aligned} p(\neg\mathrm{H} \vee \mathrm{E}\,|\,\mathrm{H}) - p(\neg\mathrm{H} \vee \mathrm{E}) \;&=\; p(\neg\mathrm{H}\,|\,\mathrm{H}) + p(\mathrm{E}\,|\,\mathrm{H}) - p(\neg\mathrm{H} \wedge \mathrm{E}\,|\,\mathrm{H}) \\ &\qquad - (1 - p(\mathrm{H}) + p(\mathrm{H} \wedge \mathrm{E})) \\ &=\; p(\mathrm{E}\,|\,\mathrm{H}) - (1 - p(\mathrm{H})) - p(\mathrm{E}\,|\,\mathrm{H})\, p(\mathrm{H}) \end{aligned}$$

$$\begin{aligned} &= p(E|H)(1-p(H)) - (1-p(H)) \\ &= (1-p(H))(p(E|H)-1) \\ &< 0. \end{aligned}$$

In other words, learning "if H, then E" (which implies $p(E|H) = 1$) always decreases the probability of H if one interprets the new information as a material conditional $\neg H \vee E$. However, there are cases where, upon learning a particular prediction, we intuitively judge the posterior probability of a hypothesis H to be greater than or equal to its prior probability. We give some examples in Section 4.2. The proposal to identify learning conditional evidence with learning the associated material conditional has trouble to account for such evidence—at least at first sight.

Alternatively, we could rethink the dynamics of Bayesian reasoning. Their main principle—Bayesian Conditionalization—is excellently suited for learning propositional evidence, but the propositional status of *conditional* evidence is very unclear. For example, the view that the truth conditions of the indicative conditional $H \rightarrow E$ depend on the truth values of H and E is much less accepted than the various non-truth-functional (suppositionalist, inferentialist, ...) accounts proposed in the literature (e.g., Edgington 1995; Douven 2016).

Therefore, one may **model learning conditional evidence by imposing constraints on the posterior distribution** p'. For instance, learning an indicative conditional "if H, then E" can be identified with learning the conditional probability $p'(E|H) = 1$. Among all distributions p' that satisfy this condition and are compatible with our background knowledge, one could choose the distribution that is in some sense closest to the prior distribution, implementing a principle of epistemic conservativeness. Such a divergence-based approach would offer a general theory of learning conditional evidence that extends beyond the special case of learning "categorical" conditional probabilities (i.e., $p'(E|H) \in \{0,1\}$).

In the introductory chapter, we have mentioned important results by Csiszár (1967, 1975) and Diaconis and Zabell (1982): minimizing an *f*-divergence between prior and posterior distribution (e.g., Kullback–Leibler divergence) agrees with Bayesian Conditionalization for the case of learning propositional evidence $p'(E) = 1$. The divergence-based approach can be seen as a **generalization of Bayesian Conditionalization:** it agrees with Bayesian Conditionalization for learning first-order propositions, and with Jeffrey Conditionalization for propositions that are learned with less than certainty. On top of this, it covers learning conditional evidence. However, also the divergence-based approach has been argued to deliver unintuitive results in a variety of cases, which we explain in Section 4.2 (Douven and Dietz 2011; Douven and Romeijn 2011). If these criticisms are correct, a convincing account of learning conditional evidence (e.g., indicative conditionals and relative frequencies) is still to be found (Douven 2012).

In this Variation, we subsume learning indicative conditionals under the more general problem of learning conditional evidence (see also Eva, Hartmann and Rafiee Rad forthcoming). We amend the methods of updating on the material conditional and minimizing an f-divergence by a causal perspective. First, we show that both methods agree for propositional evidence and deliver the same results as Bayesian Conditionalization. On top of this, we show that for conditional probability constraints of the type $p'(\mathrm{E}\,|\,\mathrm{H}) = 1$, divergence minimization agrees with conditionalizing on the material conditional $\mathrm{H} \supset \mathrm{E} = \neg\mathrm{H} \vee \mathrm{E}$ (Section 4.1). Second, we adapt the alleged counterexamples to this method to a scientific context (Section 4.2). Third, we argue that these counterexamples neglect the causal structure of the relevant propositions. Once the causal structure is properly taken into account, the divergence minimization method (and a fortiori, updating on the material conditional) yield the intuitively correct results (Section 4.3). We extend the divergence minimization method to cases where the learned relative frequencies are non-extreme, such as van Fraassen's famous Judy Benjamin case (Section 4.4). Finally, we wrap up our results and sketch questions for future research (Section 4.5).

4.1 Divergence Minimization and Bayesian Conditionalization

This section clarifies the deeper links between Bayesian Conditionalization and divergence minimization as two different modes of belief revision. It will be shown that for the case of learning propositional evidence, a large class of divergence minimization procedures agrees with Bayesian Conditionalization. More specifically, we will show that conditionalizing on the material conditional $\mathrm{H} \supset \mathrm{E} = \neg\mathrm{H} \vee \mathrm{E}$ is equivalent to divergence minimization with the constraint $p'(\mathrm{E}\,|\,\mathrm{H}) = 1$ on the posterior distribution.

We first repeat a result from the introductory chapter of this book: Bayesian (and Jeffrey) Conditionalization is equivalent to minimizing an f-divergence between prior and posterior distribution, subject to the constraints expressed by the learned evidence. Here are again the definitions of an *f*-divergence:

Definition 4.1 (*f*-divergences, discrete probability spaces). *Let the set* $\Omega = \{\omega_1, \omega_2, \ldots, \omega_N\}$ *(including the possibility* $N = \infty$*) be a discrete sample space with the maximal* σ*-algebra* 2^Ω *and probability measures* p *and* q*. An* ***f*-divergence between *q* and *p*** *is any function*

$$D_f(q, p) = \sum_{i=1}^{N} p(\omega_i) f\left(\frac{q(\omega_i)}{p(\omega_i)}\right),$$

where $f\colon \mathbb{R}^{\geq 0} \to \mathbb{R}$ *is a convex and differentiable function that satisfies* $f(1) = 0$.

For the non-discrete case (e.g., probability measures on uncountable sample spaces such as $[0;1]$, $\mathbb{R}$ or $\mathbb{R}^2$), the definition is analogous:

Definition 4.2 (*f*-divergences, general probability spaces). *For a probability space* $\Omega \subset \mathbb{R}^N$*, let* g_p *and* g_q *denote the density functions of* p *and* q *with respect to the Lebesgue-measure on* Ω*. An* ***f*-divergence between *q* and *p*** *is any function*

$$D_f(q, p) = \int_\Omega g_p(x) f\left(\frac{g_q(x)}{g_p(x)}\right) dx,$$

where $f\colon \mathbb{R}^{\geq 0} \to \mathbb{R}$ *is a convex and differentiable function that satisfies* $f(1) = 0$.

This class of functions has a number of interesting properties. An *f*-divergence cannot take non-negative values and it vanishes if and only if $p = q$ almost everywhere. This is a consequence of the convexity requirement, which also implies that f is continuous. The condition $f(1) = 0$ expresses that propositions to which p and q assign equal probability do not increase the divergence between p and q. However, unlike proper distance functions, *f*-divergences are non-symmetric and they may violate the triangle inequality. Popular *f*-divergences are the Kullback–Leibler divergence $f(x) = x \log x$, the inverse Kullback–Leibler divergence $f(x) = -\log x$, the Hellinger distance $f(x) = (1 - \sqrt{x})^2$ and the χ^2-divergence $f(x) = (x-1)^2$. Thus, the concept of an *f*-divergence subsumes a large set of divergence functions that have been used in formal epistemology and scientific applications.

It turns out that minimizing the *f*-divergence between prior and posterior distribution subject to the constraint $p'(E) = k$ is equivalent to Jeffrey Conditionalization on E (Cziszár 1967, 1975):

Theorem 4.1 (Jeffrey Conditionalization Representation Theorem). *Let $p\colon \mathcal{L} \mapsto [0;1]$ be a probability function on the sentences of a propositional language. Suppose that the probability of $\mathrm{E} \in \mathcal{L}$ shifts from its prior value $p(\mathrm{E})$ to its posterior value $p'(\mathrm{E}) = k$. Then the following two rules for defining the posterior distribution p' are equivalent:*

1. *p' is obtained from p by Jeffrey Conditionalization on E: for any proposition $\mathrm{H} \in \mathcal{L}$,*
$$p'(\mathrm{H}) = p(\mathrm{H}|\mathrm{E})\,p'(\mathrm{E}) + p(\mathrm{H}|\neg\mathrm{E})\,p'(\neg\mathrm{E}).$$
2. *p' minimizes an arbitrary f-divergence $D_f(p', p)$ between posterior distribution p' and prior distribution p, subject to the constraint $p'(\mathrm{E}) = k$.*

If $p'(\mathrm{E}) = 1$, we obtain the equivalence between Bayesian Conditionalization and divergence minimization as a special case:

Theorem 4.2 (Bayesian Conditionalization Representation Theorem). *Let $p\colon \mathcal{L} \mapsto [0;1]$ be a probability function on the sentences of a propositional language $\mathcal{L}$. Suppose that the probability of $\mathrm{E} \in \mathcal{L}$ shifts from its prior value $p(\mathrm{E})$ to its posterior value $p'(\mathrm{E}) = k$. Then the following two rules for defining the posterior distribution p' are equivalent:*

1. *p' is obtained from p by Bayesian Conditionalization on E: for any proposition $\mathrm{H} \in \mathcal{L}$, $p'(\mathrm{H}) = p(\mathrm{H}|\mathrm{E})$.*
2. *p' minimizes an arbitrary f-divergence $D_f(p', p)$ between posterior distribution p' and prior distribution p, subject to the constraint $p'(\mathrm{E}) = 1$.*

The general philosophical significance of these results should not be underestimated (see also Eva, Hartmann and Rafiee Rad forthcoming). The theorems show that Bayesian Conditionalization (and its generalization, Jeffrey Conditionalization) is equivalent to a variety of ways of belief revision where the divergence between prior and posterior distribution is minimized. Thereby they lend further support to the defense of Bayesian reasoning as a conservative and non-committal way of changing one's credences in the face of new information.

For our purposes, the case of learning conditional probabilities is especially interesting. Is there a similarly tight connection between Conditionalization and divergence-minimizing for the case of learning $p'(\mathrm{E}|\mathrm{H}) = 1$? The answer is yes: minimizing an f-divergence between prior and posterior distribution is equivalent to learning the material conditional $\mathrm{H} \supset \mathrm{E}$:

Theorem 4.3 (Material Conditional Representation Theorem). *Let* $p\colon \mathcal{L} \mapsto [0;1]$ *be a probability function on the sentences of a propositional language. Suppose that* $\mathrm{H}, \mathrm{E} \in \mathcal{L}$. *Then the following two rules for defining a probability distribution* p' *over the elements of* $\mathcal{L}$ *are equivalent:*

1. p' *is obtained from* p *by Bayesian Conditionalization on the material conditional* $\mathrm{H} \supset \mathrm{E}$*: for any proposition* $\mathrm{X} \in \mathcal{L}$, $p'(\mathrm{X}) := p(\mathrm{X} \,|\, \neg \mathrm{H} \vee \mathrm{E})$.
2. p' *is the probability distribution which minimizes any f-divergence* $D_f(p', p)$ *to the prior distribution* p*, subject to the constraint* $p'(\mathrm{E} \,|\, \mathrm{H}) = 1$.

This is a very important result: it shows that *conditionalizing on the material conditional* $\mathrm{H} \supset \mathrm{E}$ *and minimizing any f-divergence with the constraint* $p(\mathrm{E} \,|\, \mathrm{H}) = 1$ *yield the same result.* What appeared to be two fundamentally different ways of modeling the learning of conditional evidence are, upon closer inspection, just two different ways of reaching the same result.

Theorem 4.3 provides an additional philosophical justification for modeling conditional evidence by conditionalizing on the material conditional—or more generally, by minimizing divergence. The next section challenges this view by providing some specific examples that are supposed to lead both the material conditional and the divergence minimization updating procedure *ad absurdum*. We contend, however, that we can respond adequately to the alleged counterexamples.

4.2 Three Challenges for Minimizing Divergence

In a variety of papers, Richard Dietz, Igor Douven and Jan-Willem Romeijn have developed examples which challenge the divergence minimization method for learning indicative conditionals. Below, we adapt those examples to a context of scientific reasoning. Each example starts with a story that sets up the scene. Then a conditional is learned, which may prompt some previously held beliefs to change.

The Medicine Example (adapted from Douven and Romeijn 2011). A general practitioner has to choose whether or not to give drug D to a patient. She will administer D if and only if (i) D is effective against the strains of bacteria that the patient is infected with; (ii) the patient has no medical condition that makes him susceptible to serious side effects of D. The GP's assistant checks the patient's medical history and tells her boss: "If D is effective against that particular strain of bacteria, then we should administer D." Upon learning this conditional, the GP does not change her belief in the efficacy of D—rather, she reasons that the patient does not have a medical condition that makes him susceptible to side effects of D. Learning the conditional evidence should leave the probability of the efficacy of D unchanged.

The Astronomy Example (adapted from Douven and Dietz 2011). The astronomic community in the seventeenth century considers two general theories of celestial mechanics, the Ptolemaic model and the Copernican model. An astronomer observes the movements of the planets Mars, Jupiter and Saturn over an extended period and notes down his observations. He finds an agreement between periods of retrograde motion and relative brightness. Working himself through the implications of the Copernican model, he realizes: "If the Copernican model is true, then the outer planets (i.e., Mars, Jupiter and Saturn) will display retrograde motion when they are close to Earth." Already knowing that the (apparent) retrograde motion of these planets would agree with his actual observations because brightness is an indicator of spatial proximity, learning the conditional evidence should intuitively increase our confidence in the Copernican model.

The Economics Example (adapted from Douven 2012). An economist is interested in whether a country is recovering economically. During the Christmas period, she surveys the sales volume of several warehouses. It turns out to be low. She asks a colleague about the consequences of economic recovery on consumer income. Her colleague answers: "if there is an economic recovery going on, consumer income has increased", for example, because of generous year-end bonuses. Upon learning this conditional evidence, the economist thinks it is doubtful (even if not wholly excluded) whether an economic recovery is currently going on. As a result, she lowers her degree of belief in economic recovery.

These three cases describe three different ways how learning indicative conditionals may affect the probability of the antecedent: it may be lowered or increased or remain unchanged. This is bad news for the methods of minimizing divergence and updating on the material conditional because Popper and Miller's result, given at the beginning of this Variation, shows that they always lower the probability of the antecedent.

Does this mean that the project of finding a general theory of learning conditional evidence is doomed? We disagree and show that the above examples contain a lot of causal structure which is neglected in the argument that divergence minimization fails. Once this structure is taken into account by embedding the involved propositions into a causal Bayesian network, divergence minimization (and Conditionalization on the material conditional) deals successfully with the above examples.

4.3 Meeting the Challenges

To respond to the three above challenges, we adopt the following method. First, we identify all relevant variables of the problem at hand. Second, we represent causal and inferential relations between the variables by (conditional) independencies in a Bayesian network and fix the prior probability distribution p that is associated with the network. Third, we express the learned conditional evidence as a constraint on the posterior probability distribution p' and assume that the relevant independencies are not changed by learning conditional evidence. That is, they are constraints on the prior *and* the posterior probability distribution. From the story it is clear that the incoming information does not overturn causal structure; hence the posterior probability distribution should preserve it. Fourth, we obtain the posterior probability distribution p' by minimizing an arbitrary f-divergence $D_f(p', p)$ to the prior distribution p (or equivalently, by conditionalizing on the material conditional). Fifth, we check whether the results comply with our intuitions. To repeat, in comparison to minimizing an f-divergence, we have now imposed the additional constraint on the posterior distribution that causal structure (e.g., which interventions affect which variables) and inferential relations (e.g., which variables are probabilistically independent of others) remain intact.

4.3.1 The Medicine Example

We introduce three binary propositional variables to represent the medicine example. The variable E has the values E: "Drug D is effective for the bacteria the patient is infected with", and ¬E: "Drug D is not effective for these bacteria." The variable S has values S: "The patient is susceptible to serious side effects when taking drug D", and ¬S: "The patient is not susceptible to serious side effects when taking drug D." Finally, the variable A has the values A: "Administer drug D", and ¬A: "Do not administer drug D." These sentences are in the remainder identified with the propositions they express.

Before we proceed, let us show that without considering additional causal structure, conditionalizing on the material conditional leads to an intuitively wrong result. To do so, remember that the learned information is "if drug D is effective against these bacteria, then we should administer it". We interpret this as the material conditional $E \supset A$, which is equivalent to the proposition $\neg E \vee A$. Assuming $0 < p(E), p(E, \neg A) < 1$ and using the ratio analysis of conditional probability, we obtain

$$\begin{aligned} p(E \mid E \supset A) &= p(E \mid \neg E \vee A) \\ &= \frac{p(E \wedge (\neg E \vee A))}{p(\neg E \vee A)} = \frac{p(E \wedge A)}{p(\neg E \vee A)} = \frac{p(E) - p(E, \neg A)}{1 - p(E, \neg A)}. \end{aligned} \tag{4.1}$$

It is easy to verify that Equation (4.1) implies $p'(E) = p(E \mid E \supset A) < p(E)$, which conflicts with our intuitive judgment that the probability of the efficacy of the drug should remain unchanged. By Theorem 4.3, it follows that minimizing an f-divergence between p' and p, subject to the constraint $p'(A \mid E) = 1$, would lead to the same problem.

Let us now show how our suggested methodology deals with the case. The story suggests a number of dependencies and independencies between the various variables. The Bayesian network in Figure 4.1 represents the probabilistic dependencies and independencies between these variables. The arrows represent the causal relations and the effect of interventions on the variables. Proposition A is modeled as a common effect of E and S, while there is no (unconditional) interaction between drug efficacy and side effects reaction.

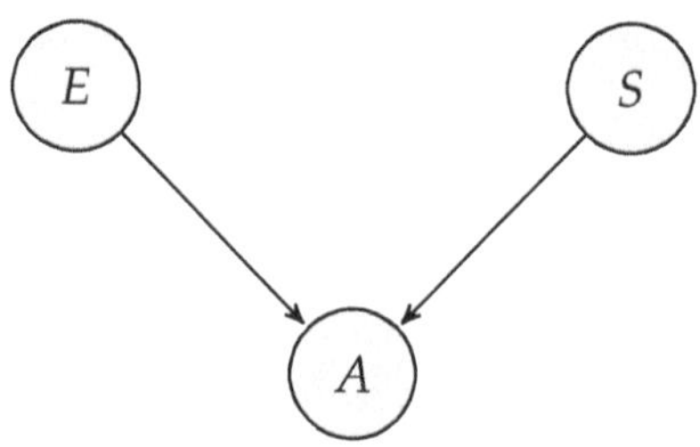

Figure 4.1: The Bayesian network representation of the medicine example.

Given the punchline of the story, we may assume that

$$p(A \mid E, \neg S) = 1. \tag{4.2}$$

In other words, if the drug is effective and the patient is not susceptible to side effects, we will administer the drug. All other conditional probabilities of the type $p(\mathrm{A} \mid \pm\mathrm{E}, \pm\mathrm{S})$ are in the open interval $(0;1)$. Let us now consider the posterior probability distribution p', which is defined over the same variables as the prior distribution. The constraint $p(\mathrm{A} \mid \mathrm{E}, \neg\mathrm{S}) = 1$ from Equation (4.2) should be preserved in that distribution too. Hence we conclude:

$$p'(\mathrm{A} \mid \mathrm{E}, \neg\mathrm{S}) = 1, \qquad p'(\neg\mathrm{A} \mid \mathrm{E}, \neg\mathrm{S}) = 0.$$

Another constraint on the posterior probability distribution is the learned conditional evidence "if the drug is effective against these bacteria, then we should administer it", which implies that

$$p'(\mathrm{A} \mid \mathrm{E}) = 1$$

and hence $p'(\neg\mathrm{A} \mid \mathrm{E}) = 0$. It follows that $p'(\neg\mathrm{A}, \mathrm{E}) = p'(\neg\mathrm{A}, \mathrm{E}, \mathrm{S}) = 0$, which in turn implies

$$0 = p'(\neg\mathrm{A}, \mathrm{E}, \mathrm{S}) = p'(\neg\mathrm{A} \mid \mathrm{E}, \mathrm{S})\, p'(\mathrm{E})\, p'(\mathrm{S}). \tag{4.3}$$

We can safely assume that $p'(\mathrm{E}) > 0$ (why should learning "if E, then A" rule out that the drug is effective?) and also that $p'(\neg\mathrm{A} \mid \mathrm{E}, \mathrm{S}) > 0$: if the patient is susceptible to side effects, it is not clear whether the GP would recommend to administer drug D. The only reasonable way to satisfy Equation (4.3) is to set $p'(\mathrm{S}) = 0$. This makes sense: the information received from the assistant suggests that the patient is not susceptible to serious side effects. We can now show the following theorem:

Theorem 4.4 (Analysis of the Medicine Example). *Let $p\colon \mathcal{L} \mapsto [0;1]$ be a probability function on the sentences of a propositional language. We assume that p is Markovian with respect to the* DAG *G in Figure 4.1 and that*

- *(i) learning conditional evidence is modeled as the constraint (4.2) on the posterior distribution $p'\colon \mathcal{L} \mapsto [0;1]$, that is, $p'(\mathrm{A} \mid \mathrm{E}, \mathrm{S}) = 1$;*
- *(i) p' preserves the conditional independence assumptions that G imposes on p; in other words, also p' is Markovian with respect to the graph in Figure 4.1;*
- *(iii) $p'(\mathrm{S}) = 0$ (as argued in the main text);*
- *(iv) p' minimizes an arbitrary f-divergence to p; or equivalently, p' is obtained from p by conditionalizing on the material conditional $\mathrm{E} \supset \mathrm{A} = \neg\mathrm{E} \vee \mathrm{A}$.*

Then $p'(\mathrm{E}) = p(\mathrm{E})$.

We observe that the proposed method yields the intuitively correct result in this case: the probability of the drug being effective is invariant under learning the conditional evidence "if E, then A". These results hold for any choice of an f-divergence, that is, they are invariant under the particular divergence minimization method that we use (e.g., Kullback–Leibler divergence, Hellinger distance, etc.). The same is, by the way, also true for our analysis of the two other cases: the astronomy and the economics example.

4.3.2 The Astronomy Example

Again, we introduce three binary propositional variables to represent the astronomy example. The variable C has the values C: "The Copernican model is true", and ¬C: "The Copernican model is false." The variable M has the values M: "The outer planets display retrograde motion when close to Earth", and ¬M: "The outer planets do not display retrograde motion when close to Earth." Finally, the variable O has the values O: "Periods of retrograde motion and relative brightness agree", and ¬O: "Periods of retrograde motion and relative brightness do not agree." The Bayesian network in Figure 4.2 represents the probabilistic dependencies and independencies between these variables: O depends on C only via M, or in other words: $O \perp\!\!\!\perp C \,|\, M$.

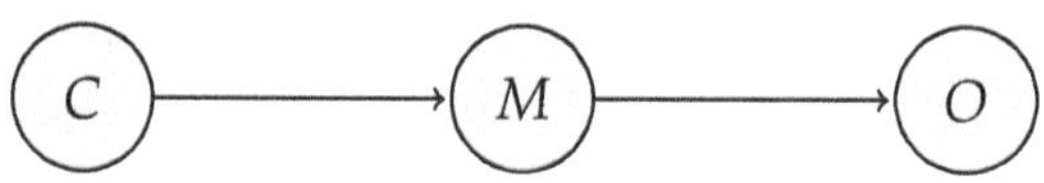

Figure 4.2: The Bayesian network representation of the astronomy example.

Next we learn two items of information. First, we learn that O obtains. Assuming that the conditional independencies depicted in Figure 4.2 do not change, this implies also $p'(\mathrm{O}) = 1$. Second, we learn the conditional evidence "if the Copernican model is true, then the outer planets will display retrograde motion when they are close to Earth", which implies that

$$p'(\mathrm{M}\,|\,\mathrm{C}) = 1. \tag{4.4}$$

We are now in a position to prove the following theorem:

Theorem 4.5 (Analysis of the Astronomy Example). *Let* $p\colon \mathcal{L} \mapsto [0;1]$ *be a probability function on the sentences of a propositional language. We assume that p is Markovian with respect to the* DAG *G in Figure 4.2 and that*

(i) *learning conditional evidence is modeled by the constraints* $p'(\mathrm{O}) = 1$ *and* $p'(\mathrm{M}|\mathrm{C}) = 1$ *on the posterior distribution* $p' \colon \mathcal{L} \mapsto [0;1]$;

(ii) p' *preserves the conditional independence assumptions that G imposes on p; in other words, also* p' *is Markovian with respect to the graph in Figure 4.2;*

(iii) p' *minimizes an arbitrary f-divergence to p subject to the constraints in (ii), or equivalently,* p' *is obtained from p by conditionalizing on* O *and* $\neg\mathrm{C} \vee \mathrm{M}$.

Then $p'(\mathrm{C}) = p(\mathrm{C})$ *if and only if*

$$p(\mathrm{M}|\mathrm{C})\, p(\mathrm{O}|\mathrm{M}) \;>\; p(\mathrm{M}|\neg\mathrm{C})\, p(\mathrm{O}|\mathrm{M}) \;+\; p(\neg\mathrm{M}|\neg\mathrm{C})\, p(\mathrm{O}|\neg\mathrm{M}). \qquad (4.5)$$

That is, our rational degree of belief in the Copernican model is increased when Equation (4.5) holds. When will this be the case? That depends on how we flesh out the details of the historical story and the background assumptions of the astronomer. The prior degree of belief in the Copernican model might have been small back in the days, since the Copernican model did not produce more accurate predictions than the Ptolemaic model and did not provide convincing explanations for many physical phenomena, such as the movement of the Earth. Hence $p(\mathrm{C})$ is small. Moreover, the observed correlation between brightness and retrograde motion is not explained by any alternative model which speaks for a small probability of

$$\begin{aligned} p(\mathrm{O}) \;&=\; p(\mathrm{C})\Big(p(\mathrm{M}|\mathrm{C})\, p(\mathrm{O}|\mathrm{M}) \;+\; p(\neg\mathrm{M}|\mathrm{C})\, p(\mathrm{O}|\neg\mathrm{M})\Big) \qquad (4.6)\\ &\quad + (1-p(\mathrm{C}))\Big(p(\mathrm{M}|\neg\mathrm{C})\, p(\mathrm{O}|\mathrm{M}) \;+\; p(\neg\mathrm{M}|\neg\mathrm{C})\, p(\mathrm{O}|\neg\mathrm{M})\Big). \end{aligned}$$

As $p(\mathrm{C})$ is small and $1-p(\mathrm{C})$ is large, we conclude that the right-hand side of Equation (4.5) must be small, too.

From the story it is also clear that $p(\mathrm{O}|\mathrm{M})$ is fairly large: given the postulated relation between a planet's position and the pattern of retrograde motion, agreement between brightness and retrograde motion is to be expected. At the same time, $p(\mathrm{O}|\neg\mathrm{M})$ will be very small as there is no reason to assume such a striking agreement if planets do not display retrograde motion when close to Earth. Finally, $p(\mathrm{M}|\mathrm{C})$ is not negligibly small. Then Equation (4.5) will be satisfied.

Hence the divergence minimization method (or equivalently, Bayesian Conditionalization on O and the material conditional $\mathrm{M} \supset \mathrm{C}$) yields the intuitively correct result. Of course, the exact result will depend of the specific details of the story, but this appears to be a very sensible feature of our approach: we have already seen before that contextual factors may determine whether learning conditional evidence raises or lowers the probability of the antecedent.

4.3.3 The Economics Example

Finally, we turn to the economics example. To represent the scenario, we introduce the following propositional variables: The variable R has the values R: "An economic recovery is going on" and ¬R: "No economic recovery is going on." The variable I has the values I: "Consumer income is increased", and ¬I: "Consumer income is not increased." The variable S has the values S: "The level of spending in warehouses is low", and ¬S: "The level of spending in warehouses is high." The Bayesian network in Figure 4.3 represents the probabilistic dependencies and independencies between these variables, as well as their causal relations. Note that the Bayesian network in Figure 4.3 has the same structure as that in Figure 4.2. Our calculation therefore proceeds as in the previous example.

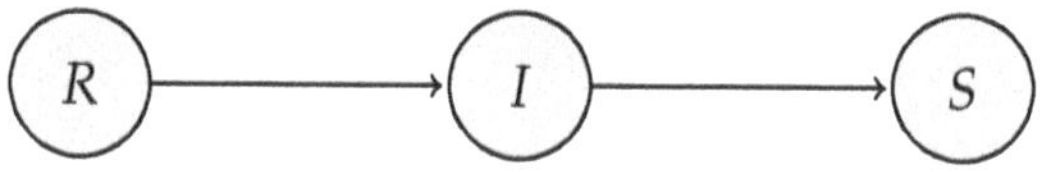

Figure 4.3: The Bayesian network representation of the economics example.

First, we learn that S obtains. Assuming that the causal structure depicted in Figure 4.3 does not change, this means that we learn that $p'(\mathrm{S}) = 1$. Second, we learn "if there is an economic recovery going on, consumer income is increased", which implies that $p'(\mathrm{I}|\mathrm{R}) = 1$. In a similar way as in the previous example, we can derive that this implies $p'(\mathrm{I}|\neg\mathrm{R}) = 1$ or $p'(\mathrm{S}|\neg\mathrm{I}) = 1$. As the first solution does not make sense—why should we be certain that consumer income is increased?—we conclude that $p'(\mathrm{S}|\neg\mathrm{I}) = 1$.

The structure of the relevant Bayesian network in the economics example, and the constraints on the posterior distribution p', are the same as in the astronomy example—see Figures 4.2 and 4.3. Hence we can apply Theorem 4.5. Whether or not the probability $p(\mathrm{R})$ of economic recovery is raised by learning these constraints depends on whether or not

$$p(\mathrm{I}|\mathrm{R})\,p(\mathrm{S}|\mathrm{I}) \;>\; p(\mathrm{I}|\neg\mathrm{R})\,p(\mathrm{S}|\mathrm{I}) \;+\; p(\neg\mathrm{I}|\neg\mathrm{R})\,p(\mathrm{S}|\neg\mathrm{I}). \tag{4.7}$$

In this case, we have evidence to the contrary. It is clear from the story that $p(\mathrm{S}|\neg\mathrm{I}) \gg p(\mathrm{S}|\mathrm{I})$: the probability of low spending is higher for equal than for increased consumer income. Hence,

$$\begin{aligned} p(\mathrm{I}|\neg\mathrm{R})\,p(\mathrm{S}|\mathrm{I}) + p(\neg\mathrm{I}|\neg\mathrm{R})\,p(\mathrm{S}|\neg\mathrm{I}) &> p(\mathrm{I}|\neg\mathrm{R})\,p(\mathrm{S}|\mathrm{I}) + p(\neg\mathrm{I}|\neg\mathrm{R})\,p(\mathrm{S}|\mathrm{I}) \\ &= p(\mathrm{S}|\mathrm{I}) \\ &\geq p(\mathrm{I}|\mathrm{R})\,p(\mathrm{S}|\mathrm{I}), \end{aligned}$$

and thus the right-hand side of Inequality (4.7) is greater than the left-hand side. We conclude that the posterior probability of economic recovery is smaller than the prior probability. Again, the proposed method yields the intuitively correct result.

4.4 Learning Relative Frequencies: The Case of Judy Benjamin

The previous examples have dealt with conditional evidence that could be expressed as an indicative conditional of the type "if H, then E". In other words, they dealt with cases where the conditional probability of E, given H, was extreme: $p'(\mathrm{E}\,|\,\mathrm{H}) = 1$. Of course, most cases of scientific inference where we learn conditional probabilities are quite different: the frequency of a certain outcome in a clinical trial will typically be a number strictly between zero and one. This section investigates how the method of minimizing f-divergences deals with **non-extreme probabilistic constraints:** $0 < p'(\mathrm{E}\,|\,\mathrm{H}) < 1$.

We already know that the f-divergence-minimization method is applicable to such cases. But does it yield unique and convincing results, like in the case of learning $p'(\mathrm{E}\,|\,\mathrm{H}) = 1$? This turns out not to be the case. As we show in the Appendix, the posterior distribution p' depends on which convex function f with $f(1) = 0$ we choose for generating an f-divergence $D_f(p', p)$.

The fact that the f-divergences deliver different results in such cases shows two things. First, it implies that the problem "How should a Bayesian learn relative frequencies?" does not have a uniquely rational solution—at least unless additional rationality constraints, such as Jaynes' maximum entropy principle, are stipulated (for a solution along those lines, see Williamson 2013). There are several admissible ways for combining (objective) relative frequencies with degrees of belief, and it is a fascinating problem for future research to find further principles for amalgamating a prior distribution with observed relative frequencies. In particular, it is an open question whether this problem has a unique solution or whether pluralism with regard to rational posterior distributions is a defensible position.

Second, it shows that the analogy with conditionalizing on a material conditional breaks down for such complicated cases. In particular, minimizing the f-divergence need not agree with Jeffrey Conditionalization on the material conditional $\neg\mathrm{H} \vee \mathrm{E}$. In the general case, divergence minimization goes significantly beyond what is contained in Bayesian Conditionalization and related principles.

The famous Judy Benjamin problem by Bas van Fraassen (1981, 1989) is a good illustration of this non-uniqueness:

The Judy Benjamin Problem A soldier, Judy Benjamin, is dropped with her platoon in a territory that is divided in two halves, Red Territory and Blue Territory, respectively, with each territory in turn being divided in equal parts, Second Company area and Headquarters Company area, thus forming four quadrants of roughly equal size. Because the platoon was dropped more or less at the center of the whole territory, Judy Benjamin deems it equally likely that she is in one quadrant as that she is in any of the others. She then receives the following radio message: "I can't be sure where you are. If you are in Red Territory, then the odds are 3 : 1 that you are in Second Company area." After this, the radio contact breaks down. Supposing that Judy accepts this message, how should she adjust her degrees of belief?

To address this question, we introduce two binary propositional variables. The variable R has the values R: "Judy lands in Red Territory", and ¬R: "Judy lands in Blue Territory." The variable S has the values S: "Judy lands in Second Company", and ¬S: "Judy lands in Headquarters." The prior probability distribution p is a uniform distribution and the posterior distribution p' has to satisfy the constraint

$$p'(\mathrm{S}|\mathrm{R}) = 3/4. \tag{4.8}$$

There has been a substantial philosophical debate about what is the rational posterior distribution to adopt in the light of (4.8) (for recent approaches, see Bovens 2010; Douven and Romeijn 2011; Titelbaum 2013). Our findings suggest that it is not surprising that this debate is still going on, because plausible rationality principles for belief change, such as divergence minimization, do not impose a uniquely rational posterior distribution. Only when a (possibly goal- and interest-relative) function f is chosen—similar to the choice of a scoring rule in prediction markets—is it possible to state which posterior distribution one should adopt. For the time being, pluralism about the Judy Benjamin case and other cases of learning non-extreme conditional evidence seems to be the appropriate default position—at least from our divergence minimization point of view.

However, all f-divergences imply some *qualitative* constraints for the Judy Benjamin case and related cases of learning non-extreme conditional probabilities:

Theorem 4.6 (f-divergences in the Judy Benjamin Case). *Let p be the uniform distribution over the algebra generated by propositions* R *and* S *(i.e., $p(\pm\mathrm{R}, \pm\mathrm{S}) = 1/4$). Suppose that we minimize the f-divergence $D_f(p', p)$ between posterior distribution p' and prior distribution p, subject to the constraint $p'(\mathrm{S}|\mathrm{R}) = k$ for $k \in (0;1)$. Then there is a $\delta \in (0;1)$ such that*

$$p'(\mathrm{R} \wedge \mathrm{S}) = \delta k, \qquad p'(\neg\mathrm{R} \wedge \mathrm{S}) = (1-\delta)/2,$$
$$p'(\mathrm{R} \wedge \neg\mathrm{S}) = \delta(1-k), \qquad p'(\neg\mathrm{R} \wedge \neg\mathrm{S}) = (1-\delta)/2.$$

This result has a couple of intuitive consequences. First, all f-divergences agree that for $k = 1/2$, the posterior and prior distribution should be equal: no new information has been learned. Second and more importantly, $p'(\neg\mathrm{R} \wedge \neg\mathrm{S}) = p'(\neg\mathrm{R} \wedge \mathrm{S})$, that is, the learning experience does not change Judy Benjamin's attitude about what would be the case if she were in Blue rather than Red Territory: it would still be equally likely for her to be in Second Company as in Headquarters. Intuitively, we can reconstruct this result by realizing that Judy's learning experience is based on *supposing* R (compare the suppositional semantics of conditionals in Edgington 1995). In the absence of information to the contrary, learning what happens if R is the case should not change our expectations on what happens if ¬R were the case. While the method of minimizing f-divergence does not single out a uniquely rational posterior distribution, all such divergences agree in the verdict that ¬R ∧ ¬S and ¬R ∧ S should remain equally likely. As evident from the proof of Theorem 4.6, this result generalizes beyond the specific case of uniform priors in the Judy Benjamin example. This shows that the proposed method can still yield non-trivial and interesting results even in complex cases.

4.5 Discussion

Minimizing an f-divergence between prior and posterior probability distribution, subject to a set of empirical constraints, is an interesting extension of the scope of Bayesian reasoning. First, whenever we perform Bayesian (or Jeffrey) Conditionalization on a first-order proposition, minimizing f-divergence will deliver the same result. Learning by minimizing f-divergence is thus a conservative extension of Bayesian Conditionalization, which does not threaten core Bayesian principles. Second, minimizing f-divergence allows for normatively attractive and computationally feasible learning of a wide range of constraints on the posterior probability distribution, such as the mean or variance of a random variable. The ability to process such evidence is an important feature of any theory of scientific inference. It is also a more general method than the well-known proposal of minimizing Kullback–Leibler divergence. Using the divergence minimization method, we can address and resolve various challenges that have been put forward against Bayesian Conditionalization and Bayesian reasoning.

This Variation has applied divergence minimization to learning conditional evidence. We have focused on evidence that imposes constraints of the type $p'(\mathrm{E}|\mathrm{H}) = 1$ on the posterior distribution. Such evidence can often be expressed in the form of indicative conditionals and learning these conditionals is an open problem for the classical Bayesian framework. We show that minimizing any *f*-divergence between prior and posterior is, for such constraints, equivalent to conditionalizing on the material conditional $\mathrm{H} \supset \mathrm{E} = \neg\mathrm{H} \vee \mathrm{E}$. Intuitively it appears that learning such information can affect the probability of H in different ways: sometimes it is raised, sometimes it is lowered, sometimes it stays equal—dependent on the context.

To deal with all three cases, we have suggested a refinement of the divergence minimization method: represent the causal and inferential relations among the relevant propositions by a Bayesian network with a set of conditional and unconditional independencies. These independencies act as constraints on both the prior and posterior distributions. After all, in the discussed examples they are not changed by learning conditional evidence. When an *f*-divergence is minimized subject to these constraints, the intuitively correct results follow. Due to the representation results in Section 4.1, this entails that conditionalizing on the material conditional yields the right result in those cases, too.

Does the proposed method also give the adequate results if more complicated scenarios are considered? We do not see a way how to answer this question in full generality. The set of possible scenarios where conditional evidence is learned is nearly unrestricted, both on the side of conditionals and on the side of relative frequencies. Typically, non-extreme conditional evidence (i.e., constraints of the type $0 < p'(\mathrm{E}|\mathrm{H}) < 1$, or learning the expectation value of a random variable) allows for different posterior distributions, dependent on which *f*-divergence is minimized. Such constraints frequently occur in scientific practice, for example when we learn observed relative frequencies. Van Fraassen's Judy Benjamin case provides another prominent example.

However, this permissiveness is no reason for pessimism: First, the method of minimizing *f*-divergences still implies interesting qualitative constraints on the posterior distribution. Second, pluralism may be a reasonable conclusion given the fact that philosophical debate in the Judy Benjamin case has failed to come up with convincing and generally accepted rationality principles that lead to a unique posterior distribution. Exploring the question of monism vs. pluralism is a worthwhile topic for future research. Finally, one may establish an argument for using f-divergences that have desirable epistemic properties. For example, if we restrict ourselves to the set of Bregman divergences, proposed in the introductory chapter in the context of measuring inaccuracy, then only the family of Kullback–Leibler divergences fits the bill. That is, Kullback–Leibler divergence is the only function that is both an f-divergence and a Bregman divergence, and this property may justify a privileged position for modeling learning experiences as minimizing Kullback–Leibler divergence. Other exciting open questions concern the nature of the implied probabilistic constraints (are they causal, evidential or still something else?) and the applicability of our analysis to reasoning with subjunctive conditionals, which arguably have a different semantics.

We are also optimistic that the proposed method will work for complicated scenarios involving more than three variables, as our examples represent diverse cases of probabilistic dependencies. The logic behind our approach is simple and intuitive: in moving from a prior to a posterior distribution, one should not only minimize the distance subject to novel evidence, but also subject to causal constraints which are not changed by learning conditional evidence. This concerns in particular the set of (conditional) independencies that the posterior distribution must satisfy. Whenever the learned evidence does not change these relations, our model provides a general and adequate method for Bayesian belief change. We conclude that the scope of evidence that Bayesian reasoners can model is much wider than that captured by Bayesian Conditionalization (i.e., first-order propositions). This observation rebuts a large number of criticisms against the Bayesian research program in philosophy of science.

Appendix: Proofs of the Theorems

Proof of Theorem 4.1: Let the probability distributions p and p' be represented by Table 4.1 and set $p'(\mathrm{E}) = k$.

H	E	p	p'
T	T	p_1	p'_1
T	F	p_2	p'_2
F	T	p_3	p'_3
F	F	p_4	p'_4

Table 4.1: The probability distributions p and p' for minimizing an f-divergence.

The distribution p' satisfies the following two constraints:

$$p'_1 + p'_3 = k, \tag{4.9}$$

$$p'_1 + p'_2 + p'_3 + p'_4 = 1. \tag{4.10}$$

Hence, we have to minimize

$$L = \sum_{i=1}^{4} p_i f(p'_i/p_i) + \lambda(p'_1 + p'_3 - k) + \mu(p'_1 + p'_2 + p'_3 + p'_4 - 1).$$

To do so, we differentiate by p'_k and obtain

$$\frac{\partial L}{\partial p'_1} = f'(p'_1/p_1) + \lambda + \mu = 0, \qquad \frac{\partial L}{\partial p'_3} = f'(p'_3/p_3) + \lambda + \mu = 0,$$

$$\frac{\partial L}{\partial p'_2} = f'(p'_2/p_2) + \mu = 0, \qquad \frac{\partial L}{\partial p'_4} = f'(p'_4/p_4) + \mu = 0.$$

These differential equations imply

$$p'_1 = \alpha p_1, \qquad p'_3 = \alpha p_3, \tag{4.11}$$

and

$$p'_2 = \beta p_2, \qquad p'_4 = \beta p_4. \tag{4.12}$$

We insert Equations (4.11) into Equation (4.9) and obtain

$$\alpha = \frac{k}{p_1 + p_3}. \tag{4.13}$$

Next, we insert Equations (4.11), (4.12) and (4.13) into Equation (4.10) and obtain

$$\beta = \frac{1-k}{p_2 + p_4}.$$

Hence,

$$p'(\mathrm{H}) = p'_1 + p'_2 = \frac{p_1}{p_1 + p_3} k + \frac{p_2}{p_2 + p_4}(1-k) = \\ p(\mathrm{H}|\mathrm{E})\,p'(\mathrm{E}) + p(\mathrm{H}|\neg\mathrm{E})\,p'(\neg\mathrm{E}).$$

From this equation it is obvious that divergence minimization yields the same result as Jeffrey Conditionalization. □

Proof of Theorem 4.2: This is a straightforward corollary of the previous theorem for the limiting case of $k=1$. □

Proof of Theorem 4.3: Let the probability distributions p and p' be represented as a function of the scalars a, b, c and a', b', c', as in Table 4.2. There,

H	E	p	p'
T	T	ab	a'
T	F	$a\bar{b}$	0
F	T	$\bar{a}c$	$\overline{a'}c'$
T	F	$\bar{a}\bar{c}$	$\overline{a'}\,\overline{c'}$

Table 4.2: The probability distributions p and p' for updating on the material conditional.

the constraint $p'(\mathrm{E}|\mathrm{H}) = b' = 1$ is already taken into account. Conventionally, we abbreviate the expression $1-a$ by $\bar{a}$, etc. We can now calculate the f-divergence between p' and p and obtain:

$$F = abf(a'/ab) + a\bar{b}f(0) + \bar{a}cf(\overline{a'}c'/\bar{a}c) + \bar{a}\bar{c}f(\overline{a'}\,\overline{c'}/\bar{a}\bar{c}). \tag{4.14}$$

We now differentiate F with respect to c' and set the resulting expression equal to zero. We obtain:

$$\overline{a'}\Big(f'(\overline{a'}c'/\bar{a}c) - f'(\overline{a'}\,\overline{c'}/\bar{a}\bar{c})\Big) = 0.$$

The convexity of f implies that f' is monotonically increasing and hence

$$\overline{a'}c'/\bar{a}c = \overline{a'}\,\overline{c'}/\bar{a}\bar{c},$$

from which we conclude that $c'=c$. Inserting this into Equation (4.14) yields

$$F = abf(a'/ab) + a\bar{b}f(0) + \bar{a}f(\overline{a'}/\bar{a}).$$

Hence,

$$\frac{\partial F}{\partial a'} = f'(a'/ab) - f'(\overline{a'}/\bar{a}) = 0,$$

from which we conclude, again using the convexity of *f*, that

$$a' = \frac{ab}{ab+\bar{a}}.$$

or equivalently,

$$p'(\mathrm{H}) = \frac{p(\mathrm{E}\wedge\mathrm{H})}{p(\mathrm{E}\vee\neg\mathrm{H})} = p(\mathrm{H}\,|\,\mathrm{E}\vee\neg\mathrm{H}) = p(\mathrm{H}\,|\,\mathrm{H}\supset\mathrm{E}).$$

This concludes the proof of the theorem. □

Proof of Theorem 4.4: We use Theorem 4.3 and represent minimizing the *f*-divergence due to the constraints $p'(\mathrm{A}\,|\,\mathrm{E},\mathrm{S}) = 1$ and $p'(\mathrm{S}) = 0$ as conditionalizing on the propositions S and $\neg(\mathrm{E}\wedge\mathrm{S})\vee\neg\mathrm{A}$. By the De Morgan rules, this is equivalent to conditionalizing on S and $\neg\mathrm{E}\vee\neg\mathrm{A}$. We then obtain:

$$\begin{aligned}
p'(\mathrm{E}) &= p(\mathrm{E}\mid \mathrm{S},\, \neg\mathrm{E}\vee\neg\mathrm{A}) \\
&= \frac{p(\mathrm{E},\,\mathrm{S},\,\neg\mathrm{E}\vee\neg\mathrm{A})}{p(\mathrm{S},\,\neg\mathrm{E}\vee\neg\mathrm{A})} \\
&= \frac{p(\mathrm{E},\mathrm{S},\neg\mathrm{A})}{p(\mathrm{E},\mathrm{S},\neg\mathrm{A}) + p(\neg\mathrm{E},\mathrm{S},\mathrm{A}) + p(\neg\mathrm{E},\mathrm{S},\neg\mathrm{A})} \\
&= \frac{p(\mathrm{E},\mathrm{S},\neg\mathrm{A})}{p(\mathrm{E},\mathrm{S},\neg\mathrm{A}) + p(\neg\mathrm{E},\mathrm{S})}.
\end{aligned}$$

We know that $p(\mathrm{E},\mathrm{S},\neg\mathrm{A}) = p(\neg\mathrm{A}\,|\,\mathrm{E},\mathrm{S})\,p(\mathrm{E},\mathrm{S}) = p(\mathrm{E},\mathrm{S})$, and also, due to the structure of the causal Bayes net, that E and S are probabilistically independent. Thus $p(\mathrm{E},\mathrm{S},\neg\mathrm{A}) = p(\mathrm{E},\mathrm{S}) = p(\mathrm{E})\,p(\mathrm{S})$. Hence,

$$p'(\mathrm{E}) = \frac{p(\mathrm{E})\,p(\mathrm{S})}{p(\mathrm{E})\,p(\mathrm{S}) + p(\neg\mathrm{E})\,p(\mathrm{S})} = p(\mathrm{E}),$$

showing the desired result. □

Proof of Theorem 4.5: We introduce some notational conventions for the joint prior probability distribution of C, M and O. We write the conditional probabilities between adjacent nodes in the Bayesian network as

$$\begin{aligned}
p_1 &:= p(\mathrm{M}|\mathrm{C}), & q_1 &:= p(\mathrm{M}|\neg\mathrm{C}), \\
p_2 &:= p(\mathrm{O}|\mathrm{M}), & q_2 &:= p(\mathrm{O}|\neg\mathrm{M}).
\end{aligned}$$

Moreover, we write $c := p(\mathrm{C})$ for the prior probability of C. After these preliminaries, we begin with the real proof.

Minimizing an f-divergence between p' and p subject to the constraints $p'(\mathrm{O}) = 1$ and $p'(\mathrm{M}|\mathrm{C}) = 1$ is, by Theorem 4.3, equivalent to conditioning on O and the material conditional $\mathrm{C} \supset \mathrm{M}$ (i.e., $\neg\mathrm{C} \vee \mathrm{M}$). This implies that all posterior probabilities of the type $p'(\neg\mathrm{O}, \pm\mathrm{C}, \pm\mathrm{M})$ vanish. Moreover, $p'(\mathrm{O}, \neg\mathrm{C}, \mathrm{M}) = 0$. Making use of the structure of the causal Bayes net and in particular the equations

$$p(\pm\mathrm{C}, \pm\mathrm{M}, \pm\mathrm{O}) \;=\; p(\pm\mathrm{C})\, p(\pm\mathrm{M}|\pm\mathrm{C})\, p(\pm\mathrm{O}|\pm\mathrm{M}),$$

this implies that

$$\begin{aligned} p'(\mathrm{C}) &= p(\mathrm{C} \mid \mathrm{O}, \neg\mathrm{C}\vee\mathrm{M}) \\ &= \frac{p(\mathrm{C}, \mathrm{O}, \neg\mathrm{C}\vee\mathrm{M})}{p(\mathrm{O}, \neg\mathrm{C}\vee\mathrm{M})} \\ &= \frac{p(\mathrm{C},\mathrm{M},\mathrm{O})}{p(\mathrm{C},\mathrm{M},\mathrm{O}) + p(\neg\mathrm{C},\mathrm{M},\mathrm{O}) + p(\neg\mathrm{C},\neg\mathrm{M},\mathrm{O})} \\ &= \frac{c p_1 p_2}{c p_1 p_2 + (1-c) q_1 p_2 + (1-c)(1-q_1) q_2}. \end{aligned}$$

The inequality $p'(\mathrm{C}) > p(\mathrm{C})$ then reads

$$\frac{c p_1 p_2}{c p_1 p_2 + (1-c) q_1 p_2 + (1-c)(1-q_1) q_2} \;>\; c;$$

and some elementary algebra (namely dividing by $c p_1 p_2$ and multiplying by the denominator of the fraction) reveals that this inequality is equivalent to

$$c \cdot 1 \;+\; (1-c)\frac{q_1 p_2 + (1-q_1) q_2}{p_1 p_2} \;<\; 1,$$

which holds if and only if

$$\frac{q_1 p_2 + (1-q_1) q_2}{p_1 p_2} \;<\; 1,$$

or equivalently,

$$q_1 p_2 + (1-q_1) q_2 \;<\; p_1 p_2.$$

Returning to full notation, this can be written as

$$p(\mathrm{M}|\neg\mathrm{C})\, p(\mathrm{O}|\mathrm{M}) \;+\; p(\neg\mathrm{M}|\neg\mathrm{C})\, p(\mathrm{O}|\neg\mathrm{M}) \;<\; p(\mathrm{M}|\mathrm{C})\, p(\mathrm{O}|\mathrm{M}),$$

and this latter inequality is identical to the one imposed in Equation 4.5, that is, the condition that we imposed at the end of Theorem 4.5. □

Proof of Theorem 4.6: The posterior distribution is denoted as follows:

$$q_1 := p'(\mathrm{R}\wedge \mathrm{S}), \qquad q_3 := p'(\neg\mathrm{R}\wedge \mathrm{S}),$$
$$q_2 := p'(\mathrm{R}\wedge\neg \mathrm{S}), \qquad q_4 := p'(\neg\mathrm{R}\wedge\neg \mathrm{S}).$$

with

$$q_1 + q_2 + q_3 + q_4 = 1. \tag{4.15}$$

Since the prior distribution p is the uniform distribution, we know that the corresponding $p_i = 1/4$. We also know that $p'(\mathrm{S}|\mathrm{R}) = k$ and therefore

$$(1-k)q_1 - kq_2 = 0. \tag{4.16}$$

Using the method of Lagrange multipliers, minimizing an f-divergence between p and p' subject to the above constraints comes down to minimizing

$$L = \sum_{i=1}^{4} p_i f(4q_i) + \lambda\big((1-k)q_1 - kq_2\big) + \mu\big(q_1+q_2+q_3+q_4-1\big).$$

The partial derivatives of L with respect to the q_i look as follows:

$$\frac{\partial L}{\partial q_1} = f'(4q_1) + \lambda(1-k) + \mu, \qquad \frac{\partial L}{\partial q_3} = f'(4q_3) + \mu,$$
$$\frac{\partial L}{\partial q_2} = f'(4q_2) - \lambda k + \mu, \qquad \frac{\partial L}{\partial q_4} = f'(4q_4) + \mu.$$

From Equation (4.16) we infer that $q_1 = k\delta$ and $q_2 = (1-k)\delta$ for some $\delta \in (0;1)$. From the two equations on the right-hand side and the convexity of f, we infer $q_3 = q_4$. With the help of Equation (4.15), we also infer $q_3 = q_4 = (1-\delta)/2$. □

Variation 5: The Problem of Old Evidence

In Variation 1, we have defined confirmation in terms of increase in firmness: evidence E confirms hypothesis H if and only if the conditional degree of belief in H given E exceeds the unconditional degree of belief in H: $p(H|E) > p(H)$. This scheme has proven to be very successful, resolving several puzzles of inductive inference (see also Variations 2–4). However, without amendments it cannot capture how already-known evidence confirms, or provides support for, a hypothesis.

The textbook example from the history of science is the precession of the perihelion of the planet Mercury (Glymour 1980; Earman 1992). For a long time, Newtonian mechanics failed to account for this phenomenon; and postulated auxiliary hypotheses (e.g., the existence of another planet within the orbit of Mercury) failed to be confirmed. Einstein realized in the 1910s that his General Theory of Relativity (GTR) accounted for the perihelion shift. According to most physicists, explaining this "old evidence" (in the sense of data known previously) conferred a substantial degree of confirmation on GTR, perhaps even more than some pieces of novel evidence, such as Eddington's 1919 solar eclipse observations. Also in other scientific disciplines, theories are commonly assessed with respect to their success at explaining away observational anomalies. Think, for example, of the assessment of global climate models against a track record of historical data, or of economic theories that try to explain anomalies in decision-making under uncertainty, such as the Allais or Ellsberg paradoxes.

We can extract a general scheme from these examples: A phenomenon E is unexplained by the available scientific theories. At some point, it is discovered that theory T accounts for E. The observation E is "old evidence": at the time when the relationship between T and E is developed, the scientist is already certain, or close to certain, that the phenomenon E is real. Indeed, in the GTR example, astronomers had been collecting data on the Mercury perihelion shift for many decades. Relative to the agent's actual degrees of belief, E is statistically independent of T. Nevertheless, E apparently confirms T because it resolves a well-known and persistent observational anomaly.

The argument can also be formalized. Proposition E confirms T if and only if $p(T|E) > p(T)$. These two probabilities are related by Bayes' Theorem:

$$p(T|E) = p(T)\frac{p(E|T)}{p(E)}.$$

When E is old evidence and already known to the scientist, her degree of belief in E is maximal: $p(E) = 1$. Because T predicts E, also $p(E|T) = 1$. It follows that the probability of T conditional on E cannot be greater than the unconditional probability:

$$p(T|E) = p(T)\,p(E|T)/p(E) = p(T)\,p(E|T) \leq p(T). \qquad (5.1)$$

Hence, E does not confirm T in the sense of being statistically relevant to T. The very idea of confirmation by old evidence, or equivalently, confirmation by accounting for well-known observational anomalies, seems impossible to describe in the Bayesian framework. This is the **Problem of Old Evidence (POE).** Since Glymour 1980, the POE has been one of the most troubling and persistent challenges for Bayesian Confirmation Theory (see also Huber 2005a; Brössel and Huber 2015).

Old evidence should not be understood literally as data that corresponds to a retrodiction of a scientific theory (e.g., in evolutionary biology, palaeoanthrology or astrophysics). After all, many things about the past are unknown and can be discovered. Rather, old evidence is already known at the moment when the confirmation relation is evaluated. The problem may also be phrased differently, as exposing that the Bayesian cannot explain how the discovery of explanatory relations between theory and evidence, or the resolution of observational anomalies by theoretical means, increases the epistemic standing of the theory. Notably, the problem does not allow for an easy fix by making assumptions on $p(T)$ and the likelihoods $p(E|\pm T)$: as long as $0 < p(T) < 1$, the Law of Total Probability implies that $p(E|T) = p(E|\neg T) = 1$ if $p(E) = 1$. Hence, T fails to be confirmed by E.

The POE has different aspects, as worked out by Ellery Eells (1985, 1990). First, there is the *static* (Eells: "ahistorical") POE: belief changes induced by discovering T, or an explanatory relationship between T and E, have already taken place. Still we would like to say that E is evidentially relevant for T: when faced with a decision between T and a competitor T′, E is a good reason for preferring T over T′. But there is also the *dynamic* (Eells: "historical") POE: it refers to the moment in time where T and its relation to E are discovered. Why does the discovery that T accounts for E raise our confidence in T? How can the discovery of an explanatory success be confirmationally relevant (see also Wenmackers and Romeijn 2016)? The dynamic POE deals with the question of confirmation relative to actual degrees of belief, whereas the static POE aims at a timeless notion of evidence.

This Variation develops new solution proposals for both the static and the dynamic POE, synthesizing the results obtained in Sprenger 2015 and Fitelson and Hartmann 2015. Section 5.1 comments on solutions of the dynamic POE proposed by Garber (1983), Jeffrey (1983), Niiniluoto (1983) and Earman (1992). On these accounts, confirmation occurs through conditionalizing on the proposition that T implies E. Section 5.2 presents an improvement on this approach. Section 5.3 analyses the static POE, while Section 5.4 explains our take on that problem. We conclude with a discussion of our results in Section 5.5.

5.1 The Dynamic Problem of Old Evidence: The Garber–Jeffrey–Niiniluoto Approach

The dynamic Problem of Old Evidence is concerned with how learning a deductive or explanatory consequence of a theory can raise our confidence in that theory. An example from the history of science may help to get this clear. In a system consisting only of the sun and a planet rotating around it, the planet's perihelion (i.e., the point where its orbit is nearest to the sun) would, according to Newtonian mechanics, be fixed. In practice, most of these perihelion points are moving, mainly due to the perturbation of the planet's orbit by other planets in the vicinity. However, for Mercury, physicists never succeeded in bringing Newtonian mechanics into agreement with the actual observations, even when these perturbations were taken into account.

Then came the General Theory of Relativity (T). It was initially not known to make better predictions for the Mercury perihelion shift (E) and it took Einstein substantial time to find out that T entails E (Brush 1989; Earman 1992). By learning the relationship $X := T \vDash E$, Einstein's confidence in T increased, since such a strong consilience of theory and data could not be expected beforehand. Thus, the inequality

$$p(T \mid X, E) > p(T \mid E)$$

seems to be a plausible representation of Einstein's degrees of belief before, and after, making the discovery that GTR explained the perihelion shift of Mercury. Consequently, the relevant piece of evidence is not E itself, but the learning of a specific relation between theory and evidence, namely that T implies, accounts for, or explains E.

However, such belief change is hard to model in a Bayesian framework. A Bayesian reasoner is assumed to be logically omniscient, and the logical fact $X = T \vDash E$ should always have been known. Hence X cannot be properly *learned* by a Bayesian: it is, and always has been, part of her background beliefs.

To solve this problem, several philosophers have relaxed the assumption of logical omniscience and enriched the set of propositions agents have degrees of belief about. New atomic sentences of the form $T \vDash E$ are added (Garber 1983; Jeffrey 1983; Niiniluoto 1983), such that Bayesian Confirmation Theory can account for our cognitive limitations in deductive reasoning. It can then be shown that, under suitable assumptions, conditionalizing on $X := T \vDash E$ confirms T.

The first models along these lines have been developed by Daniel Garber, Richard Jeffrey and Ilkka Niiniluoto in a group of papers which all appeared in 1983. Henceforth, we will refer to their family of solution proposals as the "GJN solutions". In order to properly compare our own solution proposals to the state of the art, and to assess their innovative value, we will briefly recap the achievements of the GJN models and elaborate on their limitations and weaknesses.

All GJN models aim to show that conditionalizing on the proposition X increases the posterior probability of T. Eells (1990, 211) distinguishes three steps in this endeavor: First, parting with the logical omniscience assumption and developing a formal framework for imperfect Bayesian reasoning. Second, describing which kind of relation obtains between T and E. Third, showing that learning this relation increases the probability of T. While the GJN models neglect the second step, probably in due anticipation of the diversity of logical and explanatory relations in science, they are quite explicit on Step 1 and Step 3.

Garber's model focuses on Step 1 and on learning logical truths and explanatory relations in a Bayesian framework (Garber 1983). For instance, learning logical/mathematical truths can be quite insightful and lead to great progress in science. The famous, incredibly complex proof of Fermat's Last Theorem may be a good example. Garber therefore enriches the underlying propositional language L in a way that X is one of the *atomic* sentences of the extended language L′, whose sentences are written as $\mathcal{L}'$.

Then Garber demands that the agent recognize some elementary relations in which X stands to other elements of $\mathcal{L}'$:

$$p(\mathrm{E}\,|\,\mathrm{T},\mathrm{X}) = 1, \qquad p(\mathrm{T},\mathrm{E},\mathrm{X}) = p(\mathrm{T},\mathrm{X}). \tag{5.2}$$

These constraints are an equivalent of modus ponens for a logic of degrees of belief: conditional on T and $\mathrm{X} = \mathrm{T} \vDash \mathrm{E}$, the agent should be certain that E. In other words, if an agent takes T and X for granted, then she also believes E to maximal degree. Knowledge of such elementary inference schemes sounds eminently sensible when we are trying to model the boundedly rational reasoning of a scientist. Garber then proves the following theorem: There is at least one probability function on $\mathcal{L}'$ such that *every* non-trivial atomic sentence of the form X gets a value strictly between 0 and 1. Thus one can coherently have a genuinely uncertain attitude about all propositions in the logical universe, including tautologies. Finally, Garber shows that there are infinitely many probability functions such that $p(\mathrm{E}) = 1$ and $p(\mathrm{T}\,|\,\mathrm{X},\mathrm{E}) > p(\mathrm{T}\,|\,\mathrm{E})$. A similar point is made, though with less formal detail and rigor, by Niiniluoto (1983).

While Garber's efforts are admirable, they only address the first step of solving the dynamic POE: he provides an existence proof for a solution to the POE, but he does not show that learning X confirms T for most plausible probability distributions over E, T and X. Niiniluoto (1983) sketched a solution idea without filling in the details. This lacuna was closed by Richard Jeffrey (1983), who published his solution in the same volume where Garber's paper appeared.

Jeffrey considers the meta-proposition X as an object of subjective uncertainty, but he keeps the formalism down to the standard level of Bayesian Confirmation Theory. Then he makes the following assumptions, using the notational convention $\mathrm{X}' := \mathrm{T} \vDash \neg\mathrm{E}$:

(α) $p(\mathrm{E}) = 1$.

(β) $p(\mathrm{T}), p(\mathrm{X}), p(\mathrm{X}') \in (0;1)$.

(γ) $p(\mathrm{X},\mathrm{X}') = 0$.

(δ) $p(\mathrm{T}\,|\,\mathrm{X} \vee \mathrm{X}') \geq p(\mathrm{T})$.

(ε) $p(\mathrm{T},\neg\mathrm{E},\mathrm{X}') = p(\mathrm{T},\mathrm{X}')$.

From these assumptions, Jeffrey derives $p(\mathrm{T} \mid \mathrm{X}, \mathrm{E}) > p(\mathrm{T}, \mathrm{E})$, that is, the solution to the dynamic POE.

The strength of Jeffrey's solution crucially depends on how well we can motivate Condition (δ). The other conditions are plausible: Condition (α) is just the standard presumption that at the time when confirmation takes place, E is already known to the agent. Condition (β) demands that we must not be certain about the truth of T or $\mathrm{T} \models \pm\mathrm{E}$ beforehand, in line with the typical description of the POE. Condition (γ) requires that T not entail E and $\neg$E at the same time. Finally, (ε) is a modus ponens condition similar to (5.2): the joint degree of belief in T, $\neg$E and X' is equal to the joint degree of belief in T and X', demanding that the agent recognize that the latter two propositions entail $\neg$E.

Hence, (δ) really carries the burden of Jeffrey's argument. This condition has some odd technical consequences, as pointed out by Earman (1992, 127). For instance, with plausible additional assumptions, we can derive $p(\mathrm{T} \mid \mathrm{X}) \geq 2p(\mathrm{T})$, which implies that the prior degree of belief $p(\mathrm{T})$ must have been smaller than 1/2. Jeffrey's solution of the dynamic POE does not apply to theories that were already quite probable, and this is an awkward feature.

That said, the real problem with (δ) is not technical, but philosophical. Jeffrey (1983, 148–149) supports (δ) by mentioning that Newton was, when formulating his theory of gravitation, G, convinced that it would bear on the phenomena he was interested in, namely the mechanism governing the tides. Newton did not know whether G would explain tidal phenomena or be inconsistent with them, but he knew that these phenomena fell under the scope of G, and he took this knowledge as a reason for accepting G as a working hypothesis.

To our mind, this reconstruction conflates an *evidential* virtue of a theory with a *methodological* one. Theories of which we know that they make precise predictions about interesting subject matters are worthy of further elaboration and pursuit, even if the content of their predictions is not yet known. This is basically a Popperian rationale for scientific inquiry: go for theories that have high empirical content and make precise predictions, and develop them further; they are the ones that will finally help us to solve urgent scientific problems (see also Variation 9). Newton may have followed this methodological rule when discovering that his theory of gravitation would have some implications for the tides phenomena. Making such pragmatic acceptances, however, does not entail a commitment to the thesis that the plausibility of a theory increases with its empirical content. Actually, Popper

(1959/2002, 268–269) thought the opposite: theories with high empirical content rule out more states of the world and will thus have low probability! This is just because, in virtue of making many predictions, they run a higher risk of being falsified. Indeed, it is hard to understand why increasing the empirical content of T should make T more likely to be true. Enlarging the class of potential falsifiers of T should *not* increase its plausibility. Jeffrey's Condition (δ) is therefore ill grounded and at the very least too controversial to act as a premise in a solution of the POE.

Earman (1992, 128–129) considers two alternative derivations of $p(\mathrm{T} \mid \mathrm{X}, \mathrm{E}) > p(\mathrm{T} \mid \mathrm{E})$, where assumptions different from Jeffrey's (δ) carry the burden of the argument. One of them is the inequality

$$(\varphi) \quad p(\mathrm{T} \mid \mathrm{X}) > p(\mathrm{T} \mid \neg \mathrm{X}, \neg \mathrm{X}'),$$

but it is questionable whether this suffices to circumvent the above objections. What Earman demands here is very close to what is supposed to be shown: that learning $\mathrm{T} \models \mathrm{E}$ is more favorable to T than learning that T gives no definite prediction for or against the occurrence of E. In the light of the above arguments against (δ) and in the absence of independent arguments in favor of (φ), this condition just seems to beg the question.

The second alternative derivation of $p(\mathrm{T} \mid \mathrm{X}) > p(\mathrm{T})$ relies on the equality

$$(\psi) \quad p(\mathrm{X} \vee \mathrm{X}') = 1.$$

However, as Earman admits himself, this condition is too strong: it amounts to demanding that when formulating T, the scientist was certain that it implied either E or $\neg$E. In practice, such relationships are rather discovered gradually. As Earman continues, discussing the case of GTR:

> the historical evidence goes against this supposition: ... Einstein's published paper on the perihelion anomaly contained an incomplete explanation, since, as he himself noted, he had no proof that the solution of the field equations ... was the unique solution for the relevant set of boundary conditions. (Earman 1992, 129)

Taking stock, we conclude that Garber, Jeffrey, Niiniluoto and Earman make interesting proposals for solving the dynamic Problem of Old Evidence, but that their solutions are either incomplete or based on highly problematic assumptions. We will now show how their approach to the dynamic POE can be improved.

5.2 The Dynamic Problem of Old Evidence: Alternative Explanations

A problem with the traditional GJN approaches is that they require constraints on degrees of belief (e.g., Jeffrey's (δ) or Earman's (ψ)) that are either implausibly strong or too close to the desired confirmation-theoretic conclusion $p(T \mid X, E) > p(T \mid E)$ itself. To remedy this defect, we propose to take into account whether alternatives to T adequately explain E. Let the propositions X and Y be defined as follows:

X := T *adequately explains* (or accounts for) E.

Y := *some alternative* T′ to T *adequately explains* (or accounts for) E.

Now, consider the following four ordinal constraints on the degrees of belief of a rational Bayesian agent:

$$p(T \mid X, \neg Y) > p(T \mid \neg X, \neg Y), \tag{5.3}$$

$$p(T \mid X, \neg Y) > p(T \mid \neg X, Y), \tag{5.4}$$

$$p(T \mid X, Y) \geq p(T \mid \neg X, \neg Y), \tag{5.5}$$

$$p(T \mid X, Y) > p(T \mid \neg X, Y). \tag{5.6}$$

Let's examine each of these four constraints in turn, assuming that E is very probable, perhaps even a certainty. Suppose that $\neg Y$ is the case and that no alternative to T adequately explains E. Then (5.3) asserts that T is more probable given X (i.e., T explains E) than given $\neg X$ (i.e., T does not explain E). If there is no alternative to explain E, judgments of evidential relevance translate into judgments of evidential support.

Constraint (5.4) is an even less controversial variant of the same proposition. In other words, (5.3) and (5.4) say that T's being the only adequate explanation of E confers a greater probability on T than any possibility which implies that T does not adequately explain E. These two constraints strike us as pretty uncontroversial.

The third inequality, (5.5), says that T is at least as probable, given the supposition that both T and some alternative scientific theory adequately explain E (i.e., given $X \wedge Y$), as it is given the supposition that *no* scientific theory adequately explains E (i.e., given $\neg X \wedge \neg Y$). It might even be compelling to rank $p(T \mid X, Y)$ strictly higher in one's comparative confidence ranking than $p(T \mid \neg X, \neg Y)$. After all, $X \wedge Y$ implies that T adequately explains old evidence E, whereas $\neg X \wedge \neg Y$ implies that T does *not* adequately explain E. On the other hand, one might also reasonably maintain that both suppositions place T and its alternatives on a par with respect to explaining E, and so they shouldn't confer different probabilities on T. Both of these positions are compatible with (5.5). The only thing (5.5) rules out is the claim that T is more probable given E's inexplicability ($\neg X \wedge \neg Y$) than it is given E's multiple explicability by both T and some alternative to T ($X \wedge Y$). As such, (5.5) also seems eminently reasonable.

The fourth and final constraint (5.6) also seems very plausible. If there are alternatives to T that adequately explain E, then T is more plausible if it, too, explains E than if it doesn't. This constraint mirrors the reasoning in (5.3) for the case that there are alternatives to T.

Now, the desired conclusion (5.1) follows from (5.3)–(5.5). To be precise, we can prove the following general result (see also Fitelson and Hartmann 2015):

Theorem 5.1 (Dynamic Problem of Old Evidence). *Let $p\colon \mathcal{L} \to [0;1]$ be a probability function on the sentences of a propositional language and let* T, E, X, Y $\in \mathcal{L}$ *with* $0 < p(\mathrm{T}) < 1$. *Then Conditions* (5.3)–(5.5) *jointly entail* $p(\mathrm{T}\,|\,\mathrm{X}) > p(\mathrm{T})$.

Of course this result also applies to the case where we have already conditionalized on E and $p(\mathrm{E}) = 1$:

Corollary 5.1. *Let $p\colon \mathcal{L} \to [0;1]$ be a probability function on the sentences of a propositional language and let* T, E, X, Y $\in \mathcal{L}$ *with* $0 < p(\mathrm{T}) < 1$. *Then the analogues of Conditions* (5.3)–(5.5) *for the probability distribution* $p(\cdot\,|\,\mathrm{E})$ *jointly entail* $p(\mathrm{T}\,|\,\mathrm{X},\mathrm{E}) > p(\mathrm{T}\,|\,\mathrm{E})$.

This approach has the following three distinct advantages over the traditional GJN approaches:

(i) Our approach does not require the assumption that $p(\mathrm{E}) = 1$. It may be true that our constraints (5.3)–(5.5) are most plausible given the background assumption that E is known with certainty. But we think (5.3)–(5.5) retain enough of their plausibility given only the weaker assumption that E is known with near certainty (i.e., $p(\mathrm{E}) \approx 1$).

(ii) Our approach only rests on the ordinal constraints (5.3)–(5.5), not on judgments of degrees of confirmation.

(iii) Our approach is not restricted to cases in which T (and/or alternatives T′) explains E in a deductive-nomological way. That is, our approach covers all cases in which scientists come to learn that their theory adequately explains E, not only those cases in which scientists learn that their theory entails E (or explains E deductive-nomologically).

A slight disadvantage of our approach is that Conditions (5.3)–(5.5) are themselves phrased as confirmation judgments. That is, the solution of the Problem of Old Evidence (whether X confirms T) depends on which truth-functional combinations of X and Y confirm T. Such judgments may be considered to be too close to what is supposed to be shown (for other criticisms, see Howson 2017). We think that our assumptions are plausible enough to withstand this criticism, but we also present an alternative solution, where the conditions are phrased in terms of the *likelihoods* of T and X on E. But first, we move on to the static POE.

5.3 The Static Problem of Old Evidence: A Counterfactual Perspective

The static Problem of Old Evidence is concerned with describing why old evidence E is, in a timeless sense, evidentially relevant for theory T (Eells 1985). For the Bayesian, this is a much harder nut to crack than the dynamic POE. After all, the latter was asking how a previously unknown proposition $X = T \models E$ could confirm T. In some sense it was less about the old evidence E itself than about modeling logical learning in the Bayesian framework. This challenge may be tough, but as Garber and his successors demonstrated, in principle the Bayesian has the tools to address it.

On the static POE, by contrast, the relation of evidential relevance is supposed to be independent of the moment when the evidence was observed, when a relationship between theory and evidence was discovered, and so on. It corresponds to the question, "Why is E at all—in the present as well as in the future—a reason for preferring T over its competitors?" This approach challenges the very idea of evaluating degree of confirmation with respect to actual degrees of belief, as developed in Variation 1 and applied in Variations 2–4. Similarly, the static POE replaces the conception of evidential support as probability-raising by E's power to discriminate between T and competitor T′. This is very close to the likelihoodist idea of **evidential favoring** (Hacking 1965; Edwards 1972; Royall 1997; Taper and Lele 2004; Sober 2008): a piece of evidence E favors a theory T over a competitor T′ if and only if $p(E|T) > p(E|T')$—where these probabilities are defined in a way that is stable and independent of contingent evidential learning. But how can we model such "objective", non-actual probabilities within the subjective Bayesian framework?

A general answer to this question has to be postponed to Variation 12, where we investigate the nature of conditional degree of belief in Bayesian inference, and its relation to objective chance. Here, let us first review some proposals from the literature. Christensen (1999) contends that the choice of a confirmation measure may answer this challenge. Take the measure $s^*(T, E) = p(T|E) - p(T|\neg E)$. If T entails E, as in the GTR example, then $\neg E$ also entails $\neg T$, which implies $p(T|\neg E) = 0$ and $s^*(T, E) = p(T|E) > 0$. Hence E confirms T. According to s^*, old evidence E can substantially confirm theory T, whereas the degree of confirmation is zero for measures that compare the prior and posterior probability of T, such as $d(T, E) = p(T|E) - p(T)$ or $r(T, E) = \log(p(T|E)/p(T))$. Choosing the "right" confirmation measure therefore resolves the POE.

Unfortunately, Christensen's proposal is highly problematic. First, it is questionable whether s^* is a good explicatum for degree of confirmation. In Variation 1, we have argued that s^* fails to satisfy important adequacy criteria for degree of confirmation, such as Final Probability Incrementality. Moreover, the challenge posed by the POE consists in identifying a probability distribution where E is statistically relevant to T. Christensen's proposal does not identify such a distribution, relies exclusively on the choice of a particular confirmation measure and is therefore less general than we desire.

Second, when $p(E) = 1$, $p(T|\neg E)$ may not have a clear-cut definition, since $p(T|\neg E) = p(T \wedge \neg E)/p(\neg E)$ involves a division by zero. We could solve this problem by evaluating $p(T|\neg E)$ not via the Ratio Analysis of conditional probability, but as a counterfactual degree of belief: suppose that $\neg E$ were the case, how likely would T be? But then Christensen's solution proposal is more than an appeal to a particular confirmation measure: it requires a specific approach to conditional degrees of belief which needs to be spelled out in more detail.

Such an attempt is made by Colin Howson (1984, 1985, 1991). He gives up the Bayesian explication of confirmation as statistical relevance relative to actual degrees of belief. For describing the static POE, the Bayesian should evaluate the confirmation relation with respect to a counterfactual degrees-of-belief function where E is not taken for granted:

> [T]he Bayesian assesses the contemporary support E gives T by how much the agent would change his odds on T *were he now* to come to know E .... In other words, the theory is

explicitly a theory of dispositional properties of the agent's belief-structure .... (Howson 1984, 246, original emphasis)

According to this account, conditional probabilities such as $p(\mathrm{E}|\mathrm{T})$ and $p(\mathrm{E}|\neg\mathrm{T})$ should not be understood as our actual degrees of belief in E given T or ¬T: these would be equal to 1 since E is already known and the equation

$$1 = p(\mathrm{E}) = p(\mathrm{E}|\mathrm{T})\,p(\mathrm{T}) + p(\mathrm{E}|\neg\mathrm{T})\,p(\neg\mathrm{T})$$

would imply that also $p(\mathrm{E}|\mathrm{T}) = p(\mathrm{E}|\neg\mathrm{T}) = 1$. Hence, $p(\cdot\,|\mathrm{T})$ must be a different credence function.

Howson opts for a counterfactual interpretation: it describes the degrees of belief that we would have if we did not know that E is the case. To distinguish those degrees of belief from our actual ones, given by p, denote the relevant credence function by $\tilde{p}$. For calculating $\tilde{p}(\mathrm{E}|\mathrm{T})$, we just take T for granted and calculate the probability of E conditional on this assumption. For example, suppose we are interested in whether a coin is fair (T: $\vartheta = 0.5$) or biased in favor of heads (T′: $\vartheta = 0.6$). The coin has been tossed five times (in independent and identically distributed tosses) and we have observed E = two heads, three tails. Howson's approach allows us to calculate $\tilde{p}(\mathrm{E}|\mathrm{T}) \approx 0.31$ and $\tilde{p}(\mathrm{E}|\neg\mathrm{T}) \approx 0.23$. Even if E is old evidence, we can now derive a judgment of evidential support in favor of T because $\tilde{p}(\mathrm{E}|\mathrm{T}) > \tilde{p}(\mathrm{E}|\mathrm{T}')$. In the GTR example, we could analogously conclude that $\tilde{p}(\mathrm{E}|\mathrm{T}) = 1$ because GTR implies the Mercury perihelion shift, whereas $\tilde{p}(\mathrm{E}|\mathrm{T}') \ll 1$ for Newtonian mechanics. The same inequality holds for other theories (T″, T‴, ...) that do not make definite predictions about E. Now, if $\tilde{p}$ has an "impartial" prior probability distribution, that is, $\tilde{p}(\mathrm{T}) = \tilde{p}(\mathrm{T}')$, then we infer that $\tilde{p}(\mathrm{T}|\mathrm{E}) > \tilde{p}(\mathrm{T}'|\mathrm{E})$ if and only if $\tilde{p}(\mathrm{E}|\mathrm{T}) > \tilde{p}(\mathrm{E}|\mathrm{T}')$ (proof omitted). Final Probability Incrementality then implies that E confirms T relative to $\tilde{p}$ more than it confirms T′. That is, our judgment on the conditional probability of E given T translates into a judgment of evidential support.

Of course, the counterfactual interpretation requires substantial philosophical groundwork, and we will deliver it in Variation 12. For the moment, we just ask the reader to accept that one can coherently come up with such an interpretation. The advantage is that in such a setting, we can meaningfully compare the "objective" probabilities of $\tilde{p}(\mathrm{E}|\mathrm{T})$ and $\tilde{p}(\mathrm{E}|\mathrm{T}')$ and come up with confirmation judgments that address the static POE. This is not possible if we interpret these probabilities as our actual degrees of belief in E, knowing that T. By giving up the idea that Bayesian confirmation is always relative to actual degrees of belief, Howson can model important aspects of the static POE.

However, the question is which counterfactual credence function is relevant for old evidence cases. Clearly, we need a well-defined answer in order to avoid the charge of an *ad hoc* modification. Supposing that K denotes the agent's actual background knowledge, Howson suggests the set-theoretic complement $K \setminus \{E\}$. However, this approach has been criticized for various technical and conceptual reasons, including failure of deductive closure in $K \setminus \{E\}$, and the impossibility of separating E neatly from other evidential propositions (e.g., Chihara 1987).

A possible solution to these challenges is provided by qualitative theories of belief revision, such as the AGM model (Alchourrón, Gärdenfors and Makinson 1985). It provides rules for contracting a proposition E from a set K, that is, for constructing a set $K \div E$ which does not entail E but retains as much information as possible from K. For evaluating the practical viability of Howson's proposal, it would be important to investigate whether counterfactual probability functions constructed with the help of belief revision models deliver intuitive confirmation judgments. This project is left to future research.

5.4 The Hybrid Problem of Old Evidence: Learning Explanatory Relationships

The aim of this section consists in showing that learning explanatory or deductive relationships between theory and evidence can raise our degree of belief in the theory. In other words, learning

$$X := \text{T adequately explains E}$$

raises the subjective probability of T relative to a probability function where E is taken for granted. This looks like a reformulation of the dynamic POE, but things are different from the GJN approach: we are not interested in reconstructing why X confirmed T for the actual discoverer of X (e.g., Einstein in the GTR case), but in whether X should confirm T for *all* scientists in the community. This question is related to the static POE in so far as confirmation is detached from an agent's actual degrees of belief at a particular time. We explicate evidential support by explanatory discoveries relative to a counterfactual probability function, like in the static POE. That's why we would like to call it the **hybrid POE.** It shows how discovering a deductive or explanatory relationship can be confirmatory relative to a reference probability function which scientists can reasonably use for evaluating evidential claims.

What kind of probability function $\tilde{p}$ should be chosen? Rather than just subtracting the learned proposition from our actual background knowledge, which may be deficient in various ways, $\tilde{p}(\cdot \mid \pm\mathrm{T})$ should represent the degrees of belief of a scientist who has a sound understanding of theoretical principles and their impact on observational data, conditional on the assumption that T, or ¬T, is the case (see also Earman 1992, 134). Such degrees of belief are required for making routine judgments in assessing evidence and reviewing journal articles: How probable would the actual evidence E be if T were true? How probable would E be if T were false? When T and ¬T are two definite statistical hypotheses, like in Howson's coin toss example, such judgments are immediately given by the corresponding sampling distribution. But even in more general contexts such judgments may be straightforward, or a matter of consensus in the scientific community.

We now formulate constraints on an agent's conditional degrees of belief in the hybrid POE. The first condition characterizes the elementary inferential relations between E, T and X:

$$\tilde{p}(\mathrm{E} \mid \mathrm{T}, \mathrm{X}) = 1. \qquad [5.1]$$

If T is true and T entails E, then E can be regarded as a certainty. In this scenario, X codifies a strict deductive relation between T and E. Later, we will relax this condition in order to cover more general explanatory dependencies.

To motivate the second constraint, note that learning the predictions of a refuted hypothesis is irrelevant to our assessment of the *plausibility* of the predicted events. For instance, the astrological theory on which Nostradamus based his predictions is almost certainly wrong. Upon learning the content of his predictions (e.g., the Third World War starting in 2048), we should neither raise nor lower our credence in those events. This motivates the equation $\tilde{p}(\mathrm{E} \mid \neg\mathrm{T}, \mathrm{X}) = \tilde{p}(\mathrm{E} \mid \neg\mathrm{T})$, or written differently, $\tilde{p}(\mathrm{E} \mid \neg\mathrm{T}, \mathrm{X}) = \tilde{p}(\mathrm{E} \mid \neg\mathrm{T}, \neg\mathrm{X})$. Again, these degrees of belief ought to be interpreted in a counterfactual sense: supposing that T has been disproved, would learning something about the predictions of T affect our confidence in the occurrence of E? Plausibly not, since T has ceased to be relevant for empirical forecasts. Hence we demand that if ¬T is already known, then learning X or ¬X does not change the probability of E. However, E should still be *possible* if T were false. Hence,

$$\tilde{p}(\mathrm{E} \mid \neg\mathrm{T}, \mathrm{X}) = \tilde{p}(\mathrm{E} \mid \neg\mathrm{T}, \neg\mathrm{X}) > 0. \qquad [5.2]$$

Finally, we have the following inequality:

$$\tilde{p}(\mathrm{E} \mid \mathrm{T}, \neg\mathrm{X}) < \frac{1 - \tilde{p}(\mathrm{X} \mid \neg\mathrm{T})}{1 - \tilde{p}(\mathrm{X} \mid \mathrm{T})} \, \frac{\tilde{p}(\mathrm{X} \mid \mathrm{T})}{\tilde{p}(\mathrm{X} \mid \neg\mathrm{T})}. \qquad [5.3]$$

This condition demands that the value of $\tilde{p}(E \mid T, \neg X)$ be smaller than the threshold on the right-hand side. When X and T are positively relevant to each other or probabilistically independent, [5.3] is trivially satisfied since in that case, $\tilde{p}(X|T) \geq \tilde{p}(X|\neg T)$, implying that the right-hand side of [5.3] is greater than or equal to one. But even if X and T are negatively relevant to each other, [5.3] is plausibly satisfied. When the mutual negative impact of X and T is not too strong, the two quotients in [5.3] are close to 1, and the inequality will be satisfied as long as $\tilde{p}(E \mid \neg X, T)$ is not too close to 1 itself. Given that T is assumed to be true, but that by $\neg$X, it does not fully account for E, proposition E should be far from certain for a rational Bayesian agent. Here it is essential that the conditional probabilities are interpreted in the counterfactual sense. Otherwise, we would always obtain $\tilde{p}(E \mid \cdot) = 1$ for old evidence E, regardless of which proposition stands to the right of the vertical dash. In the (plausible) case of independence of T and X, this would contradict [5.3] ($\tilde{p}(E \mid T, \neg X) < 1$).

Together with the unproblematic assumption that neither T nor $\neg$T is a certainty beforehand ($0 < \tilde{p}(T) < 1$), these three conditions are jointly sufficient to prove that X confirms T relative to E:

Theorem 5.2 (Hybrid Problem of Old Evidence, strict implication). *Let $p\colon \mathcal{L} \rightarrow [0;1]$ be a probability function on the sentences of a propositional language and let* E, T, X $\in \mathcal{L}$. *Suppose the following three conditions are satisfied:*

$$\tilde{p}(E \mid T, X) = 1; \qquad [5.1]$$

$$\tilde{p}(E \mid \neg T, X) = \tilde{p}(E \mid \neg T, \neg X) > 0; \qquad [5.2]$$

$$\tilde{p}(E \mid T, \neg X) < \frac{1 - \tilde{p}(X|\neg T)}{1 - \tilde{p}(X|T)} \, \frac{\tilde{p}(X|T)}{\tilde{p}(X|\neg T)}. \qquad [5.3]$$

Then X *confirms* T *relative to* $\tilde{p}$ *and conditional on (old evidence)* E; *that is,* $\tilde{p}(T|E, X) > \tilde{p}(T|E)$.

In other words, if E is taken for granted, learning X raises the conditional degree of belief in T if conditions [5.1]–[5.3], whose adequacy we have justified above, are accepted. Or in yet other words: if we knew little or nothing about the observational history of a discipline, and we were informed that E, then discovering X would raise our confidence in T. This seems to be a perfectly reasonable sense in which X is evidence for T, relative to E.

This theorem combines various strategies. The main idea stems from the GJN models—the confirming proposition is the discovery that T accounts for/explains E—but the relevant constraints are spelled out in terms of conditional degrees of belief, which are interpreted in a counterfactual sense, like in the static POE. Then, with the help of Bayes' Theorems, the constraints transfer to bounds on the conditional probability of T given E and X. The chosen approach does not pretend to solve the POE in full generality. What we can show, however, is that logical and explanatory relationships have confirmatory power relative to a certain (canonical?) class of counterfactual credence functions. This was one of the central intuitions that the GJN approaches tried to rescue, and we have given a rational reconstruction of this judgment.

In many cases of scientific reasoning, the condition $\tilde{p}(E \mid T, X) = 1$ appears to be too strong. It may apply well to the Mercury perihelion shift, which is deductively implied by the General Theory of Relativity, but it does not cover cases where T accounts for E in a less rigorous manner (Earman 1992; Fitelson 2004). If we allow for a weaker interpretation of X, for example, as providing some explanatory mechanism, then we are faced with the possibility that even if we are certain that T is true, and that T explains E, the conditional degree of belief in E may not be a certainty. The condition $\tilde{p}(E \mid T) < 1$ could even make sense if the relationships between T and E are deductive: the proof of X could be so complex that the scientists involved have some doubts about its soundness and refrain from assigning it maximal degree of belief. Again, Fermat's Last Theorem may be a plausible intuition pump.

For covering this case, modeled as $\tilde{p}(E \mid T, X) = 1 - \varepsilon$ for some small $\varepsilon > 0$, we prove another theorem:

Theorem 5.3 (Hybrid Problem of Old Evidence, explanatory relationship). *Let* $p\colon \mathcal{L} \to [0;1]$ *be a probability function on the sentences of a propositional language and let* E, T, X $\in \mathcal{L}$. *Suppose the following three conditions are satisfied:*

$$\tilde{p}(E \mid T, X) = 1 - \varepsilon \quad \textit{for some } 0 < \varepsilon < 1; \qquad [5.1']$$

$$\tilde{p}(E \mid \neg T, X) = \tilde{p}(E \mid \neg T, \neg X) > 0; \qquad [5.2']$$

$$\tilde{p}(E \mid T, \neg X) < (1 - \varepsilon) \frac{1 - \tilde{p}(X \mid \neg T)}{1 - \tilde{p}(X \mid T)} \frac{\tilde{p}(X \mid T)}{\tilde{p}(X \mid \neg T)}. \qquad [5.3']$$

Then X *confirms* T *relative to* $\tilde{p}$ *and conditional on (old evidence)* E; *that is,* $\tilde{p}(T \mid E, X) > \tilde{p}(T \mid E)$.

The motivations and justifications for the above assumptions are the same as in Theorem 5.2. Condition [5.1′] just accounts for lack of full certainty about the old evidence, and [5.2′] is identical to [5.2]. Moreover,

Condition [5.3] of Theorem 5.2 can, with the same line of reasoning, be extended to Condition [5.3′] in Theorem 5.3. Condition [5.3′] sharpens [5.3] by a factor of $1-\varepsilon$, but leaves the qualitative argument for [5.3] intact. As long as $\tilde{p}(\mathrm{E} \mid \mathrm{T}, \neg \mathrm{X})$ and $\tilde{p}(\mathrm{E} \mid \mathrm{T}, \mathrm{X})$ decrease by roughly the same margin, the result of Theorem 5.2 transfers to Theorem 5.3.

Thus we can extend the novel solution of POE to the case of residual uncertainty about the old evidence E—a case that is highly relevant for case studies in the history of science. If we compare this solution of the POE to Jeffrey's and Earman's proposals, we note that our assumptions [5.1], [5.2] and [5.3] are silent on whether Jeffrey's (δ)—or Earman's (φ) and (ψ), for that matter—is true or false. For a proof with the help of Branden Fitelson's PRSAT package for Mathematica (Fitelson 2008a), see Sprenger 2015. Hence we can discard Jeffrey's dubious assumption (δ) that increasing empirical content makes a theory more plausible, without jeopardizing our own results.

We have thus provided a solution of the POE that successfully tackles a hybrid version of the POE. Conceptually, this solution is anchored in the use of counterfactual (rather than actual) probability functions; and our solution makes less demanding assumptions than Jeffrey's and Earman's. We now discuss the repercussions of our results on the POE and the role of Bayesian Confirmation Theory in scientific reasoning.

5.5 Discussion

This Variation has analyzed Bayesian attempts to solve the Problem of Old Evidence (POE), and it has proposed two new solutions. We have started with a distinction between the static and the dynamic aspect of the problem. Simplifying a bit, we can say that the static POE requires finding a probability function $\tilde{p}$ where $\tilde{p}(\mathrm{E}|\mathrm{T}) > \tilde{p}(\mathrm{E}|\neg\mathrm{T})$ for old evidence E, demonstrating the evidential relevance of E for T. The dynamic problem, on the other hand, deals with the challenge to provide reasonable constraints on the actual degrees of belief p such that $p(\mathrm{T}|\mathrm{X},\mathrm{E}) > p(\mathrm{T}|\mathrm{E})$, with X denoting the proposition that T explains or deductively implies E.

We first presented our criticism of existing solutions to the dynamic POE along the lines of Garber, Jeffrey and Niiniluoto (GJN). Then we explained our own constructive proposal. Our model of the dynamic POE was based on judgments where T is confirmed by the presence or absence of alternative hypotheses accounting for old evidence E. We identified a plausible set of constraints that suffice for a resolution of the dynamic POE. By contrast, our second model targeted a hybrid version of the POE: How does logical and explanatory learning confirm T relative to a "reference probability function" $\tilde{p}$ which is normatively relevant for confirmation judgments, but different from our actual degrees of belief described by p? In other words, how can learning $\tilde{p}(\mathrm{E} \mid \mathrm{T}, \mathrm{X}) = 1$ support T?

In order to avoid that $\tilde{p}(\mathrm{E} \mid \pm\mathrm{T}, \pm\mathrm{X})$ is equal to 1 for all truth-functional combinations of T and X (and old evidence E), we conceptualized $\tilde{p}$ as a genuinely counterfactual credence function. Indeed, in scientific practice, we typically interpret $\tilde{p}(\mathrm{E}|\mathrm{T})$ and $\tilde{p}(\mathrm{E}|\neg\mathrm{T})$ as *principled* statements about the predictive import of T and ¬T on E, without referring to our complete observational record. Such judgments are part and parcel of scientific reasoning, for example, in statistical inference where theories T, T′, etc. impose definite probability distributions on the observations, and our degrees of belief $\tilde{p}(\mathrm{E}|\mathrm{T})$, $\tilde{p}(\mathrm{E}|\mathrm{T}')$, etc. follow suit. Given such a credence function and some plausible constraints on $\tilde{p}$, we showed that learning logical and explanatory relationships between T and previously known evidence E lends support to T.

Notably there are strong parallels between the POE and the prediction-vs.-accommodation debate in philosophy of science (e.g., Hitchcock and Sober 2004; Barnes 2008). Future research could investigate this relationship in more detail, and also come up with more case studies from the history of science where old evidence has confirmed a scientific theory. By describing the degree of confirmation in those historical episodes, the case studies would contribute to a better evaluation of the practical value of our solution proposals.

Other research projects that spring to mind are a more specific characterization of the counterfactual probability function for addressing the POE, integrating this function with qualitative approaches to belief revision (e.g., AGM theory), and relating an account of the POE to learning conditional evidence and relative frequencies (see Variation 4). After all, the dynamic POE can be described as learning the (strict) conditional that if T, then also E. Our theoretical account from Variation 4 could, perhaps with some modifications, be relevant for describing the conditions when learning $\mathrm{T} \models \mathrm{E}$ raises the probability of T.

Finally, a popular critique of Bayesian approaches to the POE is inspired by the view that the POE poses a principled and insoluble problem for Bayesian Confirmation Theory (see also Brössel and Huber 2015). For instance, in his monograph *Theory and Evidence*, Clark Glymour concludes his discussion of the POE as follows:

> ... our judgment of the relevance of evidence to theory depends on the perception of a structural connection between the two, and ... degree of belief is, at best, epiphenomenal. In the determination of the bearing of evidence on theory there seem to be mechanisms and stratagems that have no apparent connections with degrees of belief. (Glymour 1980, 92–93)

What Glymour argues here is not so much that a specific formal aspect of the Bayesian apparatus (e.g., logical omniscience) prevents it from solving the POE, but that these shortcomings are a symptom of a more general inadequacy of Bayesian Confirmation Theory: the inability to capture *structural relations* between evidence and theory. This criticism should not be misunderstood as claiming that confirmation has to be conceived of as an objective relation that is independent of contextual knowledge or contingent background assumptions. Rather, it suggests that solutions to the dynamic POE mistake an increase in degree of belief for a structural relation between T and E. But what makes E relevant for T is not the increase in degree of belief $p(\mathrm{T}\,|\,\mathrm{E}) > p(\mathrm{T})$, but the entailment relation between T and E. Hence Glymour's verdict that Bayesian Confirmation Theory gives "epiphenomenal" results.

To our mind, this criticism commits two oversights. First, solutions to the static POE answer Glymour's challenge by showing how the concept of evidential support can be interpreted in a way that is not tied to belief updating at a particular point in time. We have sketched such an account in Section 5.3. Second, the criticism is too fundamental to be a source of genuine concern: it is not specific to the (dynamic) POE or one of its solutions, but it attacks the entire Bayesian explication of confirmation as increase in firmness. However, as we have seen in Variation 1, Bayesian Confirmation Theory can point to many success stories: within inductive reasoning, it resolves several resilient paradoxes, such as the tacking by conjunction paradoxes, the raven paradox and the new riddle of induction. Add to this the successful applications from Variations 2–4. What we have shown here is that confirmation by old evidence is not an unsurmountable obstacle either.

The discussion of the Problem of Old Evidence concludes a set of Variations devoted to theory confirmation and various types of confirmatory arguments in science. The next group of Variations investigates and explicates, with Bayesian means, various central concepts of scientific inference: causal effect, explanatory power, intertheoretic reduction, corroboration, simplicity and scientific objectivity.

Appendix: Proofs of the Theorems

Proof of Theorem 5.1: Let

$$\mathfrak{a} := p(\mathrm{T}\,|\,\mathrm{X}\wedge\neg\mathrm{Y}), \qquad \mathfrak{b} := p(\mathrm{T}\,|\,\mathrm{X}\wedge\mathrm{Y}),$$
$$\mathfrak{c} := p(\mathrm{T}\,|\,\neg\mathrm{X}\wedge\neg\mathrm{Y}), \qquad \mathfrak{d} := p(\mathrm{T}\,|\,\neg\mathrm{X}\wedge\mathrm{Y})$$
$$x := p(\neg\mathrm{Y}|\mathrm{X}), \qquad y := p(\neg\mathrm{Y}|\neg\mathrm{X}).$$

Given these assignments, (5.3)–(5.5) translate as follows:

$$\mathfrak{a} > \mathfrak{c}, \quad \mathfrak{a} > \mathfrak{d}, \quad \mathfrak{b} \geq \mathfrak{c}, \quad \mathfrak{b} > \mathfrak{d}.$$

Suppose that $\mathfrak{a} \in (0;1]$, $\mathfrak{d} \in [0;1)$ and $\mathfrak{b}$, $\mathfrak{c}$, x, $y \in (0;1)$.[9] Then (5.3)–(5.5) jointly entail

$$\mathfrak{a}x + \mathfrak{b}(1-x) \;>\; \mathfrak{c}y + \mathfrak{d}(1-y).$$

And by the Law of Total Probability we have

$$p(\mathrm{T}|\mathrm{X}) \;=\; \mathfrak{a}x + \mathfrak{b}(1-x),$$
$$p(\mathrm{T}|\neg\mathrm{X}) \;=\; \mathfrak{c}y + \mathfrak{d}(1-y).$$

Thus (5.3)–(5.5) jointly entail $p(\mathrm{T}|\mathrm{X}) > p(\mathrm{T}|\neg\mathrm{X})$, which entails $p(\mathrm{T}|\mathrm{X}) > p(\mathrm{T})$.
□

Proof of Theorem 5.2: First, we define

$$e_1 := \tilde{p}(\mathrm{E}\,|\,\mathrm{T},\mathrm{X}), \qquad e_2 := \tilde{p}(\mathrm{E}\,|\,\neg\mathrm{T},\mathrm{X}),$$
$$e_3 := \tilde{p}(\mathrm{E}\,|\,\mathrm{T},\neg\mathrm{X}), \qquad e_4 := \tilde{p}(\mathrm{E}\,|\,\neg\mathrm{T},\neg\mathrm{X}),$$
$$t := \tilde{p}(\mathrm{T}), \qquad t' := \tilde{p}(\mathrm{T}|\mathrm{X}),$$
$$r := \tilde{p}(\mathrm{X}|\mathrm{T}), \qquad \bar{r} := \tilde{p}(\mathrm{X}|\neg\mathrm{T}).$$

By making use of [5.1] ($e_1 = 1$), [5.2] ($e_2 = e_4 > 0$) and the Extension Theorem $\tilde{p}(\mathrm{X}|\mathrm{Z}) = \tilde{p}(\mathrm{X}\,|\,\mathrm{Y},\mathrm{Z})\,\tilde{p}(\mathrm{Y}|\mathrm{Z}) + \tilde{p}(\mathrm{X}\,|\,\neg\mathrm{Y},\mathrm{Z})\,\tilde{p}(\neg\mathrm{Y}|\mathrm{Z})$, we can quickly verify the identities

$$\tilde{p}(\mathrm{E}|\mathrm{T}) \;=\; \tilde{p}(\mathrm{E}\,|\,\mathrm{T},\mathrm{X})\,\tilde{p}(\mathrm{X}|\mathrm{T}) + \tilde{p}(\mathrm{E}\,|\,\mathrm{T},\neg\mathrm{X})\,\tilde{p}(\neg\mathrm{X}|\mathrm{T})$$

---

[9]The only two conditional credences that may reasonably take extremal values here are $\mathfrak{a}$ and $\mathfrak{d}$. If T is the only theory that adequately explains E, then it may be reasonable to assign T maximal credence. And if some alternative to T (say, T′) is the only theory that adequately explains E, then it may be reasonable to assign minimal credence to T. This is why we allow $\mathfrak{a} \in (0;1]$ and $\mathfrak{d} \in [0;1)$. The other conditional credences involved in our theorem (i.e., $\mathfrak{b}$, $\mathfrak{c}$, x, y) should, in general, take non-extreme values.

$$
\begin{aligned}
&= r + e_3(1-r),\\
\tilde{p}(\mathrm{E}\,|\,\neg\mathrm{T}) &= \tilde{p}(\mathrm{E}\,|\,\neg\mathrm{T},\mathrm{X})\,\tilde{p}(\mathrm{X}\,|\,\neg\mathrm{T}) + \tilde{p}(\mathrm{E}\,|\,\neg\mathrm{T},\neg\mathrm{X})\,\tilde{p}(\neg\mathrm{X}\,|\,\neg\mathrm{T})\\
&= e_2\bar{r} + e_4(1-\bar{r})\\
&= e_2,
\end{aligned}
$$

which will be useful later. Second, we note that by Bayes' Theorem and assumption [5.1],

$$
\begin{aligned}
\tilde{p}(\mathrm{T}\,|\,\mathrm{E},\mathrm{X}) &= \tilde{p}(\mathrm{T}\,|\,\mathrm{X})\frac{\tilde{p}(\mathrm{E}\,|\,\mathrm{T},\mathrm{X})}{\tilde{p}(\mathrm{E}\,|\,\mathrm{X})}\\
&= \left(1+\frac{\tilde{p}(\neg\mathrm{T}\,|\,\mathrm{X})}{\tilde{p}(\mathrm{T}\,|\,\mathrm{X})}\frac{\tilde{p}(\mathrm{E}\,|\,\neg\mathrm{T},\mathrm{X})}{\tilde{p}(\mathrm{E}\,|\,\mathrm{T},\mathrm{X})}\right)^{-1} = \left(1+\frac{1-t'}{t'}\cdot e_2\right)^{-1}. \qquad (5.7)
\end{aligned}
$$

Third, we observe that by [5.1], [5.2] and the above identities for $\tilde{p}(\mathrm{E}\,|\,\mathrm{T})$ and $\tilde{p}(\mathrm{E}\,|\,\neg\mathrm{T})$,

$$
\begin{aligned}
\tilde{p}(\mathrm{T}\,|\,\mathrm{E}) &= \tilde{p}(\mathrm{T})\frac{\tilde{p}(\mathrm{E}\,|\,\mathrm{T})}{\tilde{p}(\mathrm{E})}\\
&= \left(1+\frac{\tilde{p}(\neg\mathrm{T})}{\tilde{p}(\mathrm{T})}\frac{\tilde{p}(\mathrm{E}\,|\,\neg\mathrm{T})}{\tilde{p}(\mathrm{E}\,|\,\mathrm{T})}\right)^{-1} = \left(1+\frac{1-t}{t}\frac{e_2}{r+e_3(1-r)}\right)^{-1}. \qquad (5.8)
\end{aligned}
$$

We also note that, by [5.3],

$$
e_3 < \frac{1-\bar{r}}{\bar{r}}\frac{r}{1-r}.
$$

Note that it is implicit in condition [5.3] that $1 > \tilde{p}(\mathrm{X}\,|\,\mathrm{T}), \tilde{p}(\mathrm{X}\,|\,\neg\mathrm{T}) > 0$, since otherwise either the expression above would be undefined (division by zero) or $\tilde{p}(\mathrm{E}\,|\,\mathrm{T},\neg\mathrm{X})$ would have to be smaller than zero, which is impossible.

This allows us to derive

$$
r+e_3(1-r) < r+\frac{1-\bar{r}}{\bar{r}}\frac{r}{1-r}(1-r) = r\cdot\left(1+\frac{1-\bar{r}}{\bar{r}}\right) = \frac{r}{\bar{r}}
$$

and consequently also

$$
\frac{1}{r+e_3(1-r)} > \frac{\bar{r}}{r}. \qquad (5.9)
$$

Moreover, note the equality

$$
\frac{1-t'}{t'} = \frac{\tilde{p}(\neg\mathrm{T}\,|\,\mathrm{X})}{\tilde{p}(\mathrm{T}\,|\,\mathrm{X})} = \frac{p(X\,|\,\neg T)}{p(X\,|\,T)}\cdot\frac{\tilde{p}(\neg\mathrm{T})}{\tilde{p}(\mathrm{T})} = \frac{\bar{r}}{r}\cdot\frac{1-t}{t}. \qquad (5.10)
$$

All this implies that

$$
\frac{\tilde{p}(\mathrm{T}\,|\,\mathrm{E},\mathrm{X})}{\tilde{p}(\mathrm{T}\,|\,\mathrm{E})} \overset{(5.7),(5.8)}{=} \left(1+\frac{1-t'}{t'}\cdot e_2\right)^{-1}\cdot\left(1+\frac{1-t}{t}\frac{e_2}{r+e_3(1-r)}\right)
$$

$$\begin{aligned}
&\overset{(5.9)}{>} \left(1+\frac{1-t'}{t'}\cdot e_2\right)^{-1}\cdot\left(1+\frac{1-t}{t}\frac{\bar{r}}{r}\cdot e_2\right)\\
&\overset{(5.10)}{=} \left(1+\frac{1-t'}{t'}\cdot e_2\right)^{-1}\cdot\left(1+\frac{1-t'}{t'}\cdot e_2\right)\\
&= 1,
\end{aligned}$$

completing the proof. The second line has also used that $e_2 > 0$, as ensured by [5.2]; and that $t, r, \bar{r} \in (0;1)$ implies $t' \in (0;1)$. □

Proof of Theorem 5.3: By means of performing the same steps as in the proof of Theorem 5.2, we can easily verify the equalities

$$\begin{aligned}
\tilde{p}(T\,|\,E,X) &= \left(1+\frac{\tilde{p}(\neg T\,|\,X)}{\tilde{p}(T\,|\,X)}\frac{\tilde{p}(E\,|\,\neg T,X)}{\tilde{p}(E\,|\,T,X)}\right)^{-1}\\
&= \left(1+\frac{1-t'}{t'}\frac{e_2}{1-\varepsilon}\right)^{-1},\\
\tilde{p}(T\,|\,E) &= \left(1+\frac{\tilde{p}(\neg T)}{\tilde{p}(T)}\frac{\tilde{p}(E\,|\,\neg T)}{\tilde{p}(E\,|\,T)}\right)^{-1}\\
&= \left(1+\frac{1-t}{t}\frac{e_2\bar{r}+e_4(1-\bar{r})}{(1-\varepsilon)r+e_3(1-r)}\right)^{-1}\\
&= \left(1+\frac{1-t}{t}\frac{e_2}{(1-\varepsilon)r+e_3(1-r)}\right)^{-1},
\end{aligned}$$

where we have made use of [5.1′] and [5.2′] = [5.2]. We also note that [5.3′] implies

$$\begin{aligned}
(1-\varepsilon)r+e_3(1-r) &< (1-\varepsilon)r+(1-\varepsilon)\frac{1-\bar{r}}{\bar{r}}\frac{r}{1-r}(1-r)\\
&= (1-\varepsilon)r\left(1+\frac{1-\bar{r}}{\bar{r}}\right)\\
&= (1-\varepsilon)\frac{r}{\bar{r}},
\end{aligned}$$

and thus we obtain

$$\frac{1}{(1-\varepsilon)r+e_3(1-r)} > \frac{\bar{r}}{(1-\varepsilon)r}. \tag{5.11}$$

This brings us to the final calculation:

$$\frac{\tilde{p}(T\,|\,E,X)}{\tilde{p}(T\,|\,E)} = \left(1+\frac{1-t}{t}\frac{1}{(1-\varepsilon)r+e_3(1-r)}e_2\right)\cdot\left(1+\frac{1-t'}{t'}\frac{e_2}{1-\varepsilon}\right)^{-1}$$

$$
\begin{aligned}
&> \left(1+\frac{1-t}{t}\frac{\bar{r}}{(1-\varepsilon)r}e_2\right)\cdot\left(1+\frac{1-t}{t}\frac{\bar{r}}{r}\frac{e_2}{1-\varepsilon}\right)^{-1} \\
&= 1,
\end{aligned}
$$

where we have simultaneously applied Equations (5.10) and (5.11) in the second line. This completes the proof. □

Variation 6: Causal Strength

Causation is a central concept in human cognition. Knowledge of causal relationships enables us to make predictions, to explain phenomena and to understand complex systems. Decisions are taken according to the effects that they are supposed to bring about. Actions are evaluated according to their causal contributions to an event. Building causal models and identifying cause and effect is a central activity in many scientific disciplines.

Since the days of Aristotle, causation has been treated primarily as a qualitative, all-or-nothing concept. A huge amount of literature has been devoted to the qualitative question "When is C a cause of E?" (e. g., Hume 1739; Suppes 1970; Lewis 1973; Mackie 1974; Woodward 2003). The comparative question "Is C or C′ a more effective cause of E?" is getting explored as well (e. g., Chockler and Halpern 2004; Halpern and Hitchcock 2015). By contrast, the quantitative question "What is the strength of the causal relationship between C and E?" has been relatively neglected. This is surprising since causal judgments regularly involve a quantitative dimension: C is a more effective cause of E than C′, the causal effect of C on E is twice as high as the effect of C′, and so on (e. g., Rubin 1974; Rosenbaum and Rubin 1983; Pearl 2001).

Principled proposals for explicating causal strength are rare and spread over different disciplines, each with its own motivation and intended context of application. This includes cognitive psychology (Cheng 1997; Icard, Kominsky and Knobe 2017), computer science and machine learning (Pearl 2000; Korb, Hope and Nyberg 2009; Korb, Nyberg and Hope 2011), statistics (Good 1961a,b; Holland 1986; Cohen 1988), epidemiology and clinical medicine (Poole 2010; Broadbent 2013), philosophy of science (Suppes 1970; Eells 1991), political philosophy and social choice theory (Braham and van Hees 2009) and legal theory (Rizzo and Arnold 1980; Hart and Honoré 1985;

Kaiserman 2017). Although these approaches use a common formalism—probability theory—the proposed explications differ substantially (see the survey of Fitelson and Hitchcock 2011). This may be due to the different purposes to which the measures are put: measuring predictive power, expressing counterfactual dependence, apportioning liability and so on. The challenge for a philosophical theory of causal strength is to characterize the various measures and to evaluate whether we should prefer one of them over its competitors, or whether we should use different measures in different contexts.

This Variation proceeds as follows. Section 6.1 specifies the sense of causal strength that we would like to explicate: the difference that causes make to their effects. We also motivate causal Bayesian networks as an appropriate formal framework for this project. Section 6.2 derives representation theorems that characterize causal strength measures in terms of the adequacy conditions that they satisfy. These theorems lend support to preferring the difference measure $\eta_d(C, E) = p(E \mid C) - p(E \mid \neg C)$ over its competitors. As the reader will notice, both the causal strength measures and the corresponding adequacy criteria often have counterparts in Bayesian Confirmation Theory (Variation 1). Section 6.3 discusses possible objections motivated by the concept of actual causal strength while Section 6.4 sketches future research questions and concludes. All proofs are contained in the Appendix. As stated at the beginning of the book, this Variation is essentially a republication of Sprenger 2018b integrating material from Sprenger and Stegenga 2017.

6.1 Interventions and Causal Bayesian Networks

Causes do not always necessitate their effects. We classify smoking as a cause of lung cancer although not every regular smoker will eventually contract lung cancer. The same is true in other fields of science, for example, when we conduct psychological experiments or choose an economic policy: interventions increase the frequency of a particular response, but they do not guarantee it. Therefore, causal relevance is often explicated as **statistical relevance** or **probability-raising:** C is a cause of E if and only if C raises the probability of E (e.g., Reichenbach 1956; Suppes 1970; Cartwright 1979; Eells 1991). A cause is the more effective the more it raises the probability of an effect. Probability-raising captures the intuition that many causes make a difference to their effects without necessitating them.

It is well known that statistical-relevance accounts of causation struggle to capture the asymmetry of causal relations. They dissolve the crucial difference between a causal inference (does bringing about C increase the probability of E?) and an observational inference (does learning C increase our confidence that E?). This is not the same: statistically associated variables, such as the number of ice cream sales and swimming pool visits on a particular day, need not be connected causally. More likely, they have common causes, such as temperatures and sunshine hours (compare Reichenbach 1956). Furthermore, unlike the cause–effect relation, statistical relevance is symmetric: if C raises the probability of E, then E also raises the probability of C.

Pearl (2000, 2011) notes that the problem is principled: causal claims go beyond the purely associational level that is encoded in probability distributions. They express how the world would change in response to interventions. Hence,

> every claim invoking causal concepts must rely on some premises that invoke such concepts; it cannot be inferred from, or even defined in, terms of statistical associations alone. (Pearl 2011, 700)

Interventionist accounts of causation offer a principled solution to this problem. The idea behind probability-raising is modified to the effect that a variable C is a cause of another variable E if and only if an intervention on C changes the probability that E takes a particular value. The probability function p can be interpreted objectively (frequencies, propensities, best-system chances, ...) or as subjective degrees of belief, dependent on the context. We contend that the Bayesian interpretation is the most natural and general one since it is often hard to define meaningful objective chances. Nonetheless, our analysis of causal strength measures can also be transferred to other interpretations of probability; in this sense, it exceeds the purely Bayesian framework.

Intervening on C removes spurious correlations by breaking the influence of the other causes of C. After intervening on the number of swimming pool visitors (e.g., by closing the pool for renovation works), learning the number of visitors (zero) does not tell us anything about temperatures or ice cream sales. By now, the interventionist account of causation is prevalent in philosophical discussions of causality (Meek and Glymour 1994; Woodward

2003, 2016) as well as in scientific applications such as causal search and discovery algorithms (Pearl 2000; Spirtes, Glymour and Scheines 2000).

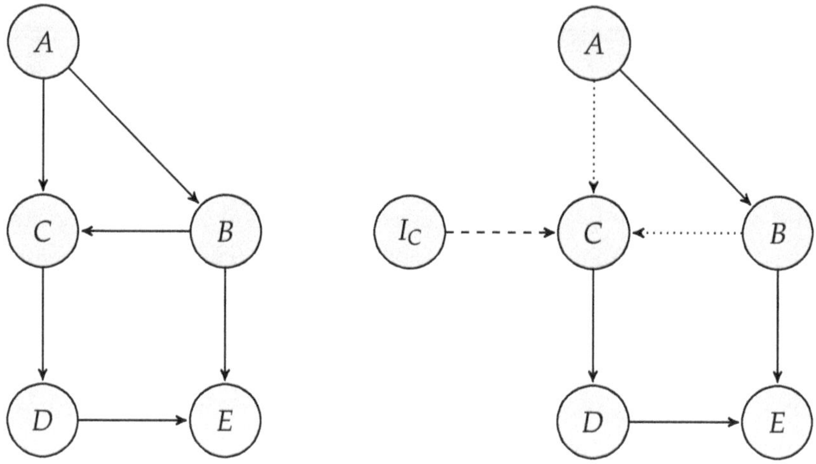

Figure 6.1: Two structurally identical DAG's with and without an intervention on the cause C. The arrows leading into C are disrupted by the intervention (the disrupted arrows are dotted); the intervention itself is represented by a dashed arrow. Reproduced from "Foundations of a Probabilistic Theory of Causal Strength" by Jan Sprenger, *Philosophical Review*, Volume 127, No. 3, pp. 371–398, with permission of Duke University Press. © 2018, doi: 10.1215/00318108-6718797.

For interventionists, causal reasoning is relative to the choice of a causal model, which we identify with a **causal Bayesian network:** a directed acyclical graph (DAG) G consisting of a set of vertices (i.e., variables) and directed edges or arrows (as in Figure 6.1). DAG's codify Pearl's "causal assumptions" in causal reasoning since the arrows represent directions of causal influence: each variable is assumed to be independent of its non-descendants, given its direct causes (i.e., its parents)—this is the famous Causal Markov Condition mentioned in the introductory chapter. Crucially, DAG's come with a probability distribution p that is Markovian relative to the variables in G, that is, it respects the probabilistic independencies which the causal reading of G implies.

In a DAG, an intervention on C amounts to breaking all arrows that lead into C, so as to control the value that C takes. This can consist in setting C to a particular value, but also in imposing a specific probability distribution on C (e.g., in medical trials, patients are randomly assigned to treatment and control groups).

The two graphs in Figure 6.1 show a causal Bayesian network before and after a (hypothetical) intervention on C, respectively. As in the previous Variations, we denote variables by italic letters (e.g., C) and use regular roman letters for particular values they take (e.g., C, ¬C, C′)—see Bovens and Hartmann 2003. The intervention on C is represented by the node I_C. Activating I_C controls the value of C and removes the influence of the parent nodes A and B. Intervening on C leads to a modified DAG G^* and probability distribution p^*, represented by the right graph of Figure 6.1. By the Causal Markov Condition and its consequence, the Manipulation Theorem (Spirtes, Glymour and Scheines 2000, 50–51), intervening on C does not alter the way C acts on causally downstream variables. Therefore we set $p^*(\cdot\,|\mathrm{C}) = p(\cdot\,|\mathrm{C})$ for all values of C.

This Variation combines the probabilistic and the interventionist perspective. We measure the causal strength of C for a target effect E as the degree to which C makes a difference to E in the post-intervention distribution p^*. This approach preserves the asymmetry of causal relations: had we intervened on E instead of C, all arrows into E in Figure 6.1 would have been cut, making C and E statistically independent. Causal strength between C and E would have been nil. Moreover, by cutting the arrows that lead into C, the intervention removes possible spurious correlations between C and E due to their common causes A and B. The technical details of this approach to measuring causal strength, including the problem of non-causal paths via common effects, are discussed in Korb, Hope and Nyberg 2009—see also Eva and Stern forthcoming and the discussion in Section 7.3.

We now add some precision to these ideas, essentially repeating our explanation from Variation 1 (see page 44). A variable C in a causal Bayesian network is characterized by its domain Ω_C and a σ-algebra $\mathcal{C}$: a set of subsets of Ω_C which contains Ω_C itself and is closed under natural set-theoretic operations such as complement and countable union. The algebra $\mathcal{C}$ describes the "structure" of Ω_C: if C is a real-valued variable, then the "natural" Borel σ-algebra for $\Omega_C = \mathbb{R}$ contains singleton sets such as $\{1\}$ or $\{2\}$, but also intervals such as $[3;4)$, $(-5;5]$ or $[0;\infty)$. Probabilities are assigned to elements of $\mathcal{C}$, corresponding to expressions such as "$C \in [0;1]$".

The causal strength of H for E depends on the features of the causal Bayesian network given by G^* and p^* that emerges from G and p by intervening on C. In particular, it is a function of the joint (post-intervention) probability distribution over C and E. A causal strength measure maps elements of $\mathcal{C} \times \mathcal{E}$ (e.g., the pair (C, E)) to a real number, denoted by $\eta(\mathrm{C}, \mathrm{E})$: it describes how C, as opposed to another value $\mathrm{C}' \in \mathcal{C}$, affects the post-intervention probability of E. This brings us to the following very general adequacy constraint, where we suppress reference to G for the sake of convenience:

Difference-Making (Causal Strength) There exists a real-valued, continuous function $f\colon [0;1]^2 \to \mathbb{R}$ such that for any putative cause $\mathrm{C} \in \mathcal{C}$ with contrast class $\mathrm{C}' \in \mathcal{C}$ (chosen as a function of C), and any putative effect $\mathrm{E} \in \mathcal{E}$, we can write the causal strength $\eta(\mathrm{C}, \mathrm{E})$ of C for E as

$$\eta(\mathrm{C}, \mathrm{E}) = f(p^*(\mathrm{E}\,|\,\mathrm{C}),\, p^*(\mathrm{E}\,|\,\mathrm{C}')),$$

where f is non-decreasing in the first argument and non-increasing in the second argument.

The idea of causal strength as difference-making is an intuition shared by counterfactual, probabilistic and interventionist accounts of causation alike. Causal strength is the higher, the more probable E is given C and the less probable E is given the contrastive value C′. Typically, we will consider $\mathrm{C}' = \neg\mathrm{C}$ or $\mathrm{C}' = \Omega_{\mathrm{C}}$. Because of this dependence between C′ and C, we suppress explicit reference to C′ in the function $\eta(\mathrm{C}, \mathrm{E})$.

Difference-Making quantifies causal strength with respect to a particular causal model. Similarly, it focuses on a **single background context,** sidestepping a substantial discussion in the field of probabilistic causation (e.g., Cartwright 1979; Dupré 1984; Eells 1991). This makes intuitive sense: when we investigate the relationship between beer consumption and obesity, causal strength depends on characteristics of the population such as age, dietary habits and general lifestyle. Compared to couch potatoes, active athletes are less likely to gain weight when they drink two or three pints a day. Causal strength claims are always relative to such a choice of context, symbolically represented by other variables that have an effect on the probability of E (e.g., A and B in the DAG of Figure 6.1). Similarly, causal strength depends on the *levels* of the cause variable that we compare (i.e., three vs. two, or five vs. zero pints per day). This choice is codified in the post-intervention distribution $p^*(\cdot)$. By contrast, we do not include external factors such as typicality, defaults and normative expectations, which have been argued to affect causal judgments (Knobe and Fraser 2008; Hitchcock and Knobe 2009; Halpern and Hitchcock 2015). We discuss objections to our approach in Section 6.3.

Notably, causal strength is blind to the presence of multiple paths leading from C to E, and to the number of mediators between C and E (see Figure 6.2). This choice is deliberate. Mediating variables are often not directly

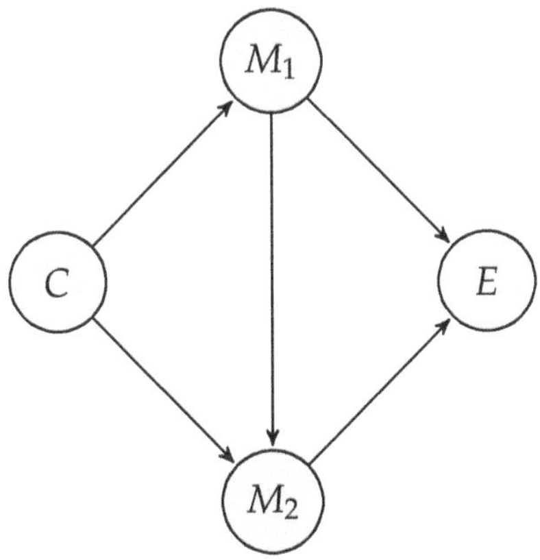

Figure 6.2: A DAG with mediators on the paths from cause C to effect E. Reproduced from "Foundations of a Probabilistic Theory of Causal Strength" by Jan Sprenger, *Philosophical Review*, Volume 127, No. 3, pp. 371–398, with permission of Duke University Press. © 2018, doi: 10.1215/00318108-6718797.

measurable. When we administer a medical drug (C) to cure migraine (E), there are numerous mediators in an appropriate causal model that includes C and E. However, the medical practitioner, who has to choose between different drugs, is mainly interested in the overall effect that C has on E (how often do migraines go away?), not in the details of causal transmission within the human body. The mapping $\eta(C, E)$ amalgamates the effects of C on E via different paths into a single number—the **total effect of C for E** (e.g., Dupré 1984; Eells 1991). This does not rule out a path-specific perspective; quite the contrary. Measures of path-specific effects supervene on elementary measures of causal strength that quantify causal strength between adjacent variables (e.g., Pearl 2001). In that sense, this chapter lays the foundations for path-specific analyses of causal strength.

6.2 Probabilistic Measures of Causal Strength

Difference-Making does not specify how C' should be chosen and how $p^*(E|C)$ and $p^*(E|C')$ should be combined. This leaves open a number of ways to explicate causal strength. Some candidate measures of causal strength that align with Difference-Making for either $C' = \neg C$ or $C' = \Omega_C$ are surveyed in Fitelson and Hitchcock 2011 and recapitulated in Table 6.1.

How should we deal with this plurality of measures of causal strength? Like in the case of confirmation measures in Variation 1, two major attitudes are possible. First, there is **monism:** there is only one adequate measure

Pearl (2000)	$\eta_{ph}(C,E) = p(E \mid C)$
Suppes (1970)	$\eta_{pr}(C,E) = p(E \mid C) - p(E)$
Eells (1991)	$\eta_{d}(C,E) = p(E \mid C) - p(E \mid \neg C)$
"Galton" (covariation)	$\eta_{ga}(C,E) = 4p(C)\,p(\neg C)\,(p(E \mid C) - p(E \mid \neg C))$
Lewis (1986)	$\eta_{r}(C,E) = \dfrac{p(E \mid C)}{p(E \mid \neg C)}$
Cheng (1997)	$\eta_{c}(C,E) = \dfrac{p(E \mid C) - p(E \mid \neg C)}{1 - p(E \mid \neg C)}$
Good (1961a,b)	$\eta_{g}(C,E) = \log \dfrac{1 - p(E \mid \neg C)}{1 - p(E \mid C)}$

Table 6.1: A list of popular measures of causal strength, adapted from Fitelson and Hitchcock 2011.

(or class of measures) of causal strength. Second, there is **pluralism:** no single measure satisfies all the conditions that an adequate measure of causal strength should possess. This is perhaps the default view. After all, intuitions about complex concepts such as causal strength may pull into different directions and lead to a set of adequacy conditions that a single measure cannot possibly satisfy. This is a lesson one might draw from the analogous projects of finding a probabilistic explication of degree of confirmation (e.g., Fitelson 1999; Meijs 2005; Brössel 2013; Crupi 2015—see also the discussion at the end of Variation 1).

The prospects for causal strength monism are, in our opinion, more promising. This monism is *qualified*: it is based on understanding causal strength as counterfactual difference-making, as informing our expectations on the efficacy of interventions on C. Whether this monism transfers to measuring causal strength as degrees of actual causation or "cause in fact" (e.g., Halpern and Pearl 2005a,b; Icard, Kominsky and Knobe 2017)—that is, as a basis for attribution and responsibility—is outside the scope of our work. However, within the explicative framework outlined by Difference-Making, we believe that there is a distinguished measure of causal strength, namely $\eta_d(C,E) = p(E \mid C) - p(E \mid \neg C)$ (Eells 1991; Pearl 2001).

In what follows, we present two different constructive arguments in favor of η_d and a negative argument against probability ratio measures. **Ordinally equivalent** measures—that is, measures that impose the same causal strength rankings on any set of cause–effect pairs—will be identified with each other. Formally, two measures η and η' are ordinally equivalent if and only if for all cause–effect pairs (C_1, E_1) and (C_2, E_2) in a causal Bayesian network,

$$\eta(C_1, E_1) > \eta(C_2, E_2) \quad \text{if and only if} \quad \eta'(C_1, E_1) > \eta'(C_2, E_2).$$

Ordinally equivalent measures can be represented as monotonically increasing functions of each other. Typical cases are addition of or multiplication with a constant, or rescalings of the type $\eta' = \log \eta$. In other words, ordinally equivalent measures may use different scales, but they agree in all comparative judgments and share most philosophically interesting properties.

The following subsections provide representation theorems for measures of causal strength and use these theorems to buttress normative arguments for a particular measure (up to ordinal equivalence). The representation theorems have independent value, too: they allow translating any normative evaluation of the adequacy conditions—including evaluations that differ from ours—into a corresponding ranking of causal strength measures.

6.2.1 Argument 1: Separability and Difference Measures

Suppose we investigate how an intervention on a class of students, such as increasing the assignment load (C), affects their exam results. The result variable R can take three values: pass with honors (R_1), regular pass (R_2) and fail (R_3). Suppose we know the causal strength of increasing the assignment load for passing with honors (i.e., $\eta(C, R_1)$) and also its causal strength for regular passes (i.e., $\eta(C, R_2)$). Since passing is just the disjunction of regular pass and pass with honors, the causal strength of C for $R_1 \vee R_2$ should exceed the causal strength for both R_1 and R_2 only if both are indeed caused, rather than prevented, by C. In other cases, causal strength for the aggregate effect should be down. This is equivalent to the following: causal strength increases under adding a disjunct to the effect when the cause is positively relevant to the disjunct, and decreases when it is negatively relevant. We obtain the following adequacy condition:

Separability of Effects Let $C \in \mathcal{C}$ and $E, E' \in \mathcal{E}$ be instantiations of the variables C and E, respectively. Suppose C' is the relevant contrast class of C, and E and E' are mutually exclusive. Then,

$$\begin{aligned}
\eta(C, E \vee E') > \eta(C, E) &\quad \text{if and only if} \quad p^*(E'|C) > p^*(E'|C'),\\
\eta(C, E \vee E') = \eta(C, E) &\quad \text{if and only if} \quad p^*(E'|C) = p^*(E'|C'),\\
\eta(C, E \vee E') < \eta(C, E) &\quad \text{if and only if} \quad p^*(E'|C) < p^*(E'|C').
\end{aligned}$$

From Separability of Effects and Difference-Making, it is possible to prove the following representation theorem:

Theorem 6.1 (Representation Theorem for Difference Measures). *All measures of causal strength that satisfy Difference-Making (Causal Strength) and Separability of Effects are of the form*

$$\eta(C,E) = p^*(E|C) - p^*(E|C').$$

This theorem implies that $\eta(C,E)$ must be the difference of the rate of E under C and a relevant contrast class C′. All such measures satisfy the equality

$$\eta(C, E \vee E') = \eta(C,E) + \eta(C,E')$$

for mutually exclusive E and E′, allowing for an easy computation of aggregate causal strength from causal strength of the disjuncts.

As mentioned before, there are two particularly natural candidates for choosing C′. First, the choice $C' = \Omega_C$ leads to a measure that quantifies how much C raises the "natural" occurrence rate of E (cf. Pearl 2011, 717):

$$\eta_{pr}(C,E) = p^*(E|C) - p^*(E).$$

By contrast, choosing $C' = \neg C$ measures the difference between the presence and absence of C (Eells 1991):

$$\eta_d(C,E) = p^*(E|C) - p^*(E|\neg C).$$

This measure captures the degree to which E depends on C. For instance, in a randomized controlled trial where we compare two levels of a drug, η_d quantifies the difference in incident rates between the treatment and the control group.

While both measures are natural and frequently cited candidates for measuring causal strength, there is a clear argument for preferring η_d. We apply causal strength in contexts where we intervene, or could hypothetically intervene, on the cause. Both measures depend, to some extent, on the post-intervention probability distribution of C, and in particular on the relative frequency of the alternative values to C—say, C_1, C_2 and C_3. This dependency is not problematic because it expresses the relevant contrast class (e.g., do we compare the efficacy of a new drug to a placebo, to the previous standard treatment or to a mixture of both?).

However, in addition to this, η_{pr} introduces a strong dependence on the base rate of C, because $p^*(E) = p^*(C)\,p^*(E|C) + p^*(\neg C)\,p^*(E|\neg C)$ and thus

$$\eta_{pr}(C,E) = p^*(E|C) - p^*(E) = (1 - p^*(C)) \cdot (p^*(E|C) - p^*(E|\neg C)).$$

This consequence is hardly acceptable. Causal strength should not depend on pragmatically motivated decisions expressed by the post-intervention frequency of C and ¬C, such as the number of patients that we allocate to the treatment and the control group. In particular, causal strength in a treatment–control experiment should not be highly sensitive to whether the treatment group consists of one hundred participants and the (possibly heterogeneous) control group of fifty participants, or the other way round. Both experiments should allow for the same kind of causal strength inferences, but for $\eta_{\text{pr}}(\text{C}, \text{E})$, the range of possible causal strength values is $[-1/3; 1/3]$ in the first case and $[-2/3; 2/3]$ in the second case. This is clearly an undesirable consequence. Since these arguments pertain to the choice of C′ and can be generalized beyond the particular function that combines $p^*(\text{E}|\text{C})$ and $p^*(\text{E}|\text{C}')$, the equality $\text{C}' = \neg\text{C}$ will be a default assumption in the remainder.

Note that both measures satisfy two important causation–prevention symmetries that will be important later on. We follow Fitelson and Hitchcock (2011) in explicating the degree to which C prevents E as the degree to which C causes ¬E, that is, the absence of E. To be able to measure causation and prevention on the same scale, we demand that the (preventive) causal strength of C for ¬E is the negative of the causal strength of C for E:

Causation–Prevention Symmetry (CPS) $\quad -\eta(\text{C}, \text{E}) = \eta(\text{C}, \neg\text{E}).$

Evidently, only measures of causal strength that take both positive and negative values can satisfy CPS. Positive causal strength indicates positive causation; negative causal strength indicates prevention; $\eta(\text{C}, \text{E}) = 0$ denotes neutral causal strength. Note that neutral strength should not be conflated with causal irrelevance. A cause can be relevant for an effect, and yet, the overall effect can be zero, for instance, when contributions via different paths cancel out. This is different from a case where there is no causal connection between C and E.

Some readers may share the basic intuition behind Causation–Prevention Symmetry, but may not want to subscribe to particular scaling properties. In that case, they could go for the following, purely ordinal version of CPS: if C is a stronger cause of E_1 than of E_2, then C prevents $\neg\text{E}_1$ more than $\neg\text{E}_2$:

Weak Causation–Prevention Symmetry (WCPS)

$$\eta(\text{C}, \text{E}_1) > \eta(\text{C}, \text{E}_2) \quad \text{if and only if} \quad \eta(\text{C}, \neg\text{E}_1) < \eta(\text{C}, \neg\text{E}_2).$$

6.2.2 Argument 2: The Multiplicativity Principle

How should causal strength combine on the single path of Figure 6.3? If causal strength is the ability of the cause to make a difference to the effect, then overall causal strength should be a function of the causal strength between the individual links. But what function $g\colon \mathbb{R}^2 \to \mathbb{R}$ should be chosen such that for a binary mediator $X \in \{\mathrm{X}, \neg\mathrm{X}\}$, both $\eta(\mathrm{C},\mathrm{E}) = g(\eta(\mathrm{C},\mathrm{X}), \eta(\mathrm{X},\mathrm{E}))$ and $\eta(\mathrm{C},\mathrm{E}) = g(\eta(\mathrm{C},\neg\mathrm{X}), \eta(\neg\mathrm{X},\mathrm{E}))$ hold?

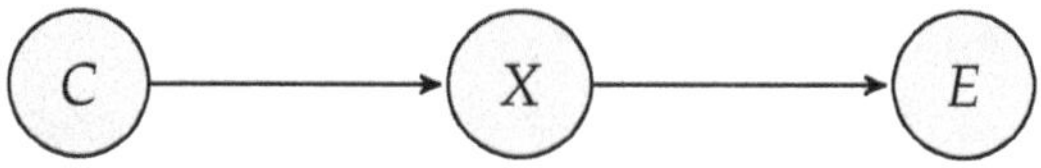

Figure 6.3: A DAG representing causation along a single path. Reproduced from "Foundations of a Probabilistic Theory of Causal Strength" by Jan Sprenger, *Philosophical Review*, Volume 127, No. 3, pp. 371–398, with permission of Duke University Press. © 2018, doi: 10.1215/00318108-6718797.

A couple of requirements suggest themselves. First, g should be symmetric: the order of mediators in a chain does not matter. Whether a weak link precedes a strong link or vice versa should not matter for overall causal strength. Second, it seems that overall causal strength is bounded by the weakest link in the chain: If C and X are almost independent, it does not matter how strongly X and E are correlated—causal strength will still be low. Similarly, if both links are weak, overall linkage will be even weaker. On the other hand, if one link is maximally strong (e.g., $\eta(\mathrm{C},\mathrm{X}) = 1$), then the strength of the entire chain will just be the strength of the rest of the chain. Perfect connections between two nodes neither raise nor attenuate overall causal strength (see also Good 1961a, 311–312).

A very simple operator that satisfies all these requirements is multiplication. This suggests the following principle:

Multiplicativity If the variables C and E are connected via a single path with a binary intermediate node X, then for $\mathrm{C} \in \mathcal{C}$, $\mathrm{E} \in \mathcal{E}$ and $\mathrm{X} \in \mathcal{X}$:

$$\eta(\mathrm{C},\mathrm{E}) = \eta(\mathrm{C},\mathrm{X}) \cdot \eta(\mathrm{X},\mathrm{E}).$$

As a corollary, we obtain that for a causal chain with multiple mediators of the type $C \to X_1 \to \cdots \to X_n \to E$,

$$\eta(\mathrm{C},\mathrm{E}) \;=\; \eta(\mathrm{C},\mathrm{X}_1) \cdot \eta(\mathrm{X}_1,\mathrm{X}_2) \cdots \eta(\mathrm{X}_{n-1},\mathrm{X}_n) \cdot \eta(\mathrm{X}_n,\mathrm{E}).$$

Multiplication may not be the only operator that fits the bill. However, it is clearly the simplest one, and *ceteris paribus*, simplicity is an added benefit for an explicatum (Carnap 1950, 5). The simple mathematical form contributes to theoretical fruitfulness, as we see in the above equation for longer causal chains.

Second, multiplicativity agrees with a lot of scientific practice. Suppose there is a linear dependency between variables E and X, modeled by the equation $E = \alpha X + i$. In those cases, the regression coefficient α is commonly interpreted as indicating the size of the causal effect that X has on E. When X depends linearly on C, too (e.g., $X = \beta C + i'$), the relation between C and E reads $E = \alpha\beta C + \alpha i' + i$, and the regression coefficient between both variables is equal to $\alpha\beta$—in agreement with Multiplicativity.

Third, suppose that in the absence of C, it is very unlikely that E: $p^*(\mathrm{E}|\neg\mathrm{C}) \approx 0$. In such circumstances, causal strength is the higher the more likely C is to bring about E. Modeling causal strength as a linear function of $p^*(\mathrm{E}|\mathrm{C})$, up to an arbitrary degree of precision, is particularly intuitive. We call this the Proportionality Principle (see the Appendix for a formal definition). It is not difficult to prove that Multiplicativity holds up to an arbitrary degree of precision when Proportionality holds. Hence, Multiplicativity also flows from a natural way to think about the strength of necessary causes. Indeed, if C is necessary for X, and X is necessary for E, then C is also necessary for E and the equation $p^*(\mathrm{E}|\mathrm{C}) = p^*(\mathrm{E}|\mathrm{X}) \cdot p^*(\mathrm{X}|\mathrm{C})$ holds, vindicating Proportionality (and Multiplicativity). These three independent arguments validate Multiplicativity as a simple, attractive and conceptually sound principle.

We can now characterize all measures that satisfy Multiplicativity alongside Difference-Making:

Theorem 6.2 (Representation Theorem for η_{d}). *All measures of causal strength that satisfy Difference-Making (Causal Strength) with the contrast class* $\mathrm{C}' = \neg\mathrm{C}$ *and Multiplicativity are ordinally equivalent to*

$$\eta_{\mathrm{d}}(\mathrm{C}, \mathrm{E}) = p^*(\mathrm{E}|\mathrm{C}) - p^*(\mathrm{E}|\neg\mathrm{C}).$$

The probability difference is a simple and intuitive quantity that measures causal strength by comparing the probability that different interventions on C impose on E. Indeed, η_{d} is straightforwardly applicable in statistical inference. For example, in clinical trials and epidemiological studies, $\eta_{\mathrm{d}}(\mathrm{C}, \mathrm{E})$ reduces to Absolute Risk Reduction, or ARR (see Section 6.2.4). Holland (1986, 947) calls η_{d} the "average causal effect" of C on E—a label that is motivated by the fact that η_{d} aggregates the strength of different causal links. Pearl (2001) uses η_{d} as the basis for developing a path-sensitive theory of causal strength.

Finally, η_d can be written as $\eta_d(C, E) = p^*(E|C) + p^*(\neg E|\neg C) - 1$. In this representation, causal strength depends linearly on two salient quantities: $p^*(E|C)$ and $p^*(\neg E|\neg C)$. They express the probability that C is *sufficient* for E and the probability that C is *necessary* for E, respectively (see also Pearl 2000). The measure η_d shares this property with Icard et al.'s (2017) measure of actual causation.

6.2.3 Argument 3: Dilution and the Ratio Measures

How strongly does C cause the conjunction of two effects—$E_1 \wedge E_2$—when C affects only one of them positively, and the other effect (say, E_2) is independent of C and of E_1? In such circumstances, we may call E_2 an "irrelevant effect". This situation is represented visually in the DAG in Figure 6.4.

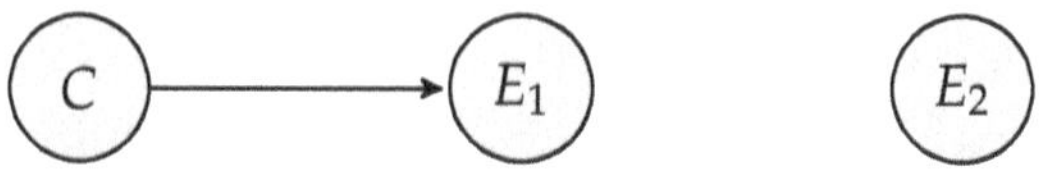

Figure 6.4: A DAG where effect E_2 is irrelevant with respect to the causal relation between cause C and primary effect E_1. Reproduced from "Foundations of a Probabilistic Theory of Causal Strength" by Jan Sprenger, *Philosophical Review*, Volume 127, No. 3, pp. 371–398, with permission of Duke University Press. © 2018, doi: 10.1215/00318108-6718797.

There are two basic intuitions about what such effects mean for overall causal strength: either causal strength is diluted when passing from E_1 to $E_1 \wedge E_2$, or it is not. **Dilution** means that adding E_2 to E_1 diminishes causal strength, that is, $\eta(C, E_1 \wedge E_2) < \eta(C, E_1)$. Contrarily, a measure is **non-diluting** if and only if in these circumstances, $\eta(C, E_1 \wedge E_2) = \eta(C, E_1)$. This amounts to the following principle:

No Dilution for Irrelevant Effects For $C \in \mathcal{C}$, $E_1 \in \mathcal{E}_1$, $E_2 \in \mathcal{E}_2$, let $E_2 \perp\!\!\!\perp C$ and $E_2 \perp\!\!\!\perp E_1$ conditional on C. Then $\eta(C, E_1 \wedge E_2) = \eta(C, E_1)$.

Incidentally, the premises of the No Dilution condition are compatible with a prima facie correlation between E_1 and E_2. However, this correlation vanishes as soon as we control for different levels of C.

Non-diluting measures of causal strength that satisfy Difference-Making can be neatly characterized. In fact, they are all ordinally equivalent to the probability ratio measure (Lewis 1986), as the following theorem demonstrates:

Theorem 6.3 (Representation Theorem for η_r and $\eta_{r'}$). *All measures of causal strength that satisfy Difference-Making (Causal Strength) with the contrast class* $C' = \neg C$ *and No Dilution for Irrelevant Effects are ordinally equivalent to*

$$\eta_r(C,E) = \frac{p^*(E|C)}{p^*(E|\neg C)}$$

and its rescaling to the $[-1;1]$ *range:*

$$\eta_{r'}(C,E) = \frac{p^*(E|C) - p^*(E|\neg C)}{p^*(E|C) + p^*(E|\neg C)}.$$

This result can be interpreted as a *reductio ad absurdum* of probability ratio measures. After all, given the lack of a causal connection between C and E_2, it is plausible that C causes $E_1 \wedge E_2$ to a smaller degree than E_1. Rain in New York on November 26, 2017 (C), affects umbrella sales in that city (E_1), but it does not affect whether FC Barcelona will win their next Champions League match (E_2). Therefore, the causal effect of rain on umbrella sales should be stronger than its causal effect on umbrella sales in conjunction with Barcelona winning their next match. This is bad news for η_r and $\eta_{r'}$.

The problems extend beyond the class of probability ratio measures. Consider the following restriction of No Dilution to the class of causal prevention:

No Dilution for Irrelevant Effects (Prevention) For $C \in \mathcal{C}$, $E_1 \in \mathcal{E}_1$, $E_2 \in \mathcal{E}_2$, let $E_2 \perp\!\!\!\perp C$ and $E_2 \perp\!\!\!\perp E_1$ conditional on C, and let C be a preventive cause of E_1. Then $\eta(C, E_1 \wedge E_2) = \eta(C, E_1)$.

Together with Weak Causation–Prevention Symmetry, this adequacy condition is sufficient to single out a particular class of measures:

Theorem 6.4 (Representation Theorem for η_{cg}). *All measures of causal strength that satisfy Difference-Making (Causal Strength) with the contrast class* $C' = \neg C$, *No Dilution for Irrelevant Effects (Prevention) and Weak Causation–Prevention Symmetry are ordinally equivalent to*

$$\eta_{cg}(C,E) = \begin{cases} \dfrac{p^*(E|C) - p^*(E|\neg C)}{1 - p^*(E|\neg C)}, & \text{if C is a positive cause of E;} \\[2ex] \dfrac{p^*(E|C) - p^*(E|\neg C)}{p^*(E|\neg C)}, & \text{if C is a preventive cause of E.} \end{cases}$$

For the case of positive causation, this measure agrees with two prominent proposals from the literature. The psychologist Patricia Cheng (1997) derived

$$\eta_c(C,E) := \frac{p^*(E|C) - p^*(E|\neg C)}{1 - p^*(E|\neg C)}$$

(that is, η_{cg} without the above case distinction) from theoretical considerations about how agents perform causal induction and called it the "causal power" of C on E. Cheng's measure is in turn ordinally equivalent to the measure

$$\eta_g(C,E) := \log \frac{p^*(\neg E|\neg C)}{p^*(\neg E|C)} = \log \frac{1-p^*(E|\neg C)}{1-p^*(E|C)}$$

that the statistician and philosopher of science I. J. Good (1961a,b) derived from a complex set of adequacy conditions. This ordinal equivalence, noted first by Fitelson and Hitchcock (2011), is evident from the equation below:

$$\eta_c(C,E) = \frac{-p^*(\neg E|C) + p^*(\neg E|\neg C)}{p^*(\neg E|\neg C)} = -e^{-\eta_g(C,E)} + 1.$$

The previous two theorems show that η_r and η_{cg} are based on the same principle: No Dilution for Irrelevant Effects. Since this property is highly suspicious, the representation results also provide evidence against η_{cg} and its cognates η_c and η_g, ruling out a prima facie attractive class of alternative measures.

6.2.4 Application: Quantifying Causal Effect in Medicine

A classical case of measuring causal strength concerns randomized controlled trials in medicine. The various outcome measures can be translated into our framework by writing observed relative frequencies of certain events as conditional probabilities under the different levels of the cause (i.e., the treatment level). For example:

$$\text{RR} = \frac{p^*(E|C)}{p^*(E|\neg C)}, \qquad \text{(Relative Risk)}$$

$$\text{ARR} = p^*(E|C) - p^*(E|\neg C), \qquad \text{(Absolute Risk Reduction)}$$

$$\text{RRR} = \frac{p^*(E|C) - p^*(E|\neg C)}{p^*(E|\neg C)}. \qquad \text{(Relative Risk Reduction)}$$

It is not difficult to relate these effect size measures to measures of causal strength. For example, RR is just the familiar probability ratio measure η_r, whereas ARR turns out to be the difference measure η_d. The measure RRR = RR − 1 is ordinally equivalent to η_r.

Normative arguments in favor of or against causal strength measures carry over to these effect size measures. Since the probability ratio measure η_r satisfies No Dilution for Irrelevant Effects, so do RR and RRR. The values of those measures do not change when irrelevant propositions are added to the effect. This can have extremely undesirable consequences. The causal effect of a painkiller on relieving headache is, according to η_r, as big as the causal effect of that drug on relieving headache *and* a completely unrelated symptom, such as lowering cholesterol levels. The measure η_r grossly misrepresents causal relevance: it conceals the fact that the high causal strength of the drug for both symptoms taken together is exclusively due to its effect on headache relief. Doctors may be misled into prescribing the drug for lowering cholesterol levels, even if it is ineffective for that purpose.

However, the defining properties of η_d, such as combining causal strength along a single path with the formula $\eta_d(C, E) = \eta_d(C, X)\, \eta_d(X, E)$, suit clinical practice very well. For example, doctors can see that overall causal strength must be weak if one of the links is tenuous. These theoretical features square nicely with decision-theoretic and epistemic arguments for preferring absolute over relative risk measures in medicine, such as the neglect of base rates in relative risk measures, and the sufficiency of η_d for identifying the most promising treatment (Stegenga 2015; Sprenger and Stegenga 2017). Briefly, the scientific application confirms our theoretical diagnosis: η_d is superior to η_r and other probability ratio measures.

6.3 Causal Contribution and Actual Causal Strength

Let us take stock. The previous section has provided three independent arguments for regarding η_d as a default measure of causal strength. The first argument was based on the Separability of Effects property, the second on the Multiplicativity Principle, and the third on the No Dilution Principle. The first two arguments showed that η_d is the only measure that satisfies those desirable properties. The third argument, by contrast, points out problems with the probability ratio family and related measures (η_r, $\eta_{r'}$, η_{cg}, η_c, η_g) based on the No Dilution property.

Each individual argument makes a good case. Cumulatively, things look even better since the three arguments operate independently from each other. Still, one may have principled doubts about uniqueness claims for causal strength measures. We will now play *advocatus diaboli* and introduce two measures that have neither the attractive properties of η_d nor the problematic properties of the No Dilution measures (e.g., η_r, η_{cg}).

Imagine, for example, that a medical drug has two side effects—diarrhea and sore throat—which are independent of each other. Both side effects are caused with the same strength t. One may want to say that the overall side effect of the medical drug is also equal to t since there is no interaction between both effects.

Conjunctive Closure For $\mathrm{C} \in \mathcal{C}$, $\mathrm{E}_1 \in \mathcal{E}_1$ and $\mathrm{E}_2 \in \mathcal{E}_2$, with $E_1 \perp\!\!\!\perp E_2$ conditional on C, the following implication holds:

$$\eta(\mathrm{C}, \mathrm{E}_1) = \eta(\mathrm{C}, \mathrm{E}_2) = t \qquad \Rightarrow \qquad \eta(\mathrm{C}, \mathrm{E}_1 \wedge \mathrm{E}_2) = t.$$

This principle facilitates calculations because we can now infer the strength of a cause C for an aggregate effect from the strength of C for the individual effects. Measures that satisfy Conjunctive Closure can be characterized neatly (compare Atkinson 2012):

Theorem 6.5 (Representation Theorem for η_{cc}). *All measures of causal strength that satisfy Difference-Making (Causal Strength) with the contrast class* $\mathrm{C}' = \neg\mathrm{C}$ *and Conjunctive Closure are ordinally equivalent to*

$$\eta_{cc}(\mathrm{C}, \mathrm{E}) = \frac{\log p^*(\mathrm{E} \,|\, \mathrm{C})}{\log p^*(\mathrm{E} \,|\, \neg\mathrm{C})}.$$

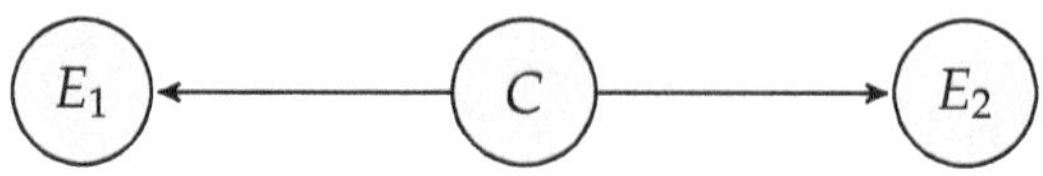

Figure 6.5: A DAG representing a simple common cause structure. Reproduced from "Foundations of a Probabilistic Theory of Causal Strength" by Jan Sprenger, *Philosophical Review*, Volume 127, No. 3, pp. 371–398, with permission of Duke University Press. © 2018, doi: 10.1215/00318108-6718797.

Although this measure fails to satisfy Separability of Effects and Multiplicativity, it may be a reasonable measure in contexts where we would like to quantify the **average strength** of a cause for a variety of independent effects. See Figure 6.5 for an illustration. For that causal structure, it is always the case that

$$\min\big(\eta_{cc}(\mathrm{C}, \mathrm{E}_1), \eta_{cc}(\mathrm{C}, \mathrm{E}_2)\big) \leq \eta_{cc}(\mathrm{C}, \mathrm{E}_1 \wedge \mathrm{E}_2) \leq \max\big(\eta_{cc}(\mathrm{C}, \mathrm{E}_1), \eta_{cc}(\mathrm{C}, \mathrm{E}_2)\big).$$

This property does not square well with the view of causal strength as difference-making, but it captures a plausible principle for averaging causal strength judgments.

Finally, one can investigate measures of **causal contribution** (e.g., Hall 2004; Kaiserman 2016; Beckers and Vennekens 2018). Suppose we ask what is the stronger cause of a car accident (E): drunk driving (C_1) or bad weather conditions (C_2)? One may answer that C_1 is a stronger cause of E than C_2 if and only if C_1 makes E more expected than C_2. In other words, one cause of an effect is stronger than another cause if it has a higher likelihood of producing the effect. This property, called Effect Production, is appealing in contexts where we want to attribute the occurrence of an event to one of its causes.

Effect Production For $C_1, C_2 \in \mathcal{C}$ and $E \in \mathcal{E}$,

$$\eta(C_1, E) > \eta(C_2, E) \quad \text{if and only if} \quad p^*(E \mid C_1) > p^*(E \mid C_2).$$

Cases where C is known with certainty suggest a further adequacy constraint. If two propositions are logically equivalent given C, it makes sense to treat them the same with respect to the causal strength that C has for them. After all, knowing that C has occurred, we cannot distinguish between them any more. Formally:

Conditional Equivalence Assume that E_1 and E_2 are logically equivalent given C. Then $\eta(C, E_1) = \eta(C, E_2)$.

It is easy to show that the Conditional Equivalence property characterizes the Pearl–Halpern measure η_{ph}:

Theorem 6.6 (Representation Theorem for η_{ph}). *All measures of causal strength that satisfy Difference-Making (Causal Strength) with contrast class $C' = \Omega_C$ and Conditional Equivalence are ordinally equivalent to*

$$\eta_{ph}(C, E) = p^*(E \mid C).$$

The Pearl–Halpern measure η_{ph} has been defended for measuring actual causal strength (Halpern and Pearl 2005a,b). It is also used in proposals for determining causal contributions among several causes of an event (Kaiserman 2016, 2017). To underscore the different angle of the discussed measures, consider a case of causal overdetermination:

> An assassin puts poison into the king's wine glass (C). If the king does not drink the wine, a (reliable) backup assassin will shoot him. The king drinks the wine and dies (E).

The Pearl–Halpern measure $\eta_{ph}(C, E) = p^*(E \mid C) \approx 1$ judges the assassin's action as a strong cause of the king's death, even if the king's fate was sealed anyway. The measure $\eta_d(C, E)$, however, disagrees (and so do other contrastive measures that compare C and $\neg$C): due to the presence of the backup assassin, poisoning the wine barely made a difference to the king's death, and $\eta_d(C, E) \approx 0$. The two groups of measures also diverge in cases where an action produces an effect, but by doing so preempts an even stronger cause.

Here is a line of argument for preferring the Pearl–Halpern explication η_{ph}: contrastive causal strength measures such as η_d judge poison in the wine as a weak cause when a backup cause is present (the second assassin—see also the discussion in Menzies 2014). But there is a sense in which poisoning is *always* a strong cause of death. Routine vaccinations are similar examples. We would not say that a vaccine is causally ineffective just because the overall risk of contracting the disease is low. By relativizing our explications to particular contexts, we seem to have lost an important aspect of strong causes: the capacity to secure an effect in a large variety of circumstances.

However, we can reconstruct this intuition in the proposed account, too. First, not each cause that secures the effect is universally strong. An umbrella is generally sufficient for protecting the person carrying it from getting wet. In the context of a desert climate, however, we would hesitate to identify umbrellas as strong causes of staying dry. Similarly, vaccinations can be more effective in some contexts and less effective in others. Think of a yellow fever vaccination, for example. If somebody travels to a region where yellow fever is endemic and gets vaccinated beforehand, contrastive causal strength measures deliver the right result: vaccination is highly efficacious ($p^*(\text{Disease Contraction} \mid \text{Vaccination, Exposure}) \gg p^*(\text{Disease Contraction} \mid \text{No Vaccination, Exposure})$). In No Exposure contexts, however, vaccination makes almost no difference to the risk of contracting yellow fever. The reason that most people don't seek yellow fever vaccination is that for them, the relevant context is No Exposure. Causal sufficiency, which is the concept explicated by the Pearl–Halpern measure η_{ph}, is different from causal strength. Crucially, it does not take into account that we only intervene on a cause if we believe the benefits to be substantial.

In general, such examples show that causal strength has a plurality of senses supported by our intuitions, not all of which are explicated by η_d. But exhausting these senses was never the goal of this chapter. Rather, we explicated causal strength as difference-making ("how would *E* change if we intervened on C?"). Within that perspective, the arguments for η_d remain compelling—at least to the degree that η_d is an excellent default measure and that the choice of other measures requires special justification. Tables

6.2 and 6.3 give an overview of which measure satisfies which adequacy condition, and how the representation theorems relate to each other.

Measure	Property										
	DM (Ω_C)	DM ($\neg C$)	CPS	WCPS	CE	EP	SE	MUL	NDIE	NDIEP	CC
Pearl–Halpern (η_{ph})	yes	(yes)	no	yes	yes	yes	no	no	no	no	no
Probability Raise (η_{pr})	yes	no	yes	yes	no	yes	yes	no	no	no	no
Difference (η_d)	no	yes	yes	yes	no	no	yes	yes	no	no	no
Probability Ratio (η_r, $\eta_{r'}$)	no	yes	no	no	no	no	no	no	yes	yes	no
Cheng/Good I (η_{cg})	no	yes	yes	yes	no	no	no	no	no	yes	no
Cheng/Good II (η_c, η_g)	no	yes	no	no	no	no	no	no	no	no	no
Conjunctive Closure (η_{cc})	no	yes	no	no	no	no	no	no	no	no	yes

Table 6.2: A classification of different measures of causal strength according to the adequacy conditions they satisfy. DM = Difference-Making (Causal Strength), CPS = Causation–Prevention Symmetry, WCPS = Weak Causation–Prevention Symmetry, CE = Conditional Equivalence, EP = Effect Production, SE = Separability of Effects, MUL = Multiplicativity, NDIE = No Dilution for Irrelevant Effects, NDIEP = No Dilution for Irrelevant Effects (Prevention), CC = Conjunctive Closure. Adapted from "Foundations of a Probabilistic Theory of Causal Strength" by Jan Sprenger, *Philosophical Review*, Volume 127, No. 3, pp. 371–398, with permission of Duke University Press. © 2018, doi: 10.1215/00318108-6718797.

Contrast Class $C' = \Omega_C$		
Separability of Effects	η_{pr}	Theorem 6.1
Conditional Equivalence	η_{ph}	Theorem 6.6
Contrast Class $C' = \neg C$		
Separability of Effects	η_d	Theorem 6.1
Multiplicativity	η_d	Theorem 6.2
No Dilution	η_r, $\eta_{r'}$	Theorem 6.3
No Dilution (Prevention) + WCPS	η_{cg}	Theorem 6.4
Conjunctive Closure	η_{cc}	Theorem 6.5

Table 6.3: An overview of all discussed causal strength measures together with the adequacy conditions that characterize the corresponding representation theorems. Adapted from "Foundations of a Probabilistic Theory of Causal Strength" by Jan Sprenger, *Philosophical Review*, Volume 127, No. 3, pp. 371–398, with permission of Duke University Press. © 2018, doi: 10.1215/00318108-6718797.

6.4 Conclusion

The goal of this Variation was to provide axiomatic foundations for a probabilistic theory of causal strength using causal Bayesian networks. Synthesizing ideas from the probability-raising and the interventionist view of causation, we have formalized causal strength as a function of the difference that interventions on a cause C make to the probability of the intended effect E.

We have characterized various measures of causal strength in terms of representation theorems, derived from a set of adequacy conditions. Such a characterization makes it possible to assess the merits of the different measures in the literature by means of assessing the plausibility of the adequacy conditions. By doing so, this chapter creates a methodological bridge to other projects in formal epistemology, such as probabilistic explications of degree of confirmation, coherence, justification and explanatory power (Fitelson 2003; Schupbach and Sprenger 2011; Shogenji 2012; Crupi 2015).

On the basis of these representation results, we have put forward arguments for using $\eta_d(C, E) = p^*(E\,|\,C) - p^*(E\,|\,\neg C)$ as a default measure of causal strength. Indeed, Holland (1986) and Pearl (2001) build their discussion of (path-specific) causal effects on the baseline measure η_d. However, they do not provide a philosophical defense of their choice—a gap that we have now closed. The theoretical analysis also agrees with practice- and decision-oriented arguments for η_d, as pointed out in the previous sections.

What remains to do? First, we aim at linking this framework to questions about the magnitude of a causal effect, such as the difference of group means (e.g., Cohen's *d* or Glass's Δ). The measure η_d might naturally be extended into this direction.

Second, this work can be connected to information-theoretic approaches to causal specificity (Weber 2006; Waters 2007; Griffiths et al. 2015). The more narrow the range of effects that an intervention is likely to produce, the more specific the cause is to the effect. How does this concept relate to causal strength, and to what extent can both research programs learn from each other?

Third, the properties of the above measures in complicated networks (e.g., more than one path linking *C* and *E*) have not been investigated. Is it possible to show, for example, how degrees of causation along different paths can be combined in an overall assessment of causal strength (compare Theorem 3 in Pearl 2001)?

Fourth, we would like to spell out how this model connects to research on actual causation and the significance of (statistical) normality and (prescriptive) norms in causal reasoning (Knobe and Fraser 2008; Hitchcock and Knobe 2009; Halpern and Hitchcock 2015; Kominsky et al. 2015; Icard, Kominsky and Knobe 2017).

Fifth, this research has implications for probabilistic explications of explanatory power (e.g., McGrew 2003; Schupbach and Sprenger 2011; Crupi and Tentori 2012). There is a tight conceptual connection between explanatory power and causal strength (e.g., Eva and Stern forthcoming). Does this mean that adequacy criteria for causal strength are also good requirements for Bayesian measures of explanatory power? An answer to this question obviously depends on one's preferred explication strategy for explanatory power. Anticipating part of the discussion in the next Variation, a prima facie difference consists in the fact that Bayesian explanatory power focuses on how C reduces (subjectively perceived) surprise in E. That is, it compares $p^*(E|C)$ and $p^*(E)$, unlike most investigated measures of causal strength. Interested readers find more detailed coverage of this question in Variation 7; and in general, exploring this relationship is a highly interesting challenge for further research. All in all, we hope that our results are promising enough to motivate further pursuit of an axiomatic theory of causal strength.

Appendix: Proofs of the Theorems

Proof of Theorem 6.1: Suppose that $p^*(E'|C) = p^*(E'|C')$ for mutually exclusive E, $E' \in \mathcal{E}$. Then we infer with the help of Generalized Difference-Making:

$$\begin{aligned} \eta(C,E) &= f(p^*(E|C),\, p^*(E|C')), \\ \eta(C, E\vee E') &= f\Big(p^*(E\vee E' \,|\, C),\, p^*(E\vee E' \,|\, C')\Big) \\ &= f\Big(p^*(E|C)+p^*(E'|C),\, p^*(E|C')+p^*(E'|C')\Big) \\ &= f\Big(p^*(E|C)+p^*(E'|C),\, p^*(E|C')+p^*(E'|C)\Big). \end{aligned}$$

Applying Separability of Effects implies $\eta(C, E\vee E') = \eta(C,E)$ and leads to the equality

$$f(p^*(E|C),\, p^*(E|C')) \;=\; f\Big(p^*(E|C)+p^*(E'|C),\, p^*(E'|C')+p^*(E'|C)\Big).$$

Since we have made no assumptions about the values of these conditional probabilities, f satisfies the formula $f(x,x') = f(x+y,\, x'+y)$ in full generality. It is then easy to see (e.g., by looking at the indifference curves of f) that there must be a function g such that $f(x,x') = g(x-x')$. Hence,

$$\eta(C,E) = f(p^*(E|C),\, p^*(E|C')) = g(p^*(E|C) - p^*(E|C')),$$

showing the desired ordinal equivalence claim. □

We also provide a formal definition of Proportionality. Suppose $f\colon [0;1]^2 \to \mathbb{R}$ is a continuous function that represents η mathematically, in agreement with Generalized Difference Making. Then, Proportionality amounts to

$$\forall \varepsilon > 0 \;\exists \delta > 0 \;\forall \alpha > 0,\; 0 < x < 1,\; y < \delta: \qquad \big|f(\alpha x, y) - \alpha f(x,y)\big| < \varepsilon.$$

Proving Multiplicativity from Proportionality is then a matter of straightforward calculus.

Proof of Theorem 6.2: By Generalized Difference-Making with $C' = \neg C$ we can focus on the function $f\colon [0;1]^2 \to \mathbb{R}$ such that $\eta(C,E) = f(p^*(E|C),\, p^*(E|\neg C))$. We would like to derive the equality

$$f(\alpha,\bar{\alpha})\cdot f(\beta,\bar{\beta}) \;=\; f\Big(\alpha\beta + (1-\alpha)\bar{\beta},\; \bar{\alpha}\beta + (1-\bar{\alpha})\bar{\beta}\Big) \qquad (6.1)$$

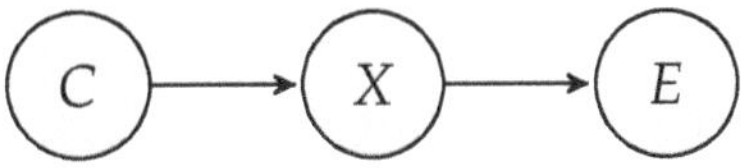

Figure 6.6: A DAG representing causation along a single path (see Figure 6.3). Reproduced from "Foundations of a Probabilistic Theory of Causal Strength" by Jan Sprenger, *Philosophical Review*, Volume 127, No. 3, pp. 371–398 (online appendix), with permission of Duke University Press. © 2018, doi: 10.1215/00318108-6718797.

for a causal-strength measure f that satisfies Multiplicativity. To this end, recall the single-path Bayesian network reproduced in Figure 6.6.

We know by Multiplicativity that for $\mathrm{C} \in \mathcal{C}$, $\mathrm{E} \in \mathcal{E}$ and $\mathrm{X} \in \mathcal{X}$,

$$\begin{aligned}
\eta(\mathrm{C},\mathrm{E}) &= \eta(\mathrm{C},\mathrm{X}) \cdot \eta(\mathrm{X},\mathrm{E}) \\
&= f(p^*(\mathrm{X}|\mathrm{C}), p^*(\mathrm{X}|\neg\mathrm{C})) \cdot f(p^*(\mathrm{E}|\mathrm{X}), p^*(\mathrm{E}|\neg\mathrm{X})) \\
&= f(p^*(\mathrm{X}|\mathrm{C}), p^*(\mathrm{X}|\neg\mathrm{C})) \cdot f(p^*(\mathrm{E}|\mathrm{X}), p^*(\mathrm{E}|\neg\mathrm{X}))
\end{aligned}$$

and at the same time,

$$\begin{aligned}
\eta(\mathrm{C},\mathrm{E}) &= f(p^*(\mathrm{E}|\mathrm{C}), p^*(\mathrm{E}|\neg\mathrm{C})) \\
&= f\Big(\sum_{\mathrm{X}=\pm\mathrm{X}} p^*(\mathrm{X}|\mathrm{C})\, p^*(\mathrm{E}|\mathrm{C},\mathrm{X}), \sum_{\mathrm{X}=\pm\mathrm{X}} p^*(\mathrm{X}|\neg\mathrm{C})\, p^*(\mathrm{E}|\neg\mathrm{C},\mathrm{X})\Big) \\
&= f\Big(\sum_{\mathrm{X}=\pm\mathrm{X}} p^*(\mathrm{X}|\mathrm{C})\, p^*(\mathrm{E}|\mathrm{X}), \sum_{\mathrm{X}=\pm\mathrm{X}} p^*(\mathrm{X}|\neg\mathrm{C})\, p^*(\mathrm{E}|\mathrm{X})\Big).
\end{aligned}$$

Combining both equations yields

$$\begin{aligned}
f(p^*(\mathrm{X}|\mathrm{C}), p^*(\mathrm{X}|\neg\mathrm{C})) \cdot f(p^*(\mathrm{E}|\mathrm{X}), p^*(\mathrm{E}|\neg\mathrm{X})) &= \\
f\Big(\sum_{\mathrm{X}=\pm\mathrm{X}} p^*(\mathrm{X}|\mathrm{C})\, p^*(\mathrm{E}|\mathrm{X}), &\sum_{\mathrm{X}=\pm\mathrm{X}} p^*(\mathrm{X}|\neg\mathrm{C})\, p^*(\mathrm{E}|\mathrm{X})\Big).
\end{aligned}$$

With the variable settings

$$\begin{aligned}
\alpha &= p^*(\mathrm{X}|\mathrm{C}), & \beta &= p^*(\mathrm{E}|\mathrm{X}), \\
\bar{\alpha} &= p^*(\mathrm{X}|\neg\mathrm{C}), & \bar{\beta} &= p^*(\mathrm{E}|\neg\mathrm{X}),
\end{aligned}$$

Equation (6.1) follows immediately.

Now we are going to show that for any extension of f to $\mathbb{R}^2$,

$$f(x,y) = -f(y-x, 0). \tag{6.2}$$

To this end, we note a couple of facts about f:

Fact 1 $f(\alpha,0)\cdot f(\beta,0) = f(\alpha\beta,0)$.

This follows immediately from Equation (6.1) with $\bar{\alpha}=\bar{\beta}=0$.

Fact 2 $f(0,1)\cdot f(\beta,\bar{\beta}) = f(\bar{\beta},\beta)$.

This follows immediately from Equation (6.1) with $\alpha=0$, $\bar{\alpha}=1$.

Fact 3 $f(1,0) = 1$.

With $\beta=1$, Fact 1 entails that $f(\alpha,0)\cdot f(1,0) = f(\alpha,0)$. Hence, either $f(1,0)=1$ or $f(\alpha,0)\equiv 0$ for all values of α. However, the latter would also imply $f\equiv 0$ and trivialize f.

Fact 4 $f(0,1) = -1$.

Equation (6.1) (with $\alpha = \beta = 0$, $\bar{\alpha} = \bar{\beta} = 1$) and Fact 3 entail that $f(0,1)\cdot f(0,1) = f(1,0) = 1$. Hence, either $f(0,1) = -1$ or $f(0,1) = 1$. If the latter were the case, then the monotonicity requirement in Generalized Difference-Making would be violated. Thus, $f(0,1) = -1$.

These facts will allow us to derive Equation (6.2). Note that (6.2) is trivial if $y=0$. So we can restrict ourselves to the case that $y>0$. We choose the variable settings

$$\alpha = (y-x)/y, \qquad \beta = 0,$$
$$\bar{\alpha} = 0, \qquad \bar{\beta} = y.$$

Then we obtain, by means of Equation (6.1) and the previously proven facts,

$$\begin{aligned} f(x,y) &= f((y-x)/y,\, 0)\cdot f(0,y) \\ &= f(y-x,\, 0)\cdot f(1/y,0)\cdot f(0,y) && \text{(Fact 1)} \\ &= f(y-x,\, 0)\cdot f(1/y,0)\cdot f(y,0)\cdot f(0,1) && \text{(Fact 2)} \\ &= f(y-x,\, 0)\cdot f(1,0)\cdot f(0,1) && \text{(Fact 1)} \\ &= -f(y-x,\, 0). && \text{(Facts 3+4)} \end{aligned}$$

This implies

$$\eta(\mathrm{C},\mathrm{E}) = f(p^*(\mathrm{E}|\mathrm{C}), p^*(\mathrm{E}|\neg\mathrm{C})) = -f\Big((-1)\cdot(p^*(\mathrm{E}|\mathrm{C}) - p^*(\mathrm{E}|\neg\mathrm{C})),\ 0\Big).$$

Hence, $\eta(\mathrm{C},\mathrm{E})$ can be represented as a function of $p^*(\mathrm{E}|\mathrm{C}) - p^*(\mathrm{E}|\neg\mathrm{C})$ only. From Generalized Difference-Making we infer that f must be non-decreasing in $p^*(\mathrm{E}|\mathrm{C}) - p^*(\mathrm{E}|\neg\mathrm{C})$. This concludes the proof of Theorem 6.2. □

Proof of Theorem 6.3: The proof relies on a move from the proof of Theorem 1 in Schupbach and Sprenger 2011. Consider three variables C, E_1 and E_2 with $E_2 \perp\!\!\!\perp C$ and $E_2 \perp\!\!\!\perp E_1 | C$. Let $\mathrm{C}\in\mathcal{C}$, $\mathrm{E}_1\in\mathcal{E}_1$ and $\mathrm{E}_2\in\mathcal{E}_2$ be propositions about the values of these variables. Then No Dilution for Irrelevant Effects implies that

$$\begin{aligned} p^*(E_1 \wedge E_2 \mid C) &= p^*(E_1|C)\, p^*(E_2|C), \\ p^*(E_1 \wedge E_2 \mid \neg C) &= p^*(E_1|\neg C)\, p^*(E_2|\neg C), \\ p^*(E_2) &= p^*(E_2|\neg C) = p^*(E_2|C). \end{aligned}$$

In particular, it follows that

$$\begin{aligned} p^*(E_1 \wedge E_2 \mid C) &= p^*(E_2)\, p^*(E_1|C), \\ p^*(E_1 \wedge E_2 \mid \neg C) &= p^*(E_2)\, p^*(E_1|\neg C). \end{aligned}$$

According to Generalized Difference-Making with $C' = \neg C$, the causal-strength measure η can be written as $\eta(C, E_1) = f(p^*(E_1|C), p^*(E_1|\neg C))$ for a continuous function f. From No Dilution and the above calculations we can infer that

$$\begin{aligned} f(p^*(E_1|C), p^*(E_1|\neg C)) &= \eta(C, E_1) \\ &= \eta(C, E_1 \wedge E_2) \\ &= f\Big(p^*(E_1 \wedge E_2 \mid C), p^*(E_1 \wedge E_2 \mid \neg C)\Big) \\ &= f\Big(p^*(E_2)\, p^*(E_1|C), p^*(E_2)\, p^*(E_1|\neg C)\Big). \end{aligned}$$

Since we have made no assumptions on the values of these probabilities, we can infer the general relationship

$$f(x, y) = f(cx, cy) \tag{6.3}$$

for all $0 < c \leq \min(1/x, 1/y)$. Without loss of generality, let $x > y$. Then, choose $c := 1/x$. In this case, Equation (6.3) becomes

$$f(x, y) = f(cx, cy) = f(1, y/x).$$

This implies that f must be a function of y/x only, that is, of the ratio $p^*(E|\neg C)/p^*(E|C)$. Generalized Difference-Making implies that all such functions must be non-increasing, concluding the proof of Theorem 6.3. □

Proof of Theorem 6.4: We write the causal-strength measure η_{cg} as

$$\eta_{cg}(C, E) = \begin{cases} \eta^+(C, E) & \text{for positive causation,} \\ \eta^-(C, E) & \text{for causal preemption.} \end{cases}$$

We know from the previous theorem that $\eta^-(\mathrm{C},\mathrm{E})$ must be ordinally equivalent to $\eta_r(\mathrm{C},\mathrm{E})$. Now we show that all $\eta^+(\mathrm{C},\mathrm{E})$-measures are ordinally equivalent to $\eta_g(\mathrm{C},\mathrm{E}) = p^*(\neg\mathrm{E}|\neg\mathrm{C})/p^*(\neg\mathrm{E}|\mathrm{C})$. Since we have already shown that η_g and η_c are ordinally equivalent, this is sufficient for proving the theorem.

Because of Generalized Difference-Making, we can represent η^+ by a function $f(x,y)$ with $x = p^*(\mathrm{E}|\mathrm{C})$ and $y = p^*(\mathrm{E}|\neg\mathrm{C})$. Suppose that there are $x > y$ and $x' > y' \in [0;1]$ such that $(1-y)/(1-x) = (1-y')/(1-x')$, but $f(x,y) \neq f(x',y')$. (Otherwise η^+ would just be a function of η_g, and we would be done.) In that case we can find a probability space such that $p^*(\mathrm{E}_1|\mathrm{C}) = x$, $p^*(\mathrm{E}_1|\neg\mathrm{C}) = y$, $p^*(\mathrm{E}_2|\mathrm{C}) = x'$, $p^*(\mathrm{E}_2|\neg\mathrm{C}) = y'$, and C screens off E_1 and E_2 (proof omitted, but straightforward). Hence $\eta^+(\mathrm{C},\mathrm{E}_1) \neq \eta^+(\mathrm{C},\mathrm{E}_2)$. By Weak Causation–Prevention Symmetry, we can then infer $\eta^-(\mathrm{C},\neg\mathrm{E}_1) \neq \eta^-(\mathrm{C},\neg\mathrm{E}_2)$.

However, since η^- is ordinally equivalent to η_r, there is a function f such that

$$\eta^-(\mathrm{C},\neg\mathrm{E}_1) = f\left(\frac{p^*(\neg\mathrm{E}_1|\mathrm{C})}{p^*(\neg\mathrm{E}_1|\neg\mathrm{C})}\right) = f\left(\frac{1-x}{1-y}\right),$$
$$\eta^-(\mathrm{C},\neg\mathrm{E}_2) = f\left(\frac{p^*(\neg\mathrm{E}_2|\mathrm{C})}{p^*(\neg\mathrm{E}_2|\neg\mathrm{C})}\right) = f\left(\frac{1-x'}{1-y'}\right).$$

By assumption,

$$\frac{1-x}{1-y} = \left(\frac{1-y}{1-x}\right)^{-1} = \left(\frac{1-y'}{1-x'}\right)^{-1} = \frac{1-x'}{1-y'},$$

and so we can infer $\eta^-(\mathrm{C},\neg\mathrm{E}_1) = \eta^-(\mathrm{C},\neg\mathrm{E}_2)$, leading to a contradiction. Hence $\eta^+(\mathrm{C},\mathrm{E})$ can be represented by a non-decreasing function of $p^*(\neg\mathrm{E}|\neg\mathrm{C})/p^*(\neg\mathrm{E}|\mathrm{C})$, completing the proof of Theorem 6.4. □

Proof of Theorem 6.5: By Generalized Difference-Making, we have that $\eta(\mathrm{C},\mathrm{E}) = f(p^*(\mathrm{E}|\mathrm{C}), p^*(\mathrm{E}|\neg\mathrm{C}))$ for some continuous function $f\colon [0;1]^2 \to \mathbb{R}$. Assume that $\eta(\mathrm{C},\mathrm{E}_1) = \eta(\mathrm{C},\mathrm{E}_2) = t$, that C screens off E_1 and E_2 and that $p^*(\mathrm{E}_1|\mathrm{C}) = p^*(\mathrm{E}_2|\mathrm{C}) = x$, $p^*(\mathrm{E}_1|\neg\mathrm{C}) = p^*(\mathrm{E}_2|\neg\mathrm{C}) = y$, for some $x, y \in \mathbb{R}$. By the Conjunctive Closure Principle, we can infer

$$\eta(\mathrm{C},\mathrm{E}_1 \wedge \mathrm{E}_2) = \eta(\mathrm{C},\mathrm{E}_1) = f(x,y).$$

Moreover, we can infer

$$\begin{aligned}\eta(\mathrm{C},\mathrm{E}_1 \wedge \mathrm{E}_2) &= f\big(p^*(\mathrm{E}_1 \wedge \mathrm{E}_2\,|\,\mathrm{C}),\ p^*(\mathrm{E}_1 \wedge \mathrm{E}_2\,|\,\neg\mathrm{C})\big)\\ &= f\big(p^*(\mathrm{E}_1|\mathrm{C})\,p^*(\mathrm{E}_2|\mathrm{C}),\ p^*(\mathrm{E}_1|\neg\mathrm{C})\,p^*(\mathrm{E}_2|\neg\mathrm{C})\big) = f(x^2,y^2).\end{aligned}$$

Taking both calculations together, we obtain

$$f(x^2, y^2) = f(x, y) \tag{6.4}$$

as a structural requirement on f, since we have not made any assumptions on x and y.

Following Atkinson (2012), we now define $u = \log x / \log y$ and define a function $g: \mathbb{R}^2 \to \mathbb{R}$ such that $g(x, u) = f(x, y)$. Equation (6.4) then implies the requirement

$$g(x^2, u) = f(x^2, y^2) = f(x, y) = g(x, u),$$

and by iterating that procedure, we obtain

$$g(x^{2n}, u) = g(x, u)$$

for some $n \in \mathbb{N}$. Due to the continuity of f and g, we can infer that g cannot depend on its first argument. Moreover, taking the limit $n \to \infty$ yields $g(x, u) = g(0, u)$. Hence, also

$$f(x, y) = g(0, u) = g\left(0, \frac{\log x}{\log y}\right),$$

and we see that

$$\eta(\mathrm{C}, \mathrm{E}) = h\left(\frac{\log p^*(\mathrm{E}|\mathrm{C})}{\log p^*(\mathrm{E}|\neg\mathrm{C})}\right)$$

for some continuous function $h: \mathbb{R} \to \mathbb{R}$. It remains to show that h is non-decreasing. Generalized Difference-Making implies that $\eta(\mathrm{C}, \mathrm{E})$ is a non-decreasing function of $p^*(\mathrm{E}|\mathrm{C})$ and a non-increasing function of $p^*(\mathrm{E}|\neg\mathrm{C})$. So it must be a non-decreasing function of $\log p^*(\mathrm{E}|\mathrm{C}) / \log p^*(\mathrm{E}|\neg\mathrm{C})$, too. This implies that h is a non-decreasing function. Hence, all measures of causal strength that satisfy Generalized Difference-Making and the Conjunctive Closure Principle are ordinally equivalent to

$$\eta_{cc}(\mathrm{C}, \mathrm{E}) = \frac{\log p^*(\mathrm{E}|\mathrm{C})}{\log p^*(\mathrm{E}|\neg\mathrm{C})}. \qquad \square$$

Proof of Theorem 6.6: We know by assumption that any measure satisfying Generalized Difference-Making with $\mathrm{C}' = \Omega_\mathrm{C}$ is of the form

$$\eta(\mathrm{C}, \mathrm{E}) = f(p^*(\mathrm{E}|\mathrm{C}), p^*(\mathrm{E})).$$

Suppose now that there are x, y, $y' \in [0; 1]$ such that $f(x, y) \neq f(x, y')$. In that case, we can choose propositions C, E_1 and E_2 and choose a probability distribution p^* such that $x = p^*(E_1|C)$, $y = p^*(E_1)$, $y' = p^*(E_2)$, $C \wedge E_1 \vDash E_2$ and $C \wedge E_2 \vDash E_1$. Then, $p^*(E_1|C) = p^*(E_1 \wedge E_2 \mid C) = p^*(E_2|C)$ and

$$\begin{aligned} \eta(C, E_1) &= f(p^*(E_1|C), p^*(E_1)) \\ &= f(p^*(E_1 \wedge E_2 \mid C), p^*(E_1)) = f(p^*(E_2|C), p^*(E_1)), \end{aligned}$$

and by Conditional Equivalence, also

$$\eta(C, E_1) = \eta(C, E_2) = f(p^*(E_2|C), p^*(E_2)).$$

Taking both equations together leads to a contradiction with our assumption $f(p^*(E_2|C), p^*(E_1)) \neq f(p^*(E_2|C), p^*(E_2))$. So f cannot depend on its second argument. Hence, all causal-strength measures that satisfy Generalized Difference-Making with $C' = \Omega_C$ and Conditional Equivalence must be ordinally equivalent to $\eta_{ph}(C, E) = p^*(E|C)$. □

Variation 7: Explanatory Power

Explanation is a central element of scientific reasoning. Physicists endorse the Big Bang theory on the basis of its ability to explain complex phenomena, such as the continuous expansion of the universe and the cosmic background radiation. Evolutionary psychologists explain gender differences in human behavior by environmental adaptations that happened in the Pleistocene (e.g., Buss and Schmitt 1993). Statisticians assess hypotheses on their basis to explain the observed data (e.g., Edwards 1972; Royall 1997). Unsurprisingly, explanatory power figures in various lists of values that describe scientists' shared criteria for assessing theories (e.g., Kuhn 1977a; McMullin 1982).

On top of this, many philosophers defend **Inference to the Best Explanation (IBE)**—that is, inferring a hypothesis on the basis of its explanatory virtues—as a rational mode of theory choice (Harman 1965; Lipton 2004). Scientists from cognitive science, artificial intelligence and computer science study IBE under the label of **abductive inference** (e.g., Hobbs et al. 1988; Bylander et al. 1991; Eiter and Gottlob 1995; Magnani 2001; Douven 2017). Finally, the concept of explanation plays an important role in psychological theories: explanation-based reasoning affects the way people learn categories, generalize properties and draw inferences (e.g., Rips 1989; Thagard 1989; Lombrozo 2006).

The large body of literature devoted to IBE and explanatory reasoning is very remarkable because there is no consensus on what constitutes a good explanation in the first place. Scientific explanations differ by discipline and context: phenomena can be subsumed under general scientific laws (Hempel and Oppenheim 1948; Friedman 1974; Kitcher 1981), predicted by statistical models (Salmon 1971), obtained as results of a causal mechanism (Machamer, Darden and Craver 2000; Bechtel and Abrahamsen 2005; Craver 2007) or explained in abstract mathematical ways, for example, by means of symmetry properties or renormalization groups (Colyvan 2001; Batterman 2002). It seems impossible to give a neat and simple answer to the question "What is a scientific explanation?" Prominent philosophers have defended the idea that all scientific explanations are ultimately of causal and/or counterfactual nature (e.g., Woodward 2003; Strevens 2009), but the recent literature leans toward pluralism about concepts of scientific explanation (e.g., Reutlinger 2016, 2017; Colombo 2017; Colombo and Wright 2017).

Bayesian models of explanatory power cannot adjudicate these debates. What they can do, however, is to measure the strength of inferential connections that an independent theory of scientific explanation has identified as explanatory (Hitchcock and Woodward 2003; Schupbach and Sprenger 2011). This is our project in this Variation. We engage in the debate about different accounts of scientific explanation only to the extent that it is necessary for motivating and defending our Bayesian explication of explanatory power. Then we argue that our Bayesian model of explanatory power is excellently suited for analyzing the conditions when IBE is a rational mode of inference.

The latter project is especially important since there is a prima facie tension between Bayesian inference and IBE. Inferring theories on the basis of their explanatory value alone may lead us to just-so stories or improbable conclusions. To return to the above example: we should not infer any implausible story about human life in the Pleistocene just on its capacity to explain features of current behavior (e.g., Gould and Lewontin 1979). Explanatory value and high (subjective) probability can diverge. Bas van Fraassen (1989) sharpens this criticism and contends that an explicit role of explanatory power in Bayesian inference leads straight into incoherence. After all, Bayesians assess the plausibility of a scientific hypothesis on the basis of its posterior probability. This probability is a function of its prior probability and the conditional probability of the data, given the competing hypotheses. Explanationists want to base their assessment of a scientific hypothesis not only on posterior probabilities, but also on its explanatory power. Perhaps the most natural way to accommodate these explanatory considerations in the Bayesian paradigm is to assign a probability bonus to explanatorily powerful hypotheses. Van Fraassen shows, however, that such a procedure is inconsistent with Conditionalization and allows for a diachronic Dutch book against the explanationist. Bayesian rationality and IBE turn out to be incompatible. To defend the thesis that "the Bayesian and the explanationist can be friends" (Lipton 2001), one needs to find a place for explanatory power *within* the Bayesian paradigm, which is a much more challenging task. It requires, in the first place, an account of what explanatory power could mean when expressed in terms of probabilities (e.g., Okasha 2000; Salmon 2001; Schupbach 2011b).

This Variation takes up the challenge. First, we engage with the view that explanations are essentially causal, and that successful explanations are based on a relationship of actual causation (Section 7.1). Second, we motivate why, and under which circumstances, explanatory power can be explicated in terms of statistical relevance: as reducing the degree to which an explanandum is surprising (Section 7.2). Third, we derive Bayesian explications of explanatory power by means of several representation theorems (Section 7.3). We compare the advantages and drawbacks of various explanatory power measures (Section 7.4) and conclude by sketching projects for future research on the integration of probabilistic and explanatory inference (Section 7.5).

7.1 Causal Theories of Explanatory Power

An explication of explanatory power in Bayesian terms requires some motivation. After all, most modern theories of scientific explanation focus on the explanatory role of causal processes and interactions leading to the explanandum (e.g., Salmon 1984; Cartwright 1989; Dowe 2000; Machamer, Darden and Craver 2000; Craver 2007; Strevens 2009). On those accounts, phenomena are explained by their causes, such as mechanisms that describe how and why the explanandum was produced: Failure of the brakes explains a car accident. Wine poisoning explains the death of the king. Exposure to violent video games explains aggressive behavior. In all of these examples, causal efficacy grounds explanatory power. So perhaps it is not surprising that until the early twentieth century, the concept of explanation was subordinate to the concept of causation, for instance in the writings of David Hume or Immanuel Kant. Even if we concede that the causal view of explanation may not be universally applicable (see Reutlinger 2017), it might, with an eye on applications, be a promising line of research to identify explanatory power with a suitable measure of causal strength, examined in the previous Variation.

Following this research program, Joseph Halpern and Judea Pearl, two renowned AI researchers specializing in causal inference, have proposed an explication of explanatory power with causal Bayesian networks (Halpern and Pearl 2005a,b). They postulate that "the role of explanation is to provide the information needed to establish causation" (Halpern and Pearl 2005b, 897). Halpern and Pearl use the interventionist account of causality, presented in the previous Variation, to redefine the notion of explanatory power: "we view an explanation as a fact that is not known for certain but, if found to be true, would constitute a genuine cause of the explanandum, regardless of the agent's initial uncertainty" (ibid.). Thereby Halpern and Pearl relativize an explanation to the epistemic state of an agent and introduce a pragmatic, subject-dependent component. They contend that the value of a certain variable $C = C_1$ counts as an explanation of some fact E roughly if and only if (i) the agent regards E as true; (ii) $C = C_1$ is a cause of E; (iii) there are possible contexts where C_1 is false. The last clause serves to rule out vacuous explanations. The goodness of an explanation is then quantified by the probability $p(E \mid do(C = C_1))$, that is, the conditional probability of E given that we force the value C_1 on the variable C.

Halpern and Pearl's line of reasoning also connects to experimental work on **probabilistic measures of actual causation:** We often assign higher causal strength to statistically abnormal causes. Suppose the occurrence of an event E depends on two distinct and independent causes A and C, such as in the following vignette:

> Alex is playing a board game. Every turn of the game involves simultaneously rolling two six-sided dice and flipping a fair coin. Alex will either win or lose the game on his next turn.
>
> Alex will only win the game if the total of his dice rolls is greater than 2 AND the coin comes up heads. It is very likely that he will roll higher than 2, and the coin has equal odds of coming up heads or tails.
>
> Alex flips the coin and rolls his dice at exactly the same time. The coin comes up heads, and he rolls a 12, so just as expected, he rolled greater than 2. Alex wins the game. (Kominsky et al. 2015, 205, Experiment 4)

This vignette is contrasted with a case where the second paragraph is modified as follows:

> Alex will only win the game if the total of his dice rolls is greater than 11 AND the coin comes up heads. It is very unlikely that he will roll higher than 11, but the coin has equal odds of coming up heads or tails. (ibid.)

Thus, the two cases differ only regarding the statistical normality of the dice roll. And this makes a difference for the participants' responses: when the outcome of the dice roll (A) is abnormal, the coin flip (C) is less often judged to be a cause of winning the game (E) than in cases where the dice roll has a normal outcome. Kominsky and his collaborators call this effect

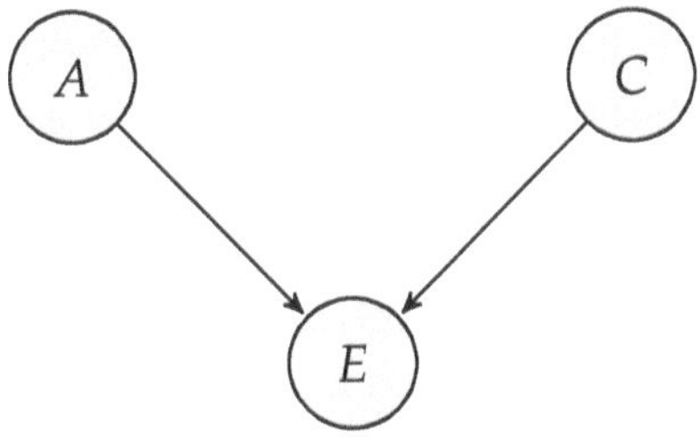

Figure 7.1: The causal structure of the vignette by Kominsky et al. (2015). A = total number of dice rolls, C = outcome of the toin coss, E = win or lose game. The structural equation is $E \equiv (C = \text{"heads"} \wedge A > 2)$. In the modified case the last condition is replaced with $A = 12$.

causal superseding. Inspired by this work, Icard, Kominsky and Knobe (2017) propose the following measure of actual causation that takes into account the normality of the putative cause:

$$\eta_{\mathrm{A}}(\mathrm{C}, \mathrm{E}) \;=\; p(\neg\mathrm{C})\, p(\neg\mathrm{E} \,|\, do(\neg\mathrm{C}, \mathrm{A})) \;+\; p(\mathrm{C})\, p(\mathrm{E} | \mathrm{C}).$$

In the coin-tossing example, we obtain, by filling in the relevant information from the vignettes,

$$\eta_{\mathrm{A}}(\mathrm{C}, \mathrm{E}) \;=\; 1 - p(\mathrm{C}) + p(\mathrm{A})\, p(\mathrm{C}).$$

This measure of actual causation quantifies the causal strength of C in the presence of a second, competing cause (A). In other words, it measures the extent to which we can attribute the occurrence of E to one of its causes, namely C. It is monotonically increasing in $p(\mathrm{A})$ and thus in line with the outcome of the experiments: the more expected the occurrence of A, the higher the (actual) causal strength that C has on E.

One could now argue that this is exactly what explanatory power consists in. Indeed, judgments of explanatory power are often highly correlated with those of (actual) causation, as witnessed by the experiments of Colombo, Postma and Sprenger (2016) and Colombo, Bucher and Sprenger (2017a,b). In the same sense that generic causal strength answers prediction-seeking questions ("what would happen if we intervened on C?"), actual causation answers explanation-seeking questions. Thus we could identify explanatory power with degree of actual causation, thereby connecting two central concepts in philosophy of science.

This idea runs into several problems, however. For starters, we can encounter probability-lowering explanations. For actual causation, probability-lowering is not necessarily an issue: if a football player shoots on goal from a difficult angle and the ball ends up in the goal nonetheless (e. g., because the goalie blunders), his shot has definitely caused the goal—even if it lowered the probability of a goal, compared to passing the ball to a fellow player. But we would hesitate to say that the player's shot *explains* the goal.

Furthermore, the actual causation account of explanatory power does not do justice to the variety of explanatory reasoning in science, for example, functional explanations. Bas van Fraassen (1980) illustrates this point with the famous flagpole story from Bromberger 1965: We can explain the length of the shadow of a flagpole by the height of the flagpole, conditional on the angle of the sun. Prima facie, the reverse explanation does not work: the length of the shadow does not explain the height of the flagpole. Effects do not explain their causes. Van Fraassen points out, however, that such judgments depend on pragmatic factors: in a specific context, the height of the flagpole could be explained by the fact that it was manufactured to cast a shadow of a certain length at a certain time of the day. Sundials work in this way, for example. When we aim to explain phenomena, causal reasoning is not automatically privileged over functional reasoning.

For this reason, we do not adopt a strictly causal theory of explanatory power. Rather, we will conceptualize explanatory power as the strength of a probabilistic argument, which may (but need not) have causal support. On this view, to be expounded in the next section, the power of an explanation expresses the degree to which the explanans rationalizes the explanandum. To which extent such a proposal covers the wide variety of explanatory reasoning in the sciences is of course a topic for future debate.

7.2 Statistical Relevance and Explanatory Power

The view of explanations as arguments connects explanatory reasoning to logical inference, and in particular, abduction: inferring a hypothesis H on the basis of its explanatory qualities for a given explanandum E. The first characterization of abductive inference dates back to the American pragmatist C. S. Peirce:

> Long before I first classed abduction as an inference it was recognized by logicians that the operation of adopting an explanatory hypothesis—which is just what abduction is—was subject to certain conditions. Namely, the hypothesis cannot be admitted, even as a hypothesis, unless it be supposed that it would account for the facts or some of them. The form of inference, therefore, is this:

> The surprising fact, E, is observed;
>
> But if H were true, E would be a matter of course.
>
> Hence, there is reason to suspect that H is true. (Peirce 1931–1935, V.189)

Peirce's characterization contains two crucial premises: First, the phenomenon E is surprising: $p(\mathrm{E})$ is small. Second, given H, the effect E is "a matter of course", that is, $p(\mathrm{E}|\mathrm{H})$ is close to unity. If these premises are satisfied, Peirce concludes that "there is reason to suspect that H is true"—not necessarily a conclusive reason, but at least *some* reason to accept H.

Philosophers working on explanatory inference have adopted Peirce's analysis of abductive reasoning. Indeed, this is the reason why abductive arguments are often called "explanatory arguments" or "inferences to the best explanation": H explains E, H makes E a "matter of course". Like Peirce, Carl G. Hempel is explicit about this connection:

> [T]he [explanatory] argument shows that, given the particular circumstances and the laws in question, the occurrence of the phenomenon *was to be expected*; and it is in this sense that the explanation enables us to *understand why* the phenomenon occurred. (Hempel 1965a, 337, original emphasis)

And, one page later:

> the explanatory information must provide good grounds for believing that X [the explanandum] did in fact occur; otherwise, that information would give us no adequate reason for saying: "That explains it—that does show why X occurred." (Hempel 1965a, 368)

Explanation thus has a central epistemic function—namely to resolve the epistemic puzzle surrounding the explanandum, to make it a matter of course given the explanans. Good explanations have predictive value for the explanandum—a classical view that is also defended in recent papers (e.g., Douglas 2009a).

Note the subtle difference between Peirce and Hempel. Peirce stresses that E must have been surprising beforehand, Hempel doesn't—at least not explicitly. This difference corresponds to the choice between two different types of probabilistic explications of explanatory power: one that focuses on the statistical relevance of H for E (e.g., by comparing $p(\mathrm{E}|\mathrm{H})$ and $p(\mathrm{E})$), and another that focuses solely on the degree to which E is probable given H. This, by the way, is the same choice that we have already faced in Variations 1 and 6. There, we distinguished confirmation as firmness from confirmation as *increase* in firmness, and difference-making measures of causal strength such as η_d from production-oriented measures such as η_{ph}. In the remainder, we will opt for the difference-making approach to explanatory power, based on the concept of statistical relevance, and provide an axiomatic characterization of various measures of explanatory power.

The Peirce–Hempel approach to explanatory power as the degree to which the explanans rationalizes the explanandum is also congenial to James Woodward's counterfactual analysis of scientific explanation:

> an explanation ought to be such that it can be used to answer what I call a *what-if-things-had-been-different question*: the explanation must enable us to see what sort of difference it would have made for the explanandum if the factors cited in the explanans had been different in various possible ways. (Woodward 2003, 11)

While the Peirce–Hempel approach is not committed to the counterfactual nature of explanatory reasoning, it shares Woodward's sentiment that a good explanation must make a difference for the phenomenon to be explained. Evidently, it is not meant to apply to each aspect of explanatory reasoning in science, but we believe that it is a general and non-committal view of explanatory power that has a wide range of applications. In the following section, we make it precise.

7.3 Representation Theorems for Measures of Explanatory Power

We begin our explication of explanatory power with the definition of an appropriate framework. It is almost identical to the one explained in the introductory chapter and used in Variations 1 and 6. Explanans H and explanandum E are represented as instantiations of variables H and E on measurable spaces $(\Omega_H, \mathcal{H})$ and $(\Omega_E, \mathfrak{E})$. Ω_H denotes the possible set of values that H can take, and $\mathcal{H} \subset \mathcal{P}(\Omega_H)$ denotes the σ-algebra of sentences about the values of H. For instance, if H is a real-valued variable, then elements of $\mathcal{H}$ are specific claims about the value of H such as "$H = 0$" or "$H = 1$", but also more general sentences such as "H takes a positive value" (i.e., $H \in (0; \infty)$), or "the absolute value of H is between -1 and 1" (i.e., $H \in [-1; 1]$). Ditto for E. As before, we do not draw a sharp line between sentences and the propositions they express.

These variables are embedded into a causal Bayesian network: a directed acyclical graph (DAG) G with a probability distribution p that is Markovian relative to the variables in G. Explanatory power is explicated as a function of the features of such a distribution. More precisely, we explicate explanatory power in a contrastive way, as we did with causal strength: it varies as a function of how the explanans H raises the probability of E. This brings us to the following general adequacy constraint, which is analogous to the explication of evidential support and causal strength measures in Variations 1 and 6.

Difference-Making (Explanatory Power) There is a real-valued, continuous function f: $[0;1]^2 \to \mathbb{R}$ such that for a putative explanans $\mathrm{H} \in \mathcal{H}$ and a putative explanandum $\mathrm{E} \in \mathfrak{E}$, we can write the explanatory power $\mathcal{E}(\mathrm{H},\mathrm{E})$ of H on E as

$$\mathcal{E}(\mathrm{H},\mathrm{E}) = f(p(\mathrm{E}|\mathrm{H}), p(\mathrm{E})),$$

where f is non-decreasing in the first argument and non-increasing in the second argument.

Difference-Making implies that among two competing explanations of the same phenomenon, we should prefer the one which rationalizes the explanandum to a higher degree. Note that this is not entirely uncontroversial since degree of explanatory power may depend on the goodness of an explanation *and* its plausibility. Difference-Making focuses on the first component only.

Moreover, not all statistically relevant hypotheses provide a satisfactory causal explanation. Causes usually explain their effects, but many scientists would be reluctant to claim that effects explain their causes. Common cause scenarios lead to similarly misleading judgments. Returning to the example from the introductory chapter, suppose that Alice and Bob have heavy stomach pain. We know that they have collected mushrooms and eaten a risotto prepared from them. None of them is an expert in distinguishing edible from poisonous mushrooms. A plausible causal representation of the story is $A \leftarrow M \rightarrow B$, where the particular values A and B stand for Alice's and Bob's illness, respectively, and M stands for eating poisonous mushrooms. Now, B is quite probable given A (after all, we know that they have eaten the same dish), and vice versa. But we would not want to say that Bob's stomach pain *explains* Alice's stomach pain, although it renders it less surprising. Rather, they seem both to be explained by the common cause, namely eating the poisonous mushrooms.

There are various roads to solving these problems. An evident proposal is to copy the move from Variation 6 (Causal Strength) and to evaluate $\mathcal{E}(\mathrm{H},\mathrm{E})$ with respect to a distribution p^* where one has *intervened* on the explanans.

This would remove the non-explanatory correlation between A and B in the above example: the intervention cuts the causal connection between A and B, and Bob's stomach pain ceases to have explanatory power for Alice's stomach pain. Eating poisonous mushrooms, however, turns out to be an excellent explanation of both explananda since the causal arrows point in the right direction. Intervening on the putative explanans also precludes effects from explaining their causes.

However, this proposal does not solve the problem of non-explanatory correlations induced by common effects of E and H: Suppose I would like to explain that my car battery is empty (E). I know that the car fails to start (K). Suppose that the car fails to start if the battery is empty or if there is no fuel in the tank (¬F). Represent this situation by the collider structure $E \rightarrow K \leftarrow F$ and the equation $\mathrm{K} \equiv \mathrm{E} \vee \neg \mathrm{F}$. Relative to the distribution $p_K(\,\cdot\,)$, which includes knowledge that the car did not start, F renders E a matter of course: given that there is fuel in the tank, the car battery must have been empty. The inference is sound, but it just does not feel like a good (causal) explanation.

Following the proposal of Korb, Hope and Nyberg (2009, 250–251), we could eliminate common effects of explanans and explanandum (or causal paths involving them) from the causal Bayesian network because they are explanatorily inert. A different proposal is made by Eva and Stern (forthcoming): they suggest intervening on the explanans H as well as on the known background variables (fixing the latter at their actual values) and to calculate explanatory power with respect to the resulting probability distribution. We leave it to the reader to evaluate both approaches and to choose an appropriate probability distribution for calculating explanatory power. For the rest of our analysis, the choice does not matter as long as spurious explanations are taken care of.

We shall now present three major proposals for measuring explanatory power based on the concept of statistical relevance. We provide axiomatic characterizations of the measures and then engage in a normative comparison. We omit Popper's (1959/2002) proposal $\mathcal{E}(\mathrm{H},\mathrm{E}) = \big(p(\mathrm{E}\,|\,\mathrm{H}) - p(\mathrm{E})\big)/\big(p(\mathrm{E}\,|\,\mathrm{H}) + p(\mathrm{E})\big)$, since he provides no independent motivation, and the phrase "explanatory power" is used in a heuristic sense only, in the context of explicating a measure of degree of corroboration.

According to a classical research tradition (e.g., Hempel and Oppenheim 1948), explanatory power is structurally similar to predictive and confirmatory value. In line with this approach, one may propose the following link between explanatory and confirmatory value: H explains E_1 better than E_2 if and only if E_1 raises the probability of H to a higher level than E_2 does.

Confirmatory Value

$$\mathcal{E}(H, E_1) > \mathcal{E}(H, E_2) \quad \text{if and only if} \quad p(H|E_1) > p(H|E_2).$$

According to Confirmatory Value, explanatory power is structurally similar to degree of confirmation in so far as a candidate explanans H performs best on those phenomena that confirm it the most. It is perhaps the purest probabilistic implementation of Hempel and Oppenheim's (1948) thesis that explanation is structurally similar to prediction. We can now characterize the measures that satisfy Confirmatory Value as well as the Difference-Making property (for similar representation theorems, see Crupi, Chater and Tentori 2013; Cohen 2016):

Theorem 7.1 (Representation Theorem for $\mathcal{E}_{GMG}$). *All measures of explanatory power $\mathcal{E}(H, E)$ that satisfy Difference-Making (Explanatory Power) and Confirmatory Value are ordinally equivalent to the Good–McGrew measure $\mathcal{E}_{GMG}$, proposed by I. J. Good (1960) and Timothy McGrew (2003):*

$$\mathcal{E}_{GMG}(H, E) = \log \frac{p(E|H)}{p(E)}. \tag{7.1}$$

The Good–McGrew measure $\mathcal{E}_{GMG}$ is structurally identical to the log-ratio measure of confirmation $r(H, E)$ from Variation 1. This identity, and the general rationale behind Confirmatory Value, suggest further adequacy conditions. For instance, for advocates of Inference of the Best Explanation, "observations support the hypothesis precisely because it would explain them" (Lipton 2000, 185). This reductionist claim can be cast into the following quantitative form that identifies explanatory power and degree of confirmation:

Confirmation–Explanation Reduction For any $E \in \mathfrak{E}$ and $H \in \mathcal{H}$, $\mathcal{E}(H, E) = c(H, E)$.

However, Crupi (2012) provides a convincing argument why Confirmation–Explanation Reduction does not fly. First, consider the following symmetries regarding the relation between positive and negative explanatory power:

Positive–Negative Symmetry (I)

$$\mathcal{E}(H_1, E) > \mathcal{E}(H_2, E) \quad \text{if and only if} \quad \mathcal{E}(H_1, \neg E) < \mathcal{E}(H_2, \neg E).$$

Positive–Negative Symmetry (II)

$$\mathcal{E}(H, E_1) > \mathcal{E}(H, E_2) \quad \text{if and only if} \quad \mathcal{E}(H, \neg E_1) < \mathcal{E}(H, \neg E_2).$$

That is, the more H_1 is able to explain E, compared to competing hypothesis H_2, the less is it able to explain $\neg$E. This principle is a plausible constraint on any logic of explanatory power. Also the second principle—if H explains E_1 better than E_2, then it is less of a failure in explaining $\neg E_2$ than in explaining $\neg E_1$—certainly makes sense.

Crupi shows that Positive–Negative Symmetry (II) and Confirmation–Explanation Reduction are jointly incompatible with Final Probability Incrementality—the intuitive principle that confirmation rankings between two pairs (H, E) and (H, E') depend only on $p(H|E)$ versus $p(H|E')$ (see Variation 1). Since there is no apparent reason for giving up Final Probability Incrementality as an adequacy criterion for confirmation measures, Crupi's result demonstrates that degree of confirmation and explanatory power should be seen as distinct concepts, and why we should search for independent explications of explanatory power.

Consider now a case where irrelevant evidence E' is added to the explanandum E which is explained by hypothesis H. On this case, Schupbach and Sprenger (2011, 115, notation changed) write:

> [A]s the evidence becomes less statistically relevant to some explanatory hypothesis H (with the addition of irrelevant propositions), it ought to be the case that the explanatory power of H relative to that evidence approaches the value at which it is judged to be explanatorily irrelevant to the evidence ($\mathcal{E} = 0$).

Schupbach and Sprenger transfer this property to the case of negative statistical dependence: addition of statistically independent evidence *dilutes* (negative) explanatory power and brings it closer to the neutral value of zero. Crupi and Tentori (2012, 370), on the other hand, think that this property would allow us "to indefinitely relieve a lack of explanatory power, no matter how large, by adding more and more irrelevant explananda, simply at will". In line with this reasoning, they propose the following condition:

Explanatory Justice If E' is statistically independent from E, H and their conjunction $E \wedge H$, then

(i) if $\mathcal{E}(H, E) > 0$, then $\mathcal{E}(H, E \wedge E') < \mathcal{E}(H, E)$; and

(ii) if $\mathcal{E}(H, E) \leq 0$, then $\mathcal{E}(H, E \wedge E') = \mathcal{E}(H, E)$.

When Explanatory Justice and Positive–Negative Symmetry (II) replace the Confirmatory Value in Theorem 7.1, this suffices for demonstrating another representation result (see also Crupi and Tentori 2012; Cohen 2016):

Theorem 7.2 (Representation Theorem for $\mathcal{E}_{CT}$). *All measures of explanatory power $\mathcal{E}(H, E)$ that satisfy Difference-Making (Explanatory Power), Positive–Negative Symmetry (II) and Explanatory Justice are ordinally equivalent to the Crupi–Tentori measure $\mathcal{E}_{CT}$:*

$$\mathcal{E}_{CT}(H, E) = \begin{cases} \dfrac{p(E|H) - p(E)}{1 - p(E)} & \text{if } p(E|H) \geq p(E), \\ \dfrac{p(E|H) - p(E)}{p(E)} & \text{if } p(E|H) < p(E). \end{cases}$$

The third and last candidate measure which we review is a proposal by Schupbach and Sprenger (2011):

$$\mathcal{E}_{SS}(H, E) = \frac{p(H|E) - p(H|\neg E)}{p(H|E) + p(H|\neg E)},$$

which is ordinally equivalent to the ratio of posterior probabilities $r := p(H|E)/p(H|\neg E)$: dividing the numerator and denominator of $\mathcal{E}_{SS}$ by $p(H|\neg E)$ leads to $\mathcal{E}_{SS} = (r-1)/(r+1)$. Since

$$r(H, E) = \frac{p(H|E)}{p(H|\neg E)} = \frac{p(E|H)\,(1 - p(E))}{p(E)\,(1 - p(E|H))}$$

is equivalent to the betting odds on $E|H$, divided by the betting odds on E, $\mathcal{E}_{SS}$ satisfies Difference-Making, too. One can therefore interpret $\mathcal{E}_{SS}$ as the ratio to which conditioning on H raises the odds that a rational agent would be willing to lay on E.

Schupbach and Sprenger's justification of $\mathcal{E}_{SS}$ is based on the idea that the values of $p(H)$ do not affect $\mathcal{E}(H, E)$ when there is a deductive relationship between H and E, or $\neg$H and E. Instead of copying their approach, we will present an adequacy condition that allows for a simple representation theorem when taken together with Difference-Making:

Independent Background Theories Suppose there is a theory T such that

$$p(H\,|\,E, T) = p(H|E) \quad \text{and} \quad p(H\,|\,\neg E, T) = p(H|\neg E).$$

Then $\mathcal{E}(H, E) = \mathcal{E}_T(H, E)$, where $\mathcal{E}_T$ refers to explanatory power calculated with respect to a probability distribution conditional on T.

Informally, this amounts to the following: if a theory T is irrelevant to the interaction between explanans H and explanandum E (and its negation ¬E), then conditionalizing on T does not affect the degree of explanatory power. For example, if H is a particular hypothesis in genetics (e.g., a mechanism for transferring individual traits from one generation to the next) and E is an observed pattern of trait inheritance, then the explanatory power of H for E should not depend on whether or not we take the General Theory of Relativity (= T) for granted. Using this condition, we obtain a representation theorem for $\mathcal{E}_{SS}$:

Theorem 7.3 (Representation Theorem for $\mathcal{E}_{SS}$). *All measures of explanatory power* $\mathcal{E}(H,E)$ *that satisfy Difference-Making (Explanatory Power) and Independent Background Theories are ordinally equivalent to the posterior ratio* $p(H|E)/p(H|\neg E)$, *and thus to the Schupbach–Sprenger measure* $\mathcal{E}_{SS}$.

$\mathcal{E}_{SS}$ also satisfies Positive–Negative Symmetry and the first clause of the Explanatory Justice condition, but neither its second clause nor Confirmatory Value. Indeed, the second clause of Explanatory Justice is a major point of contention in the debate about different measures of explanatory power, as evidenced by Crupi and Tentori 2012, Cohen 2015, 2016 and Cohen 2018. These papers also offer different representation theorems for $\mathcal{E}_{SS}$. For instance, Cohen (2015) proposes three defining conditions for $\mathcal{E}_{SS}$: (i) all tautological hypotheses receive constant explanatory power; (ii) a strong symmetry condition $\mathcal{E}(\neg E,H) = -\mathcal{E}(E,H)$; (iii) a somewhat stronger independence of prior probabilities.

Table 7.1 gives an overview of the various properties of the three candidate measures. It is notable that their mathematical structure corresponds to confirmation measures from Variation 1, with the roles of evidence and hypothesis reversed: $\mathcal{E}_{GMG}(H,E)$ is identical to $r(E,H)$, $\mathcal{E}_{CT}(H,E)$ takes the same values as the partial entailment measure $z(E,H)$ and $\mathcal{E}_{SS}(H,E)$ equals the likelihood ratio measure $l(E,H)$ (see Table 1.3).

Measure	Property						
	DM	CV	S1	S2	EJ^+	EJ^-	IBT
Good–McGrew ($\mathcal{E}_{GMG}$)	yes	yes	yes	no	no	yes	no
Crupi–Tentori ($\mathcal{E}_{CT}$)	yes	no	yes	yes	yes	yes	no
Schupbach–Sprenger ($\mathcal{E}_{SS}$)	yes	no	yes	yes	yes	no	yes

Table 7.1: A classification of different measures of explanatory power according to the adequacy conditions they satisfy. DM = Difference-Making (Explanatory Power), CV = Confirmatory Value, S1 = Positive–Negative Symmetry (I), S2 = Positive–Negative Symmetry (II), EJ^+ = Explanatory Justice, first clause (positive explanatory power), EJ^- = Explanatory Justice, second clause (negative explanatory power), IBT = Independent Background Theories.

7.4 Comparison of the Measures

We now proceed to a **normative comparison of the three measures** and begin with a critique of the Good–McGrew measure $\mathcal{E}_{GMG}$. For starters, we note that $\mathcal{E}_{GMG}$ allows for the **conjunction of irrelevant evidence** (Schupbach and Sprenger 2011, 114–115). Suppose that H has positive explanatory power for E and that $p(E' \mid E, H) = p(E' \mid H)$ for some other piece of evidence E′. In that case, $\mathcal{E}_{GMG}(E \wedge E', H) = \mathcal{E}_{GMG}(E, H)$. Schupbach and Sprenger consider this property—possessed by neither their measure $\mathcal{E}_{SS}$ nor the Crupi–Tentori measure $\mathcal{E}_{CT}$—as problematic and illustrate their objection with an example: Let E be an observed Brownian motion, let H be an appropriate physical explanation of that motion, and let E′ be a proposition about the mating season of the American tree frog. Clearly, H explains E much better than it explains $E \wedge E'$—the Brownian motion *and* the tree frog mating season proposition. A substantial part of $E \wedge E'$ stays unexplained. This phenomenon exemplifies failure to satisfy the first clause of the Explanatory Justice condition, as described in the Schupbach and Sprenger quote on page 196.

The above criticism echoes the paradox of irrelevant conjunctions that we have encountered in Variation 1, applied to the log-ratio measure $r(H, E) = \log(p(H \mid E)/p(H))$. In that case, r allowed for tacking irrelevant conjunctions to the confirmed hypothesis without lowering the degree of confirmation. On this basis, r was ruled out as an appropriate confirmation measure. The same argument pattern applies here: tacking irrelevant conjunctions to the explanandum should *lower* the degree of explanatory power, not leave it constant.

In defense of $\mathcal{E}_{GMG}$, Cohen (2016) notes that the bite of Schupbach and Sprenger's objection depends on whether E′ is *meant* to be explained by H or not. For example, if E′ is collected for practical or procedural reasons only (e.g., certain demographic data in a psychological survey), then it seems that H should not be penalized for failing to explain E′. Whether irrelevant conjunctions are problematic depends on the focus of the explanation: is H supposed to explain all of the evidence or just the part that we consider crucial?

We are not sure that this proposal manages to save $\mathcal{E}_{GMG}$. After all, if data E′ do not stand in need of an explanation, they should not be modelled as part of the explanandum. Rather, one should calculate explanatory power relative to a probability distribution conditional on E′ (compare the Modularity requirement in Variation 1). We find it problematic that explanatory power is invariant under adding data that are *part of the explanandum* but fail to be rationalized by the hypothesis.

What about the other two measures? Should $\mathcal{E}_{SS}$ or $\mathcal{E}_{CT}$ be preferred? Of course we are not completely unbiased in answering this question: one of the authors of this monograph (J. S.) developed the $\mathcal{E}_{SS}$-measure together with Jonah Schupbach. We will now advance two arguments in favor of $\mathcal{E}_{SS}$, both of them due to Cohen (2018).

The first criticism of $\mathcal{E}_{CT}$ concerns the **scaling properties** of both measures. It is based on a simple coin-flipping example: Suppose there are two identical-looking coins, one of which is fair while the other is biased (say, with a 70/30 bias in favor of heads). To block the influence of prior probability, we suppose that a coin is chosen at random, with equal probability. Then we start tossing it. Consider the hypothesis H that the tested coin is biased, and the observation E_N that all N tosses of the coin turn out to be heads. Certainly H explains E_N to a certain degree—primarily because this course of events would be a truly extraordinary chance under the alternative ¬H.

However, the Crupi–Tentori measure disagrees: as N increases, the measure $\mathcal{E}_{CT}(H, E_N)$ quickly approaches zero (e.g., $\mathcal{E}_{CT}(H, E_{10}) = 0.014$). In other words, $\mathcal{E}_{CT}$ treats a statistically highly relevant hypothesis as if it were independent of the explanandum. E is surprising under H, but it is much more surprising under ¬H—a fact that is not reflected by $\mathcal{E}_{CT}$. By contrast, Schupbach and Sprenger's measure $\mathcal{E}_{SS}$ converges to a reasonable, but not very high value ($\mathcal{E}_{SS}(H, E_N) \longrightarrow 0.33$ for $N \to \infty$), indicating that H outperforms ¬H while being a far-from-perfect explanation. In other words, $\mathcal{E}_{SS}$ captures the contrastive nature of scientific explanations (van Fraassen 1980) better than $\mathcal{E}_{CT}$.

The second criticism is based on how $\mathcal{E}_{CT}$ deals with **irrelevant evidence.** If (negative) explanatory power remains constant under the addition of irrelevant evidence, as the downward clause of Explanatory Justice demands, then Crupi and Tentori should also believe that $\mathcal{E}_{CT}$ remains constant under the addition of irrelevant disjunctions to the hypothesis. Hence they should require that $\mathcal{E}_{CT}(H, E) = \mathcal{E}_{CT}(H \vee H', E)$ whenever H' is statistically independent of E, H and $E \wedge H$, and $p(E|H) < p(E)$. However, Cohen (2018, Claim 1) shows that in this case, $\mathcal{E}_{CT}(H, E) < \mathcal{E}_{CT}(H \vee H', E)$. This leads the entire motivation of the second, negative clause of Explanatory Justice *ad absurdum*: explanatory power is unchanged for the explanandum $E \wedge E'$, but it is increased for the explanans $H \vee H'$. This can be construed as an argument in favor of the Schupbach–Sprenger measure $\mathcal{E}_{SS}$. While Schupbach and Sprenger consistently reject the intuition behind the negative clause of Explanatory Justice, Crupi and Tentori need to explain why the case of disjunction on the explanans level needs to be handled differently than the case of conjunction of the explanandum level. The entire debate is still quite young, though, and future arguments may overturn our verdict in favor of $\mathcal{E}_{SS}$.

7.5 Discussion

This Variation motivated and compared various Bayesian accounts of explanatory power. Models of explanatory power are not meant to decide the debate on what constitutes a scientific explanation. However, once there is consensus on the explanatory character of a hypothesis–phenomenon relationship, they provide a means of quantifying the strength of that relationship. This is not only of great use for studying explanatory reasoning in science—it also provides a formal basis for studying the conditions under which an IBE, that is, an inference to the most powerful explanation, leads to hypotheses that we can rationally accept.

To motivate a Bayesian approach to measuring explanatory power, we have first expounded two grand traditions for conceiving of scientific explanation—the causal-interventionist view and the view of explanations as arguments. Probably none of them captures completely what scientific explanations are about; yet, both traditions retain important features of scientific reasoning that can be used for an explication of explanatory power. To this purpose, we have set up a Bayesian framework for measures of explanatory power that conceptualizes explanatory power as argument strength, but keeps track of causal relations between the relevant variables.

The core of this Variation has been the comparison of three different Bayesian measures of explanatory power and their characterization by means of representation theorems. The Schupbach–Sprenger measure $\mathcal{E}_{SS}$ emerges as a particularly apt candidate. To contest this claim, its competitors—the Good–McGrew measure $\mathcal{E}_{GMG}$ and the Crupi–Tentori measure $\mathcal{E}_{CT}$—need to address objections related to their functional form.

Even so, any of the three measures can help to investigate IBE from a Bayesian perspective. Recent empirical work has shown that people accept hypotheses rather on the basis of their explanatory power than on the basis of objective chances (Douven and Schupbach 2015a,b). Indeed, assessing explanatory power can be cognitively easier than applying Bayes' Theorem and calculating posterior probabilities. This motivates further empirical research into the circumstances under which people's reasoning conforms to IBE. From a normative point of view, Schupbach (2011b, 2018) has used computer simulations in order to show that IBE—understood as inference to the hypothesis with the highest explanatory power—is a reliable mode of inference. Peirce's inference scheme

$$\frac{\begin{array}{c}\text{E}\\ \text{H explains E}\end{array}}{\text{H}}$$

is replaced by the scheme

$$\frac{\begin{array}{c}\text{E}\\ \mathcal{E}(\text{H},\text{E}) > \mathcal{E}(\text{H}_i,\text{E}) \text{ for all alternatives } \text{H}_i\end{array}}{\text{H.}}$$

This sophisticated form of IBE approximates Bayesian reasoning very well, and in Schupbach's simulations, the explanatorily most valuable hypothesis matches the true hypothesis in an overwhelming number of cases. Trpin and Pellert (forthcoming) also conduct simulations where they study the behavior of Bayesian inference in cases of uncertain evidence, and they come to the conclusion that IBE-based inference often outperforms Jeffrey Conditionalization. More research along these lines may help to determine the conditions under which IBE is a sound form of scientific reasoning, and shed light on issues where IBE takes a prominent role, such as the ongoing debate between realists and anti-realists (see also Variation 3).

Another dimension of investigating measures of explanatory power consists in empirical work. In an experiment that transfers the design of Crupi, Tentori and González 2007 to the case of explanatory power, Schupbach

(2011a) finds that $\mathcal{E}_{SS}$ provides the best description of participants' judgments of explanatory power. The statistical analysis has, however, been criticized by Clark Glymour (2015). Recent experiments on explanatory power (e.g., Colombo, Bucher and Sprenger 2017a,b) confirm that explanatory judgment is sensitive to statistical relevance, lending empirical support to the Bayesian research program on explanatory reasoning and explanatory power. The above studies also reveal a strong link between judgments of explanation, confirmation and rational acceptability. Furthermore, Lombrozo (2007) has investigated how the simplicity of an explanation affects its perceived value. Future studies could transfer this design from Lombrozo's artificial and idealized scenario, involving inhabitants of an alien planet, to a more familiar setting and obtain parochially more valid results.

Finally, there is ample room for combining empirical and theoretical research on explanatory inference in the Bayesian paradigm. A particularly salient issue concerns the role and interplay of causal and probabilistic factors in explanatory reasoning. One could, for example, envision a systematic comparison of measures of actual causation and explanatory power. Can they be studied in a similar way? Does the formal apparatus presented above also help to understand features of actual causation, such as sensitivity to normality of causes? And what are the relevant differences? More generally, how do explanatory, causal and probabilistic forms of reasoning interact (Lombrozo 2009, 2011, 2012; Sloman and Lagnado 2015)? Do empirically observed differences have correlates on the level of a formal Bayesian analysis?

We hope that our contribution will stimulate further research on the nature of explanation. In particular, we hope that our results will help to promote "the prospects for a naturalized philosophy of explanation" (Lombrozo 2011, 549), where philosophical theorizing about the nature of explanation is constrained and informed by empirical evidence about the psychology of explanatory power and where, on the other hand, philosophical research stimulates empirical investigations into explanatory reasoning.

Appendix: Proofs of the Theorems

In general, the proofs in this Variation are very similar to the analogous results from Variation 1 and 6, with the roles of E and H replaced by other propositions. In particular, Difference-Making (Explanatory Power) is just the same as Prior–Posterior Dependence, with $p(\mathrm{H}|\mathrm{E})$ and $p(\mathrm{H})$ being replaced by $p(\mathrm{E}|\mathrm{H})$ and $p(\mathrm{E})$. A similar claim can be made about Difference-Making (Explanatory Power) and Difference-Making (Causal Strength)—see below.

For this reason, we will restrict ourselves to sketching the proof ideas and explaining why the proofs are structurally identical to those given in earlier chapters.

Proof of Theorem 7.1: From a structural point of view, Difference-Making (Explanatory Power) corresponds to Prior–Posterior Difference, and Explanation–Confirmation Symmetry corresponds to Law of Likelihood, with the places of E and H reversed (i.e., H, E_1 and E_2 correspond to E, H_1 and H_2, respectively). Thus we can follow the proof of the first statement of Theorem 1.2 (i.e., the representation theorem for the log-ratio measure $r(\mathrm{H},\mathrm{E})$). Since any confirmation measure $c(\mathrm{H},\mathrm{E})$ that satisfies Prior–Posterior Difference and Modularity must be ordinally equivalent to

$$r(\mathrm{H},\mathrm{E}) = \log \frac{p(\mathrm{H}|\mathrm{E})}{p(\mathrm{H})},$$

any measure of explanatory power $\mathcal{E}(\mathrm{H},\mathrm{E})$ that satisfies Difference-Making and Explanation–Confirmation Symmetry must be ordinally equivalent to the Good–McGrew measure

$$\mathcal{E}_{\mathrm{GMG}}(\mathrm{H},\mathrm{E}) = \log \frac{p(\mathrm{E}|\mathrm{H})}{p(\mathrm{E})}. \qquad \square$$

Proof of Theorem 7.2: First, we transfer the proof of Theorem 6.4: When we replace $\mathrm{E}|\mathrm{C}$ with $\mathrm{E}|\mathrm{H}$ and $\mathrm{E}|\neg\mathrm{C}$ with E, then Difference-Making (Causal Strength) just becomes Difference-Making (Explanatory Power). Similarly, No Dilution for Irrelevant Effects (Prevention) takes the form of the second clause of Explanatory Justice, and Weak Causation–Prevention Symmetry takes the form of Positive–Negative Symmetry (II). Thus, we can conclude that all measures that satisfy these three conditions (for explanatory power) are ordinally equivalent to the measure that satisfies the corresponding conditions for causal strength, namely

$$\eta_{cg}(C,E) = \begin{cases} \dfrac{p^*(E|C) - p^*(E|\neg C)}{1 - p^*(E|\neg C)} & \text{if C is positively relevant for E,} \\ \dfrac{p^*(E|C) - p^*(E|\neg C)}{p^*(E|\neg C)} & \text{if C is negatively relevant for E.} \end{cases}$$

By means of the above replacement operation ($E|C \rightarrow E|H$ and $E|\neg C \rightarrow E$), we infer that any such measure must have the form of the Crupi–Tentori measure $\mathcal{E}_{CT}$. □

Proof of Theorem 7.3: Like in the proof of Theorem 7.1, Difference-Making (Explanatory Power) corresponds to Prior–Posterior Difference, with the places of E and H reversed. Independent Background Theories corresponds in the same way to Modularity. Thus, we can follow the proof of the second statement of Theorem 1.2 (i.e., the representation theorem for the likelihood ratio measure $l(H,E)$ and the Kemeny–Oppenheim measure $k(H,E)$). Since any confirmation measure $c(H,E)$ that satisfies Prior–Posterior Difference and Modularity must be ordinally equivalent to

$$k(H,E) = \frac{p(E|H) - p(E|\neg H)}{p(E|H) + p(E|\neg H)},$$

any measure $\mathcal{E}(H,E)$ of explanatory power that satisfies Difference-Making and Independent Background Theories must be ordinally equivalent to the Schupbach–Sprenger measure

$$\mathcal{E}_{SS}(H,E) = \frac{p(H|E) - p(H|\neg E)}{p(H|E) + p(H|\neg E)}. \qquad \square$$

Variation 8: Intertheoretic Reduction

Establishing relations between different theories, by unification or by reduction, is an important goal of science. As to unification, theories with a wide scope and a small number of basic postulates have been found attractive by scientists at all times. Take, for example, Newtonian mechanics: it explains terrestrial as well as celestial motion, and it unifies Galilei's invariance principle with Kepler's Laws of planetary motion. Or consider Maxwell's theory of electrodynamics: it provides a unified account of electric and magnetic forces, and the laws governing their interaction.

Relations of **intertheoretic reduction** are more complex, and their cognitive value is less obvious. The most famous example from physics is statistical mechanics: micro-level laws about the motion of molecules provide the foundations for a macro-level theory about the behavior of gases and fluids, namely thermodynamics. What exactly is involved in a reduction is a matter of philosophical controversy (see van Riel and Van Gulick 2014), but in any case, it involves accounting for the behavior of a system by describing the behavior of its constituents. In the case of statistical mechanics, the concepts and laws of a **phenomenological theory** $\mathbf{T}_P$, such as thermodynamics, are "reduced" to the concepts and laws of a more **fundamental theory** $\mathbf{T}_F$, such as statistical mechanics. Often this reduction is executed by means of deriving the laws of $\mathbf{T}_P$ from those of $\mathbf{T}_F$ (Nagel 1961; Schaffner 1967)—more on this below. Following standard terminology, we say that $\mathbf{T}_P$ is the "reduced" theory and that $\mathbf{T}_F$ is the "reducing" theory. Other examples of (putative) intertheoretical reductions are chemistry to atomic physics, rigid-body mechanics to particle mechanics, psychology to neuroscience, and game theory to agent-based models.

Reductions, if successful, are celebrated by scientists because they allow for a unified theoretical framework in which one can investigate the phenomenological as well as the fundamental theory. They also allow for precise predictions on the phenomenological level motivated by assumptions on the fundamental level. They may provide some deep understanding and explanation of the nature of central concepts of the involved disciplines. For instance, the thermodynamic concept of heat is identified with the energy transfer by a disordered, microscopic action on a system of molecules, described by statistical mechanics. For these reasons, intertheoretic reductions make large contributions to the cognitive advancement of science.

In this Variation, we show how the establishment of intertheoretic reductions boosts the cognitive value of the theories involved. More specifically, we show that if there is a reductive relation between two theories, then confirmation flows both from the phenomenological to the fundamental theory and from the fundamental to the phenomenological theory. For instance, evidence that exclusively confirms statistical mechanics before the reduction also confirms (though perhaps to a lower degree) thermodynamics after the reduction, and vice versa. In this sense, we obtain a more coherent and unified theoretical picture.

Section 8.1 sets the scene by outlining the Generalized Nagel–Schaffner (GNS) model of reduction, which serves as the foil for our Bayesian analysis. Section 8.2 contains the main argument: We consider the confirmation of $\mathbf{T}_F$ and $\mathbf{T}_P$ in two scenarios, one with and one without intertheoretic reduction. We conclude that confirmation is stronger in the post-reduction scenario. Section 8.3 discusses various implications of this result. In Section 8.4, we sum up our results and outline a number of open problems. Further detail is contained in the articles "Who is Afraid of Nagelian Reduction?" and "Confirmation and Reduction" by Dizadji-Bahmani, Frigg and Hartmann (2010, 2011), on which this chapter is based.

8.1 The Generalized Nagel–Schaffner Model

The idea behind the GNS model of reduction is best illustrated by the familiar case of thermodynamics and statistical mechanics. Thermodynamics describes systems like gases and solids in terms of macroscopic properties such as volume, pressure, temperature and entropy, and gives a correct description of the behavior of such systems. Statistical mechanics, by contrast, aims to account for the laws of thermodynamics in terms of dynamical laws governing the microscopic constituents of macroscopic systems (Frigg 2008). In particular, statistical mechanics aims to show that the Second Law of Thermodynamics is a consequence of the mechanical motion of the gas molecules. Consider a container divided in two halves by a partition wall. The left half is filled with a gas, while the right half is empty. If we remove the partition, the gas will spread and soon be evenly distributed throughout the entire container; as the gas spreads, its entropy increases. This is an instance of a process obeying the Second Law of Thermodynamics. Roughly speaking, the Second Law says that the entropy of a closed system cannot decrease, and usually increases when the system is left on its own in a non-equilibrium state. The aim of statistical mechanics is to account for the Second Law in terms of equations governing the motion of gas molecules and some probabilistic assumptions; that is, it aims to show that the Second Law is a consequence of its basic postulates—or approximately so.

That analogues of the laws of the phenomenological theory (here: thermodynamics) should follow from the laws of the fundamental theory (here: statistical mechanics) is the basic idea of GNS. Consider a phenomenological theory $\mathbf{T}_\mathrm{P}$ and a fundamental theory $\mathbf{T}_\mathrm{F}$, which are each identified with a set of empirical propositions. So let

$$\mathbf{T}_\mathrm{P} := \left\{\mathrm{T}_\mathrm{P}^{(1)}, \ldots, \mathrm{T}_\mathrm{P}^{(n_\mathrm{P})}\right\} \quad \text{and} \quad \mathbf{T}_\mathrm{F} := \left\{\mathrm{T}_\mathrm{F}^{(1)}, \ldots, \mathrm{T}_\mathrm{F}^{(n_\mathrm{F})}\right\}.$$

The reduction of $\mathbf{T}_\mathrm{P}$ to $\mathbf{T}_\mathrm{F}$ consists of the following three steps (Schaffner 1967):

1. Adopt auxiliary assumptions describing the particular setup under investigation. Here, these are assumptions about the mechanical properties of the gas molecules. Then derive from these and $\mathbf{T}_\mathrm{F}$ a restricted version of each proposition $\mathrm{T}_\mathrm{F}^{(i)}$. Denote these by $\mathrm{T}_\mathrm{F}^{*(i)}$ and the corresponding set by
$$\mathbf{T}_\mathrm{F}^{*} := \left\{\mathrm{T}_\mathrm{F}^{*(1)}, \ldots, \mathrm{T}_\mathrm{F}^{*(n_\mathrm{F})}\right\}.$$

2. $\mathbf{T}_\mathrm{P}$ and $\mathbf{T}_\mathrm{F}$ are formulated in different vocabularies. In our example, statistical mechanics talks about trajectories in phase space and probability measures while thermodynamics talks about macroscopic properties such as pressure and temperature. In order to connect the two theories, we adopt bridge laws. These connect terms of one theory with terms of the other, for instance mean kinetic energy in statistical mechanics with temperature in thermodynamics. Substituting the terms in $\mathbf{T}_\mathrm{F}^{*}$ with terms from $\mathbf{T}_\mathrm{P}$ as per the bridge laws yields
$$\mathbf{T}_\mathrm{P}^{*} := \left\{\mathrm{T}_\mathrm{P}^{*(1)}, \ldots, \mathrm{T}_\mathrm{P}^{*(n_\mathrm{P})}\right\}.$$

3. Show that each element of $\mathbf{T}_P^*$ is strongly analogous to the corresponding element of $\mathbf{T}_P$.

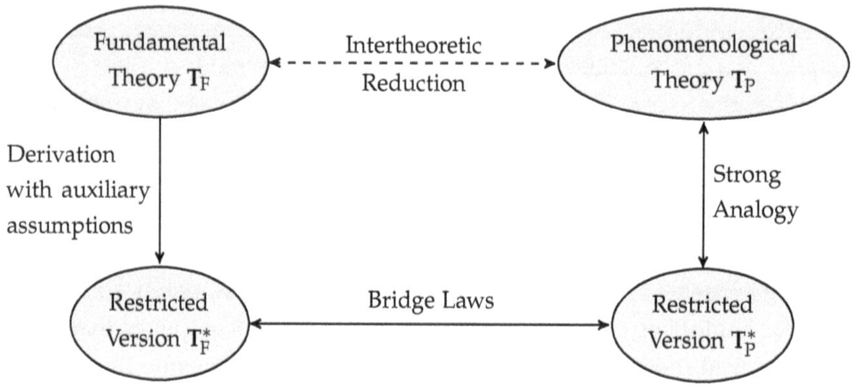

Figure 8.1: The Generalized Nagel–Schaffner (GNS) model of intertheoretic reduction.

If these conditions obtain, we say that $\mathbf{T}_P$ **is reduced to** $\mathbf{T}_F$. See Figure 8.1 for a graphical illustration.

We now explain two central notions that occur in the GNS model of intertheoretic reduction.

First, the notion of **strong analogy,** which may appear inappropriately vague. After all, Nagel himself stressed the importance of logical and mathematical relations that hold between the reducing and the reduced theory. These strong links between $\mathbf{T}_P$ and $\mathbf{T}_F$ seem to be watered down by introducing a concept that introduces a great deal of subjective judgment on behalf of the scientist. However, it is often impossible to derive the *exact* laws of $\mathbf{T}_P$. For instance, we cannot derive the exact Second Law of Thermodynamics from statistical mechanics, which is a probabilistic theory, whereas the Second Law is supposed to hold without exception. Thus *exact* derivability is too stringent a requirement. It suffices to deduce laws that are *approximately the same* as the laws of $\mathbf{T}_P$. For the case of statistical mechanics and thermodynamics, we derive a probabilistic law that is strongly analogous to the Second Law of Thermodynamics: the proposition that entropy is highly likely to increase over time, which is known as Boltzmann's Law. This revision of the original model has been developed in a string of publications by Schaffner (1967, 1969, 1976, 1977, 1993) and, indeed, by Nagel (1979) himself.

In sum, an intertheoretic reduction subsumes a strongly analogous version $\mathbf{T}_P^*$ of the phenomenological theory $\mathbf{T}_P$ under the fundamental theory $\mathbf{T}_F$.

This involves deriving a restricted version $\mathbf{T}_F^*$ of $\mathbf{T}_F$ with the help of boundary conditions and auxiliary assumptions, and using bridge laws to obtain $\mathbf{T}_P^*$ from $\mathbf{T}_F^*$.

This brings us to the second point: the notion of a **bridge law.** While Nagel himself remains relatively non-committal about the exact form and nature of bridge laws, Schaffner (1976, 614–615) offers a concise characterization of bridge laws, which he calls *reduction functions*. For Schaffner, a reduction function is a statement to the effect that a term y_P of $\mathbf{T}_P^*$ and a term y_F of $\mathbf{T}_F^*$ are coextensional. For example, the terms "temperature" and "mean kinetic energy" are coextensional when applied to a gas. Moreover, physical properties usually have magnitudes: a gas does not have a temperature *simpliciter*; it has a temperature of so and so many degrees Kelvin. Thus, a bridge law does not only establish coextensionality of two terms; it also specifies the functional relationship between their magnitudes and the units of measurement. Stated more precisely, the bridge law contains a function f such that $\tau_P = f(\tau_F)$, where τ_P and τ_F are the values of y_P and y_F, respectively. So we can give the following tentative definition of bridge laws (we will qualify this statement below): A bridge law is a statement to the effect that (i) y_P applies if, and only if, y_F applies, and (ii) $\tau_P = f(\tau_F)$.

Both the concept and the epistemology of strong analogy and bridge laws have served as the basis for criticism of the GNS model of reduction. For instance, the so-called New Wave Reductionists (e.g., Churchland 1979, 1985; Bickle 1998) deny that bridge laws play an important role in the discovery of reductive relations. However, we do not want to engage (again) in a debate about the merits and drawbacks of the GNS model, but to show how it can be used for demonstrating the confirmatory value of reductive relations. Therefore we refer the interested reader to Dizadji-Bahmani, Frigg and Hartmann 2010, where these and similar criticisms are addressed and, to our mind, convincingly rebutted.

8.2 Reduction and Confirmation

Consider how theories are supported by evidence. With regard to our two theories $\mathbf{T}_P$ and $\mathbf{T}_F$, there are three kinds of evidence: evidence that confirms only the phenomenological theory, evidence that confirms only the fundamental theory and evidence that confirms, to some degree, both. Thermodynamics and statistical mechanics provide salient examples for all three kinds of confirmation (see also Dizadji-Bahmani, Frigg and Hartmann

2011, 323–324). For the first case, consider what is known as the Joule–Thomson process: There are two chambers of different dimensions, filled with a gas and connected to each other by a permeable membrane. At the end of each chamber, there is a piston which allows the pressure and volume for the gas in that chamber to be varied by applying a force. The pressure in the first chamber is higher than the pressure in the second. Now push the gas from the first chamber into the second, but so slowly that the pressure remains constant in both chambers and no heat is exchanged with the environment. Then the gas in the second chamber cools down. The expected amount of cooling can be calculated using the principles of thermodynamics, and is found to coincide with experimental values. So we have a confirmation of thermodynamics, but not of statistical mechanics, since no statistical mechanics assumptions have been used in the argument.

For the second case, consider the dependence of a metal's electrical conductivity on temperature. From statistical mechanics one can derive an equation that relates change in conductivity of certain metals to a change in temperature. This equation is also confirmed in experiments, whereas thermodynamics is entirely silent about this phenomenon.

Third, consider again the gas confined to the left half of the box which spreads evenly when the dividing wall is removed. It follows from thermodynamics that the thermodynamic entropy of the gas increases; at the same time, it is a consequence of statistical mechanics that the Boltzmann entropy increases in that process. So the spreading of the gas confirms both statistical mechanics and thermodynamics. We shall now explicate this intuition in a Bayesian model (Dizadji-Bahmani, Frigg and Hartmann 2011, Section 4).

8.2.1 Before the Reduction

Let us examine the situation before a reduction is established. To simplify things, we assume that $\mathbf{T}_\mathrm{P}$ and $\mathbf{T}_\mathrm{F}$ have only one element each, viz. T_F and T_P, respectively. The generalization to more than one element is conceptually straightforward. Furthermore, E confirms both T_F and T_P, E_F confirms only T_F, and E_P confirms only T_P. Introducing corresponding propositional variables T_F, T_P, E, E_F and E_P, we can represent the situation before the attempted reduction in the Bayesian network depicted in Figure 8.2.

Following our methodology, we have to specify the probabilities of T_F and T_P (i.e., of all root nodes) and the conditional probabilities of E, E_F and E_P (i.e., of all child nodes), given their parents. We denote:

$$t_\mathrm{F} := p(\mathrm{T}_\mathrm{F}), \qquad\qquad t_\mathrm{P} := p(\mathrm{T}_\mathrm{P}),$$

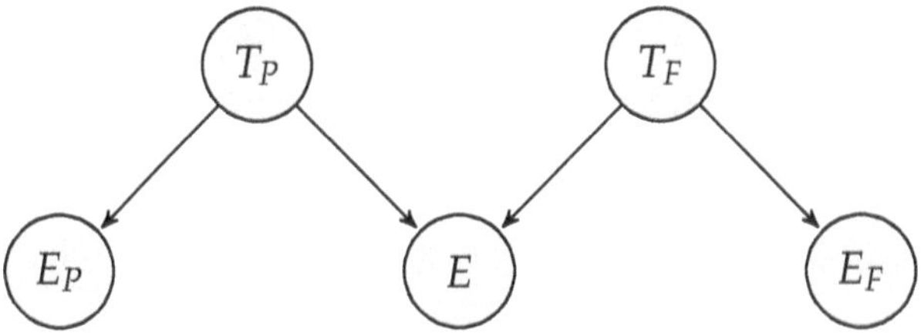

Figure 8.2: The Bayesian network representation of the situation *before* an intertheoretic reduction. Adapted under the terms of the Creative Commons Attribution Noncommercial (CC BY-NC 3.0) license from "Confirmation and Reduction: A Bayesian Account" by Foad Dijadzi-Bahmani, Roman Frigg and Stephan Hartmann, *Synthese*, Volume 179, No. 2, pp. 321–338. © 2010, doi: 10.1007/s11229-010-9775-6.

$$\begin{aligned} p_F &:= p(E_F \mid T_F), & q_F &:= p(E_F \mid \neg T_F), \\ p_P &:= p(E_P \mid T_P), & q_P &:= p(E_P \mid \neg T_P), \\ \alpha &:= p(E \mid T_F, T_P), & \beta &:= p(E \mid T_F, \neg T_P), \\ \gamma &:= p(E \mid \neg T_F, T_P), & \delta &:= p(E \mid \neg T_F, \neg T_P). \end{aligned} \tag{8.1}$$

These parameters cannot be freely chosen, as we assume that the following conditions hold: First, E_F confirms T_F, hence $p_F > q_F$. Second, E_P confirms T_P, hence $p_P > q_P$. Third, E confirms T_F, and fourth, E confirms T_P. The last two conditions entail the following constraints on α, β, γ and δ (all proofs are in the Appendix):

$$(\alpha - \beta)t_F + (\gamma - \delta)\bar{t}_F > 0, \tag{8.2}$$

$$(\alpha - \gamma)t_P + (\beta - \delta)\bar{t}_P > 0. \tag{8.3}$$

These inequalities hold, for example, if $\alpha > \beta, \gamma > \delta$, which seems to be a natural condition. One may also want to require that $p(T_F \mid E, E_F) > p(T_F)$ and $p(T_P \mid E, E_P) > p(T_P)$. Note, however, that both inequalities follow from the above four conditions (proof omitted).

Given this network structure and the conditional independences encoded in it, it is easy to see, for example, that the variable E_F is independent of T_P given T_F and that E_P is independent of T_F given T_P. In symbols:

$$E_F \perp\!\!\!\perp T_P \mid T_F, \qquad E_P \perp\!\!\!\perp T_F \mid T_P. \tag{8.4}$$

Hence, E_F does not confirm (or disconfirm) T_P, and E_P does not confirm (or disconfirm) T_F:

$$p(T_P \mid E_F) = p(T_P), \qquad p(T_F \mid E_P) = p(T_F).$$

We conclude that there is no flow of confirmation from one theory to the other. The intuitive reason for this is that there is no chain of arrows from E_F to T_P. Note also that the variables T_F and T_P are probabilistically independent before the reduction:

$$p(\mathrm{T_F}, \mathrm{T_P}) = p(\mathrm{T_F})\, p(\mathrm{T_P}) = t_F t_P. \tag{8.5}$$

All this, however, may look different in practice. The two theories could be more intimately connected: on the level of the mathematical formalism, or by some evidence E that supports both theories. In this case, scientists will attempt to reduce one theory to the other. Let us now model this situation.

8.2.2 After the Reduction

Recall the three steps involved in reducing one theory to another, set out in Section 8.1: First, derive $\mathrm{T_F^*}$ from the auxiliary assumptions and $\mathrm{T_F}$. Second, introduce bridge laws and obtain $\mathrm{T_P^*}$ from $\mathrm{T_F^*}$. Third, show that $\mathrm{T_P^*}$ is strongly analogous to $\mathrm{T_P}$.

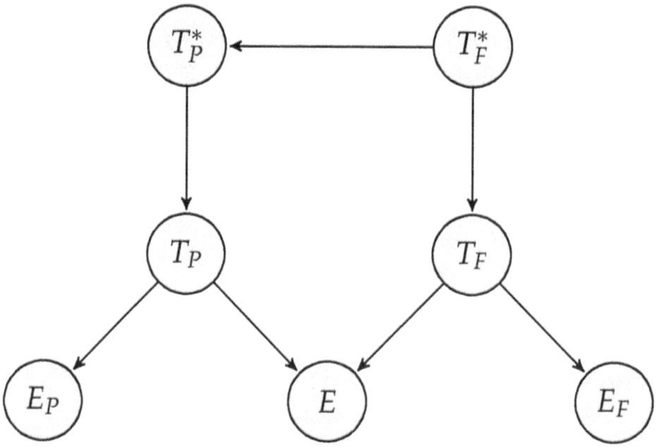

Figure 8.3: The Bayesian network representation of the situation *after* an intertheoretic reduction. Adapted under the terms of the Creative Commons Attribution Noncommercial (CC BY-NC 3.0) license from "Confirmation and Reduction: A Bayesian Account" by Foad Dijadzi-Bahmani, Roman Frigg and Stephan Hartmann, *Synthese*, Volume 179, No. 2, pp. 321–338. © 2010, doi: 10.1007/s11229-010-9775-6.

The post-reduction relationship between $\mathrm{T_F}$ and $\mathrm{T_P}$ is depicted in the Bayesian network in Figure 8.3. Let us denote the post-reduction probabilities by the function p_R. Note that when we speak of prior and posterior probabilities in the remainder of this chapter, we do not refer to p and p_R but to $p^*(\cdot)$ and $p^*(\cdot\,|\mathrm{E})$, or to $p_R(\cdot)$ and $p_R(\cdot\,|\mathrm{E})$.

To complete the network, we specify the following conditional probabilities:

$$p^*_P := p_R(T_P\,|\,T^*_P), \qquad q^*_P := p_R(T_P\,|\,\neg T^*_P), \tag{8.6}$$

$$p^*_F := p_R(T^*_F\,|\,T_F), \qquad q^*_F := p_R(T_F\,|\,\neg T_F). \tag{8.7}$$

Note that Equation (8.6) replaces the second equation in the first line of Equation (8.1). We have to represent the bridge law in probabilistic terms, too. Naturally, we require

$$p_R(T^*_P\,|\,T^*_F) = 1 \quad \text{and} \quad p_R(T^*_P\,|\,\neg T^*_F) = 0. \tag{8.8}$$

All other probability assignments hold as in the case of Equation (8.1). Requiring this condition makes sure that we can later compare the two scenarios, that is, the situations before and after the reduction.

Three remarks about the steps in the reduction are in order. First, T^*_F may be more or less good. How good it is depends on the context (i.e., the application in question and the auxiliary assumptions made) and on the judgment of the scientists involved. In line with our Bayesian approach, we assume that the judgment of the scientists can be expressed in probabilistic terms. Second, the move from T^*_F to T^*_P in virtue of the bridge laws may be controversial amongst scientists. Whilst bridge laws are non-conventional factual claims, different scientists may assign different credences to them. Third, what counts as strongly analogous will also depend on the specific context and on the judgment of the scientists. For example, whether entropy fluctuations can be neglected or not cannot be decided independently of the specific problem at hand, see Callender 2001. All this fits well with our Bayesian account.

Note that, in the Bayesian network in Figure 8.3, there is now a direct sequence of arrows from T_F to T_P: the path through T^*_F to T^*_P. Hence we expect that E_F is now probabilistically relevant for T_P and that E_P is now probabilistically relevant for T_F. And this is indeed what we find: the independencies formulated in Equation (8.4) do not hold any more. We state our results in the following two theorems:

Theorem 8.1 (Confirmation Transmission, I). E_F *confirms* T_P *if and only if* $(p_F - q_F)(p^*_F - q^*_F)(p^*_P - q^*_P) > 0$.

This theorem entails that E_F confirms T_P if the following three conditions hold: (i) E_F confirms T_F (i.e., $p_F > q_F$), (ii) T_F confirms T^*_F (i.e., $p^*_F > q^*_F$), and (iii) T_P^* confirms T_P (i.e., $p_P^* > q_P^*$). These conditions are immediately plausible. Condition (i) was assumed from the beginning, and Conditions (ii) and (iii) make sure that there is a positive flow of confirmation from T_F to $T_F^* \equiv T_P^*$ (*qua* bridge law) and from T_P^* to T_P.

Theorem 8.2 (Confirmation Transmission, II). E_P *confirms* T_F *if and only if* $(p_P - q_P)(p_F^* - q_F^*)(p_P^* - q_P^*) > 0$.

This theorem is analogous to the previous theorem. It entails that E_P confirms T_F if the following three conditions hold: (i) E_P confirms T_P (i.e., $p_P > q_P$), (ii) T_F confirms T_F^* (i.e., $p_F^* > q_F^*$), and (iii) T_P^* confirms T_P (i.e., $p_P^* > q_P^*$).

In our representation, the bridge law states a *perfect correlation* between T_F^* and T_P^*. A bridge law is posited by scientists working in a particular field, and it may happen that not everybody in that community is convinced of it. Thus, different scientists may assign different credences to a particular bridge law. In cases where a lower probability is assigned to a bridge law, the reduction may still be epistemically valuable—the flow of confirmation will just be less. The exact amount of evidential support depends, of course, on the values of the relevant probabilities.

For future reference, let us calculate the prior probability of the conjunction of both theories. We obtain

$$p_R(T_F, T_P) = t_F(p_F^* p_P^* + \bar{p}_F^* q_P^*).$$

In a similar way, we may calculate the posterior probability of both theories given the total evidence, that is, the expression $p_R(T_F, T_P \mid E, E_F, E_P)$—see Section 8.4.

Finally, let us remark on the specific representation we have chosen in the Bayesian Network in Figure 8.3. Clearly, having a sequence of arrows from T_F to T_P ensures that confirmation can flow from one theory to the other. However, this sequence of arrows is not just driven by our wish to establish a flow of confirmation from the reducing to the reduced theory: it also makes scientific sense. First, T_F^* is an approximation of T_F. It follows from it and depends on it, hence the direction of the arrow. Second, we have drawn an arrow from T_F^* to T_P^* although the propositional variables in question are, *qua* the bridge law, interchangeable with each other. This is modeled by assigning appropriate conditional probabilities. The arrow could have also been drawn from T_P^* to T_F^*. In this case we would have to demand $p(T_F^* \mid T_P^*) = 1$ and $p(T_F^* \mid \neg T_P^*) = 0$. These conditions are, however, equivalent to Equations (8.8) for non-extreme priors. Third, it may look strange that we have drawn an arrow from T_P^* to T_P to model the relation of strong analogy as a symmetrical relation. We would like to reply to this objection that, then, "analogy" is perhaps not the right word, as T_P^* is indeed stronger than T_P, and so it makes sense to draw an arrow from T_P^* to T_P. We conclude that the chain of arrows from T_F to T_P is indeed plausible.

8.3 Why Accept a Purported Reduction?

Under what conditions should we accept a proposed reduction? More specifically, given everything we know about the domains of the two theories, when should we accept a proposed reduction and when should we reject it? In the Bayesian framework, theories are accepted on the basis of their probabilities and their confirmation track record. But which probabilities are relevant? The previous section focused on the probabilities of T_F and T_P individually. But perhaps one is interested in the "package" as a whole, that is, the *conjunction* of T_F and T_P. If so, should we look at the *prior* probability of the conjunction of T_F and T_P after the reduction (that is, without accounting for the total evidence)? Or at the *posterior* probability of the conjunction of T_F and T_P, that is, the probability of T_F and T_P given the total evidence (i.e., E, E_F and E_P)? We examine these proposals in turn.

Let us first compare the prior probabilities of the conjunction of T_F and T_P before and after the reduction. Before the reduction, the two theories are independent, as expressed in Equation (8.5). For convenience, let us restate the condition:

$$p(T_F, T_P) = t_F t_P. \tag{8.9}$$

We calculate the prior probability of the conjunction of T_F and T_P after the reduction and obtain

$$p_R(T_F, T_P) = t_F\left(p_F^* p_P^* + \bar{p}_F^* q_P^*\right). \tag{8.10}$$

While the expression in Equation (8.9) is an explicit function of t_P, the expression in Equation (8.10) is not. This is because, after the reduction, T_P is no longer a root node, and so it is not assigned a prior probability. In order to meaningfully compare the situations before and after the reduction, we have to level the playing field. In other words, we should not only assume that $p_R(E_P\,|\,T_P) = p(E_P\,|\,T_P)$, and so on, but also that $p_R(T_P) = p(T_P)$. Let us therefore calculate:

$$\tilde{t}_P := p_R(T_P) = t_F^* p_P^* + \bar{t}_F^* q_P^* \tag{8.11}$$

with

$$t_F^* = p_R(T_F^*) = p_R(T_P^*) = p_F^* t_F + q_F^* \bar{t}_F. \tag{8.12}$$

Alternatively, we have:

$$\tilde{t}_P := (p_F^* p_P^* + \bar{p}_F^* q_P^*) t_F + (q_F^* p_P^* + \bar{q}_F^* q_P^*) \bar{t}_F. \tag{8.13}$$

This equation follows if we insert Equation (8.12) into Equation (8.11) or by direct calculation from the Bayesian network depicted in Figure 8.3. We now require $p_R(T_P) = p(T_P)$, i.e.,

$$t_P = \tilde{t}_P, \tag{8.14}$$

and replace t_P in Equation (8.9) by the expression for $\tilde{t}_P$ given in Equation (8.13).

Now we calculate the difference between the joint probabilities of T_F and T_P before and after the reduction:

$$\Delta_0 := p_R(T_F, T_P) - p(T_F, T_P) \tag{8.15}$$

For this expression, we obtain the following result:

Theorem 8.3 (Independence Preservation).

$$\Delta_0 = (p_F^* - q_F^*)(p_P^* - q_P^*) t_F \bar{t}_F, \tag{8.16}$$

and $\Delta_0 = 0$ *if and only if* $p_F^* = q_F^*$ *or* $p_P^* = q_P^*$. *Analogously,* $\Delta_0 > 0$ *if* $p_F^* > q_F^*$ *and* $p_P^* > q_P^*$.

The first part of the theorem says that if either T_F and T_F^* are independent or T_P^* and T_P are independent, then T_F and T_P remain independent after the reduction. The second part says that the conjunction of T_F and T_P is more likely after the reduction if T_F confirms T_F^* and if T_P^* confirms T_P.

Next, let us compare the posterior probabilities of the conjunction of T_F and T_P before and after the reduction. To do so, we calculate the difference

$$\Delta_1 := p_R(T_F, T_P \mid E, E_F, E_P) - p(T_F, T_P \mid E, E_F, E_P)$$

and obtain:

$$\Delta_1 = (p_F^* - q_F^*)(p_P^* - q_P^*) t_F \bar{t}_F \cdot \alpha \tilde{\Delta}_1. \tag{8.17}$$

The explicit expression for $\tilde{\Delta}_1$ is given in the Appendix. Equation (8.17) then entails the following theorem:

Theorem 8.4 (Joint Posterior Probabilities, I). $\Delta_1 = 0$ *if* $p_F^* = q_F^*$ *or* $p_P^* = q_P^*$. *That is, the posterior probability of* $T_F \wedge T_P$ *equals the prior probability if one of the two equalities above is satisfied.*

This result has an intuitive interpretation: If either T_F and T_F^* or T_P and T_P^* are independent, then the flow of confirmation from T_F to T_P (and vice versa) is stopped and the epistemic situations before and after the reduction are the same. We can then infer:

Theorem 8.5 (Joint Posterior Probabilities, II). $\Delta_1 > 0$ *if the following three conditions hold: (i)* $\beta, \gamma > \delta$, *(ii)* $0 < x_F, x_P < 1$ *and (iii)* $(p_F^* - q_F^*)(p_P^* - q_P^*) > 0$. *That is, the posterior probability of* $T_F \wedge T_P$ *exceeds the prior probability if the above inequalities are satisfied.*

Condition (i) seems natural in the light of inequalities (8.2) and (8.3). In fact, it is a rather weak condition which also holds, for example, for Set 2 below. Condition (ii) makes sure that E_F confirms T_F and E_P confirms T_P; we have assumed this throughout. Condition (iii) is our usual condition on the dependency between T_P and T_P^* as well as that between T_F and T_F^*. Hence none of these conditions is in any way problematic. Given the total relevant evidence, the joint probability of the fundamental theory T_F and the phenomenological theory T_P is higher *ceteris paribus* when a reductive relationship holds between them.

8.4 Discussion

The Generalized Nagel–Schaffner model of reduction (GNS) is a template for describing the conceptual and epistemic impact of intertheoretic reductions. Embedding GNS into the Bayesian framework, we have formulated criteria that help us assess proposed reductions epistemically, and we have shown how a reduction facilitates the flow of confirmation from the reducing theory to the reduced theory and back.

A GNS reduction between two theories, such as thermodynamics and statistical mechanics, is epistemically advantageous in virtue of our main results, Theorems 8.1 and 8.2. Specifically, we have shown that a reduction ensures that evidence which prior to reduction supported only one of the theories comes to support the other theory as well. Moreover, a successful reduction increases both the prior and the posterior probability of the conjunction of both theories (Theorem 8.5).

Our Bayesian account also shows to what extent the various judgments depend on the probabilistic judgments of the scientists, connecting—or so we argue—our account to scientific practice. Disagreement about the epistemic value of a reduction can be traced back to disagreement about the assignment of the relevant prior probabilities. This need not be a disagreement about exact numbers and may also take the form of qualitative (e.g., ordinal) plausibility judgments.

As usual, we finish the Variation with a series of proposals for follow-up projects.

First, one might propose that a possible reduction be accepted if the conjunction of T_F and T_P is better confirmed by the evidence after the reduction, compared to the situation before the reduction. Determining whether this is the case requires an analysis in terms of *degree of confirmation*. That is, one has to choose one of the various measures of confirmation as increase in firmness (see Variation 1). Dizadji-Bahmani, Frigg and Hartmann (2011) conduct such an analysis for the difference measure $d(H, E)$ and come to the conclusion that the degree of confirmation is usually greater if a reduction has taken place than if not. Several other confirmation measures have to be checked and the stability of these results has to be explored.

The previous observation suggests that strong coherence between the fundamental and the phenomenological theory may be confirmation-conducive (Dietrich and Moretti 2005; Moretti 2007). So, second, it would be interesting to compare degrees of coherence before and after an intertheoretic reduction has taken place (Bovens and Hartmann 2003). Here, one might want to focus on the two theories in question, or on the conjunction of the theories and all available evidence. It might be reasonable to focus on the latter, as the evidence is also uncertain and one might, in the end, be interested in the coherence of the entire package, comprising all available theories and all available evidence. Should coherence considerations play a role when it comes to deciding whether a theory should be accepted?

Third, one may want to examine the situation where evidence for, say, the fundamental theory *disconfirms* the phenomenological theory. How ought one to assess the value of a reduction in these situations?

Fourth, our Bayesian models open a path for studying other types of intertheoretic dependence. Plausible topics are singular limit relations between high- and low-level theories (Batterman 2002), or qualitative "stories" that establish an explanatory link between a fundamental theory and an idealized model capturing specific aspects of that theory (e.g., models in quantum chromodynamics, Hartmann 1999). For making such a project succeed, we imagine that formally minded philosophers who build Bayesian models join forces with philosophers studying specific scientific disciplines (e.g., physics, biology, economics) who calibrate those models with detailed case studies.

Finally, our work on intertheoretic reduction is also related to debates about the (dis-)unity of science. The Bayesian models indicate how one can argue for specific hypotheses about the structure of science: for example, whether it consists of several disconnected clusters of theories and models, or whether it has a tight and integrated network structure.

Appendix: Proofs of the Theorems

Theorem 8.1, preliminary calculations: Let us start with the situation before the reduction and the Bayesian network represented in Figure 8.2. The joint distribution $p(T_F, T_P, E, E_F, E_P)$ is given by the expression

$$p(T_F)\, p(T_P)\, p(E\,|\,T_F, T_P)\, p(E_F|\,T_F)\, p(E_P|\,E_P).$$

Using the methodology described in Bovens and Hartmann 2003, Chapter 3, we obtain:

$$p(\mathrm{T_F}, \mathrm{E}) \;=\; \sum_{T_P, E_F, E_P} p(\mathrm{T_F}, T_P, \mathrm{E}, E_F, E_P) \;=\; t_F(t_P\alpha + \bar{t}_P\beta). \tag{8.18}$$

Similarly, we calculate

$$p(\mathrm{T_P}, \mathrm{E}) \;=\; t_P(t_F\alpha + \bar{t}_F\gamma), \tag{8.19}$$

$$p(\mathrm{E}) \;=\; t_F(t_P\alpha + \bar{t}_P\beta) + \bar{t}_F(t_P\gamma + \bar{t}_P\delta) \tag{8.20}$$

$$\;=\; t_P(t_F\alpha + \bar{t}_F\gamma) + \bar{t}_P(t_F\beta + \bar{t}_F\delta). \tag{8.21}$$

To prove Equation (8.2), we note, using the definition of conditional probability, that $p(\mathrm{T_P}|\mathrm{E}) > p(\mathrm{T_P})$ iff $p(\mathrm{T_P}, \mathrm{E}) - p(\mathrm{T_P})\, p(\mathrm{E}) > 0$. Using Equations (8.19) and (8.21), we obtain

$$p(\mathrm{T_P}, \mathrm{E}) - p(\mathrm{T_P})\, p(\mathrm{E}) \;=\; t_P\bar{t}_P\Big((\alpha-\beta)t_F + (\gamma-\delta)\bar{t}_F\Big), \tag{8.22}$$

from which Equation (8.2) immediately follows. The proof of Equation (8.3) proceeds accordingly using Equations (8.19) and (8.20).

Next, we calculate the prior probability of the two theories:

$$p(\mathrm{T_F}, \mathrm{T_P}) \;=\; \sum_{E, E_F, E_P} p(\mathrm{T_F}, \mathrm{T_P}, E, E_F, E_P) \;=\; p(\mathrm{T_F})\, p(\mathrm{T_P}) \;=\; t_F t_P.$$

Similarly, we obtain for the posterior probability $P_1^* := p(\mathrm{T_F}, \mathrm{T_P}\,|\,\mathrm{E}, \mathrm{E_F}, \mathrm{E_P})$:

$$\begin{aligned} P_1^* &= \frac{p(\mathrm{T_F}, \mathrm{T_P}, \mathrm{E}, \mathrm{E_F}, \mathrm{E_P})}{p(\mathrm{E}, \mathrm{E_F}, \mathrm{E_P})} \\ &= \frac{t_F t_P p_F p_P \alpha}{t_F t_P p_F p_P \alpha + t_F \bar{t}_P p_F q_P \beta + \bar{t}_F t_P q_F p_P \gamma + \bar{t}_F \bar{t}_P q_F q_P \delta} \\ &= \frac{t_F t_P \alpha}{t_F(t_P\alpha + \bar{t}_P x_P \beta) + \bar{t}_F x_F (t_P\gamma + \bar{t}_P x_P \delta)}, \end{aligned} \tag{8.23}$$

with the probability ratios $x_F := q_F/p_F$ and $x_P := q_P/p_P$. □

Proof of Theorem 8.1: Let us now turn to the situation after the reduction and the Bayesian network represented in Figure 8.3. The joint distribution $p_R(T_F, T_P, T_F^*, T_P^*, E, E_F, E_P)$ is given by

$$p_R(T_F)\, p_R(E \mid T_F, T_P)\, p_R(E_F \mid T_F)\, p_R(E_P \mid E_P)\, p_R(T_P \mid T_P^*)\, p_R(T_P^* \mid T_F^*)\, p_R(T_F^* \mid T_F).$$

To simplify our notation, we introduce the following abbreviations:

$$\varphi_\alpha := p_F^* p_P^* + \bar{p}_F^* q_P^*, \qquad \varphi_\beta := p_F^* \bar{p}_P^* + \bar{p}_F^* \bar{q}_P^*,$$
$$\varphi_\gamma := q_F^* p_P^* + \bar{q}_F^* q_P^*, \qquad \varphi_\delta := q_F^* \bar{p}_P^* + \bar{q}_F^* \bar{q}_P^*.$$

For later use, we note that $0 < \varphi_\alpha, \varphi_\beta, \varphi_\gamma, \varphi_\delta < 1$ and

$$\varphi_\alpha - \varphi_\gamma = \varphi_\delta - \varphi_\beta = (p_F^* - q_F^*)(p_P^* - q_P^*), \tag{8.24}$$
$$\varphi_\alpha + \varphi_\beta = \varphi_\gamma + \varphi_\delta = 1. \tag{8.25}$$

We then obtain for the prior probability of the conjunction of both theories after the reduction:

$$p_R(\mathrm{T_F}, \mathrm{T_P}) = t_F \varphi_\alpha. \tag{8.26}$$

For the posterior $P_2^* := p_R(\mathrm{T_F}, \mathrm{T_P} \mid \mathrm{E}, \mathrm{E_F}, \mathrm{E_P})$, we obtain:

$$P_2^* = \frac{t_F \alpha \varphi_\alpha}{t_F(\alpha \varphi_\alpha + x_P \beta \varphi_\beta) + \bar{t}_F x_F(\gamma \varphi_\gamma + x_P \delta \varphi_\delta)}. \tag{8.27}$$

Similarly, we calculate

$$p_R(\mathrm{T_P}) = t_F \varphi_\alpha + \bar{t}_F \varphi_\gamma, \tag{8.28}$$
$$p_R(\mathrm{T_P} \mid \mathrm{E_F}) = \frac{t_F \varphi_\alpha + \bar{t}_F x_F \varphi_\gamma}{t_F + \bar{t}_F x_F}, \tag{8.29}$$
$$p_R(\mathrm{T_F} \mid \mathrm{E_P}) = \frac{t_F(\varphi_\alpha + x_P \varphi_\beta)}{t_F(\varphi_\alpha + x_P \varphi_\beta) + \bar{t}_F(\varphi_\gamma + x_P \varphi_\delta)}. \tag{8.30}$$

We now calculate

$$\begin{aligned} p_R(\mathrm{T_P} \mid \mathrm{E_F}) - p_R(\mathrm{T_P}) &= \frac{t_F \bar{t}_F (\varphi_\alpha - \varphi_\gamma)(1 - x_F)}{t_F + \bar{t}_F x_F} \\ &= \frac{t_F \bar{t}_F (p_F - q_F)(p_F^* - q_F^*)(p_P^* - q_P^*)}{p_F(t_F + \bar{t}_F x_F)}. \end{aligned}$$

This proves Theorem 8.1, since the sign of the above expression is decided by the sign of

$$(p_F - q_F)(p_F^* - q_F^*)(p_P^* - q_P^*).$$

□

Proof of Theorem 8.2: This theorem is the mirror result of Theorem 8.1, and also the proof runs analogously. Following the same steps as above, we calculate

$$\begin{aligned} p_R(T_F|E_P) - p_R(T_F) &= \frac{t_F\bar{t}_F(\varphi_\alpha - \varphi_\gamma)(1-x_P)}{t_F(\varphi_\alpha + x_P\varphi_\beta) + \bar{t}_F(\varphi_\gamma + x_P\varphi_\delta)} \\ &= \frac{t_F\bar{t}_F(p_P - q_P)(p_F^* - q_F^*)(p_P^* - q_P^*)}{p_P\Big(t_F(\varphi_\alpha + x_P\varphi_\beta) + \bar{t}_F(\varphi_\gamma + x_P\varphi_\delta)\Big)}, \end{aligned}$$

and again, we discover that the sign of this expression corresponds to the sign of

$$(p_P - q_P)(p_F^* - q_F^*)(p_P^* - q_P^*),$$

as requested by the theorem. □

Proof of Theorem 8.3: To prove Equation (8.16), we note that, by Equation (8.26),

$$\Delta_0 = (\varphi_\alpha - t_P)t_F.$$

We now use Equations (8.14) and (8.28) and obtain

$$\Delta_0 = (\varphi_\alpha - t_F\varphi_\alpha - \bar{t}_F\varphi_\gamma)t_F = (\varphi_\alpha - \varphi_\gamma)t_F\bar{t}_F.$$

Equation (8.16) then follows using Equation (8.24). The conditions on the sign of Δ_0 stated in Theorem 8.3 immediately follow from Equation (8.16). □

Proof of Theorem 8.4 and 8.5: Both theorems concern the same quantity — Δ_1—so it is convenient to prove them together. From Equations (8.23) and (8.27), we conclude that we can write Δ_1 as follows:

$$\Delta_1 = (\varphi_\alpha - \varphi_\gamma)t_F\bar{t}_F \cdot \alpha\tilde{\Delta}_1,$$

with

$$\tilde{\Delta}_1 = N_1^{-1}N_2^{-1} \cdot \tilde{\Delta}_1'$$

and

$$\begin{aligned} N_1 &= t_F(t_P\alpha + \bar{t}_P x_P\beta) + \bar{t}_F x_F(t_P\gamma + \bar{t}_P x_P\delta), \\ N_2 &= t_F(\alpha\varphi_\alpha + x_P\beta\varphi_\beta) + \bar{t}_F x_F(\gamma\varphi_\gamma + x_P\delta\varphi_\delta). \end{aligned}$$

Theorem 8.4 follows by observing that $\Delta_1 = 0$ if $\varphi_\alpha - \varphi_\gamma = 0$. By Equation (8.24), this will be the case if either $p_F^* = q_F^*$ or $p_P^* = q_P^*$. This concludes the proof of Theorem 8.4.

For proving Theorem 8.5, note first that $N_1, N_2 > 0$. We are therefore most interested in $\tilde{\Delta}'_1$, which is given by

$$\tilde{\Delta}'_1 \;=\; t_{\mathrm{F}} x_{\mathrm{F}} (\varphi_\alpha - \varphi_\gamma)(\gamma - \delta x_{\mathrm{P}}) \;+\; t_{\mathrm{F}} x_{\mathrm{P}} (\beta - \delta x_{\mathrm{F}}) \;+\; \gamma \varphi_\gamma x_{\mathrm{F}} \;+\; \delta \bar{\varphi}_\gamma x_{\mathrm{F}} x_{\mathrm{P}}.$$

From Conditions (i) and (ii) of Theorem 8.5, we conclude that $\gamma > \delta x_{\mathrm{P}}$ and $\beta > \delta x_{\mathrm{F}}$. Hence $\tilde{\Delta}'_1 > 0$, and Theorem 8.5 follows from the same considerations about the sign of $\varphi_\alpha - \varphi_\gamma$ that we have made above, when proving Theorem 8.4. □

Variation 9: Hypothesis Tests and Corroboration

Scientific reasoning often proceeds by coming up with innovative hypotheses, subjecting them to experimental testing, and appraising how well they have stood up to the test. For Critical Rationalists such as Karl R. Popper (1959/2002), the critical attitude that we express by repeatedly testing our best scientific theories constitutes the basis of rational inquiry about the world. Arguably, such tests have already been conducted in antiquity: think of Eratosthenes' test of different hypotheses about the circumference of the Earth, conducted by comparing the height of the sun in two different places at the same time. However, the design and interpretation of hypothesis tests was formalized very late: in the first half of the twentieth century. The emergence of statistics—the science of analyzing and interpreting data—played a crucial role in this process. It provided science with probabilistic tests, above all **null hypothesis significance tests (NHST),** which have acquired a predominant role in scientific reasoning. NHST belong to the toolbox of **frequentist statistics,** where inferential procedures such as point estimation, confidence intervals and hypothesis tests are justified by their favorable long-run properties.

NHST are based on testing a precise hypothesis H_0—the "null" or default hypothesis—against an unspecific alternative H_1. Usually the null denotes the absence of a causal effect in an experimental manipulation, for example, that a medical drug is no better than a placebo treatment. The rationale of NHST is negative: before inferring to the presence of an effect, or taking it for granted in practical decisions we make (e.g., prescribing the drug to a patient), we need to have strong evidence against the null hypothesis that an effect is absent. NHST are applied across all domains of science, but are especially prominent in psychology and medicine.

Typically, the null hypothesis is not considered a candidate for literal truth. After all, most experimental manipulations have *some* minuscule (positive or negative) effect, though it may be practically meaningless. A diet cure that reduces weight on average by half a kilo has a positive effect, but is unlikely to be followed in practice. NHST aim at finding out whether H_0 is a *good proxy* for the general statistical model, or whether we should replace it by a specific alternative postulating an effect.

Most null hypotheses H_0: $\mu = \mu_0$ postulate that an unknown parameter μ, such as the mean of a Normal (i.e., Gaussian) distribution $N(\mu, \sigma^2)$ takes a precise value μ_0. By contrast, the alternative H_1: $\mu \neq \mu_0$ corresponds to the logical disjunction of all other precise hypotheses about μ (e.g., Neyman and Pearson 1933, 1967; Fisher 1956).

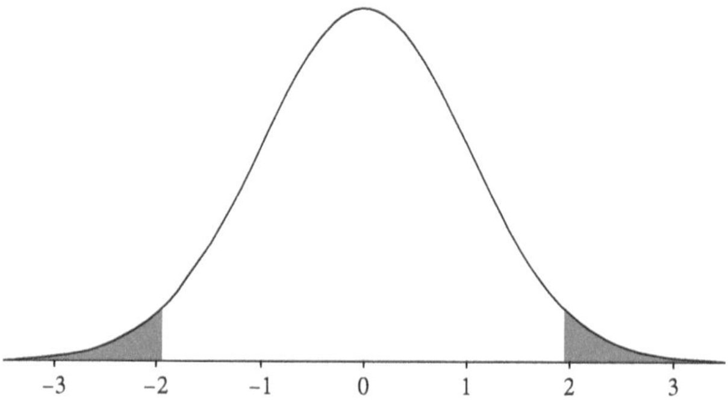

Figure 9.1: Visualization of the p-value $p = .05$ for testing the mean of a Normal distribution, with H_0: $\mu = 0$ and H_1: $\mu \neq 0$. The curve denotes the probability density function of H_0. The p-value corresponds to the size of the shaded area, divided by the size of the total area under the graph of the probability density function.

NHST quantify evidence against the null hypothesis by means of ***p*-values:** the probability of obtaining a result that diverges from the null hypothesis at least as much as the actual data (e.g., Fisher 1956, 38–43). The p-value is the smaller, the more the result falls into one of the tails of the distribution. In general, for a random variable X with realization x and a function $z(x)$ that measures divergence from the null hypothesis (e.g., the difference between the mean under H_0 and the actual sample mean), the p-value can be defined as

$$p := p_{H_0}\Big(\big|z(X)\big| \geq \big|z(x)\big|\Big). \tag{9.1}$$

Another name for p-values is "observed significance levels". Conventionally, p-values smaller than .05 count as "statistically significant evidence" against the null hypothesis, p-values smaller than .01 count as "highly significant evidence", and so on. Since the null hypothesis could explain such low p-values only by reference to a very unlikely event, and since chance is not accepted as a good explanation in scientific reasoning, NHST count a statistically significant p-value as evidence for the alternative hypothesis H_1—the presence of a causal effect.

Notwithstanding the popularity of NHST in scientific inference, they are a highly contested method. The rationale behind NHST has often been argued to be flawed, and their use in scientific practice has promoted various misunderstandings such as the base rate fallacy (e.g., Hacking 1965; Spielman 1974; Cohen 1994; Royall 1997; Ziliak and McCloskey 2008; Romeijn 2014). Partly this is because modern NHST are a hybrid of two different schools of statistical inference: the behavioral approach by Jerzy Neyman and Egon Pearson, which focuses on the control of error in inference (Neyman and Pearson 1933, 1967; Mayo 1996, 2018), and R. A. Fisher's idea of judging hypotheses solely on the basis of their fit with the data (Fisher 1935/74, 1956).

NHST and p-values are also problematic from a practical perspective. In the behavioral sciences, systematic replications of published experiments often yield substantially lower effect sizes and fail to be statistically significant (Ioannidis 2005b; Open Science Collaboration 2015; Camerer et al. 2016). Similar findings have been obtained in cancer biology (Nosek and Errington 2017). From the point of view of meta-scientists such as John Ioannidis, NHST contribute to this **replication crisis** by promoting **publication bias** and the "file drawer effect" (Rosenthal 1979; Bakker, Wicherts and van Dijk 2012): since only statistically significant results are perceived as interesting and publishable, a lot of valuable research supporting the null hypothesis ends up in the proverbial file drawer and is not disseminated to the scientific community.

In particular, there is barely any methodological guidance on **how we should interpret a non-significant result,** that is, a result where we fail to reject the null. p-values greater than .05 do not have an evidential interpretation, or license a judgment of support in favor of the null hypothesis—even when they are close to unity. This feature is not a bug of NHST, but hard-wired into their methodology:

> [I]t should be noted that the null hypothesis is never proved or established, but is possibly disproved, in the course of experimentation. Every experiment may be said to exist only in

> order to give the facts a chance of disproving the null hypothesis. (Fisher 1935/74, 16)

Encyclopedias and statistics textbooks (e. g., Chase and Brown 2000; Wasserman 2004) mirror this attitude when they classify p-values greater than .05 by phrases such as "the evidence is insufficient to support a conclusion" (Wikipedia), or by analogies to legal reasoning: A verdict of 'not guilty' does not imply that the defendant is innocent. He may in effect be guilty; the verdict only means that the forensic evidence is not strong enough to justify a conviction. Similarly, non-significant p-values do not imply support for the null hypothesis.

However, not all non-significant outcomes are evidentially on a par. Consider a Binomial model where we are testing the hypothesis that a coin is fair (H_0: $\mu = 1/2$) and toss the coin a hundred times. Neither $x = 52$ nor $x = 58$ qualifies as significant evidence against H_0 at the $p = 0.05$ level, but there is certainly a difference in the performance of the null hypothesis on both data sets. While the former observation aligns well with the null hypothesis, the latter could be interpreted as moderate evidence against it. The classical NHST methodology, which refuses to interpret p-values greater than 0.05, fails to quantify this difference. From that standpoint, p-values of .13 and .76 (i.e., the values from the above Binomial example with $x = 58$ and $x = 52$) have the same practical meaning: they are above the range where results are statistically significant, and therefore neither evidence for nor against the null.

Since NHST cannot express evidence in favor of the null, they contribute to publication bias: when only statistically significant results with p-values smaller than .05 are evidentially interpretable, it is not easy to relate bigger p-values to general scientific conclusions and to package the experimental results into a convincing narrative. Such experiments are therefore more likely to end up in the proverbial file drawer and not to be shared with the scientific community, although their methodological quality is not inferior to experiments with statistically significant outcomes. This effect is reinforced by the common tendency to identify statistically significant results with scientifically relevant results, which are often the ones that draw attention, spark novel hypothesis and get published in high-impact venues (Ziliak and McCloskey 2008; Francis 2014). As a consequence, effect sizes in published research tend to be inflated and null hypotheses are underestimated with respect to their empirical support.

The indifference of NHST toward quantifying support in favor of the null also neglects that null hypotheses are, due to their precision, predictive value and high testability, often of significant theoretical and practical importance (Gallistel 2009; Morey et al. 2014). They may express the independence of two factors in a causal mechanism, postulate that performance difference between two groups is due to chance, or claim that a generic medical drug is equally effective as the originally patented drug. Due to their salience in theoretical inference and decision-making, it is imperative that a conceptual framework for scientific hypothesis testing be able to quantify evidence in their favor.

This means that we require an answer to the question whether the observed results **corroborate** the null hypothesis. Did it stand up to the test? Should we prefer it to the alternative? These questions were already discussed by Karl R. Popper in his writings on hypothesis testing and scientific method. It is useful to recall Popper's approach:

> By the **degree of corroboration** of a theory I mean a concise report evaluating the state (at a certain time t) of the critical discussion of a theory, with respect to the way it solves its problems; its degree of testability; the severity of tests it has undergone; and the way it has stood up to these tests. Corroboration (or degree of corroboration) is thus an evaluating report of past performance. Like preference, it is essentially comparative. (Popper 1979, 18, our emphasis)

We would like to highlight some of the characteristics that Popper points out. First, corroboration judgments appraise the performance of the null hypothesis in a severe test, rather than just stating a failure to find significant evidence against it. Second, high degrees of corroboration need not guide us to truth (Popper 1979, 21). Instead, the function of corroboration is comparative and pragmatic: it guides our practical preferences over competing hypotheses, for example, the choice of the hypothesis on which we base the next experiment (Popper 1959/2002, 416). This is exactly what most scientists are after when testing a complex set of hypotheses. More generally, in Popper's perspective, science proceeds through the successive elimination of hypotheses, and degrees of corroboration guide practical preferences over the competing hypotheses. Notably, corroboration does not inspire any *confidence* in the tested hypothesis: it is just the statement that a hypothesis has survived severe tests.

This Variation, which partly builds on results from Sprenger 2018d, explores the prospects for a corroboration-based epistemology of hypothesis tests. We first discuss attempts to reduce degree of corroboration to Bayesian degree of confirmation (Section 9.1). Then, we describe Popper's own explication of corroboration (Section 9.2) and address the more general question of whether testability and past performance may be synthesized into a statistical-relevance measure of corroboration (Section 9.3). The answer, which we derive using the axiomatic methods from Variations 1, 6 and 7, is negative: no such measure can simultaneously satisfy a certain set of desirable constraints. This result, recently published in Sprenger 2018d and reproduced in this Variation, seems to create insurmountable problems for Bayesian explications of corroboration. However, they can be solved by moving to a different conceptual framework. We construct a measure of corroboration that fruitfully applies Popperian thinking to hypothesis tests and that can be understood as a generalization of Bayesian inference (Section 9.4). Finally, we compare this measure to p-values in NHST and standard Bayesian inference (Section 9.5). While the practical merits of the new corroboration measure are still to be evaluated, it demonstrates two important theoretical insights: First, non-significant results in NHST can be given a valid evidential interpretation. Second, Popper and Bayes may be less fierce opponents than the popular picture—and Popper himself—has it. Indeed, a large part of this Variation is devoted to expounding Popper's ideas and to making them consistent with a Bayesian perspective on scientific inference.

9.1 Confirmation versus Corroboration

The point of measuring corroboration is to quantify the extent to which a (null) hypothesis has stood up to attempts at refuting it. Degree of corroboration evaluates past performance. For deterministic hypotheses, corroborating evidence simply validates their predictions. The more specific the evidence, the more it corroborates the hypothesis. This rationale essentially corresponds to the hypothetico-deductive model of theory confirmation (e.g., Gemes 1998). While this model may be adequate as a qualitative theory of corroboration, it is not applicable to statistical hypotheses and NHST. Here, a different, quantitative model has to be developed (see also Popper 1959/2002, 265–266).

However, do we really need the concept of corroboration to interpret non-significant results in NHST? Can't we just describe them in terms of Bayesian confirmation? We saw in Variation 1 that evidence E confirms hypothesis H (in the sense of increase in firmness) if and only if $p(\mathrm{H}|\mathrm{E}) > p(\mathrm{H})$, that is, if E is statistically relevant to H. Or equivalently, $p(\mathrm{E}|\mathrm{H}) > p(\mathrm{E}|\neg\mathrm{H})$, that is,

E is more expected under H than under its negation ¬H. Before explicating a new and complex concept—corroboration—we should first argue why it is not coextensive with confirmation as increase in firmness.

In other words, we have to address the **Monist Thesis:** The epistemic function of the concept of corroboration can be taken over by the Bayesian concept of confirmation. The monist replaces a judgment of corroboration by a judgment of increase in firmness. The Monist Thesis can be aligned with authors such as Howson and Urbach (2006) and Lee and Wagenmakers (2014), who argue that NHST should be abandoned and be replaced by Bayesian hypothesis testing.

We shall now present three objections to the Monist Thesis. This does not rule out that explications of corroboration and confirmation may agree numerically; rather, the point is to show that the two *concepts* are not redundant and need different explication strategies.

> **Objection 1:** Corroboration does not aim at inferring probable hypotheses, or at raising our degree of belief in the tested hypothesis.

This objection contends that scientific hypotheses and models are idealizations of the external world, which are judged by their ability to capture relevant causal relations and to predict future events (see the survey in Frigg and Hartmann 2012). The epistemic function of corroboration consists in determining whether the data are consistent with the tested hypothesis, or whether the results agree "well enough" with the null hypothesis H_0 that we may use it as a proxy for a more general statistical model.

Consequently, corroborated hypotheses should not be regarded as true or empirically adequate—not even conditional on assuming the truth of the overarching statistical model. Instead, they are useful and tractable idealizations (Bernardo 2012; Gelman and Shalizi 2012, 2013)—see also Variation 12 for more detail on this point. Corroboration is a guide to practical preference over competing hypotheses; it does not ground confidence in the truth of the tested hypothesis (Popper 1959/2002, 281–282). To sharpen this point, note that Popper rejected the subjective interpretation of probability while defending the epistemic role of corroboration (op. cit., Chapter 8 and 10, Appendix vii).

Degree of confirmation, by contrast, is related to change of confidence in a hypothesis. Characteristically, all confirmation measures $c(H, E)$ possess the Final Incrementality Property from Variation 1:

$$c(H, E) > c(H, E') \qquad \text{if and only if} \qquad p(H|E) > p(H|E'). \qquad (9.2)$$

This condition demands that E confirms H more than E′ if and only if E raises the probability of H to a higher level than E′ does (Festa 2012; Crupi 2015). For a Popperian, however, corroboration should not measure (subjective) belief change, but (objective) past performance. Indeed, even if we did not have subjective degrees of belief in the tested hypothesis or were unable to elicit them, we should still be able to assess the past performance of the null hypothesis by a judgment of corroboration.

> **Objection 2:** On a Bayesian account, hypotheses with prior probability $p(\mathrm{H}) = 0$ cannot be confirmed. Yet they are perfectly acceptable candidates for being corroborated.

This point was first raised by Karl R. Popper (1959/2002, Appendix vii). As a consequence of Bayes' Theorem, any hypothesis with prior probability $p(\mathrm{H}) = 0$ also has posterior probability $p(\mathrm{H}|\mathrm{E}) = p(\mathrm{H})\,p(\mathrm{E}|\mathrm{H})/p(\mathrm{E}) = 0$. No such hypothesis can be confirmed in the sense of increase in firmness. But they certainly occur in scientific practice and we should be able to *corroborate* them: scientists often deal with an uncountable set of candidate hypotheses where each singleton hypothesis receives zero weight (e. g., different values of a physical parameter). Testing whether such hypotheses are good proxies for more general statistical models certainly makes sense. Confirmation as increase in firmness does not capture this practice, but corroboration does.

> **Objection 3:** The notion of corroboration exhibits more asymmetries than the notion of confirmation.

To appreciate this point, note that the logic of NHST is asymmetric: unlike the null hypothesis, the alternative is usually not a precise hypothesis. Remember our introductory example of testing the null hypothesis H_0: $\mu = \mu_0$ against the alternative H_1: $\mu \neq \mu_0$. NHST aim at finding out whether H_0 is a good proxy for the more general model represented by the alternative. Finding H_0 highly corroborated is a precise conclusion in its favor, whereas a "rejection" of the null leaves open which of the alternatives is corroborated. Confirmation judgments, however, are symmetric: disconfirmation of H amounts to confirmation of ¬H, up to the extent that confirmation measures satisfy symmetries such as $c(\neg\mathrm{H}, \mathrm{E}) = -c(\mathrm{H}, \mathrm{E})$ (Crupi, Tentori and González 2007). While confirmation measures $c(\mathrm{H}, \mathrm{E})$ pitch a hypothesis H against its negation ¬H, measures of corroboration $\zeta(\mathrm{H}_0, \mathrm{E})$ pitch a null hypothesis H_0 against a set of distinct alternatives. Section 9.4 will elaborate this idea in more detail.

These objections undermine the Monist Thesis sufficiently to motivate that the concept of corroboration stands in need of an independent explication and cannot be reduced to degree of confirmation. We begin by reviewing Popper's classical proposal for a measure of corroboration (see also Sprenger 2018d, 141–144).

9.2 Popper on Degree of Corroboration

Popper's first writings on degree of corroboration, in Chapter 10 of *The Logic of Scientific Discovery* (1934/2002), do not engage in a quantitative explication. Apparently this task is deferred to a scientist's common sense (see, e.g., Popper 1959/2002, 265–267). However, this move makes the entire concept of corroboration vulnerable to the charge of subjectivism: without a quantitative criterion, it is not clear which corroboration judgments are sound and which aren't (Good 1968b, 136). Especially if we aim at gaining objective knowledge from hypothesis tests, we need a precise explication of degree of corroboration.

Popper faces this challenge in a couple of articles in *The British Journal of Philosophy of Science* (Popper 1954, 1957, 1958) that form, together with a short introduction, Appendix ix of *The Logic of Scientific Discovery*. In these articles, Popper develops and defends a measure of degree of corroboration. Popper argues that this measure cannot correspond to a probability, that is, the plausibility of the tested theory (or hypothesis) conditional on the observed evidence:

> ... the probability of a statement ... simply does not express an appraisal of the severity of the tests a theory has passed, of the manner in which it has passed these tests. (Popper 1959/2002, 411)

In particular, logical content and informativity contribute to the testability of a theory and to its degree of corroboration:

> The main reason for this is that the *content* of a theory—which is the same as its *improbability*—determines its *testability* and *corroborability*. (ibid., original emphasis)

Recall that testability, identified with the empirical content or informativity of a hypothesis, is an essential cognitive value for Popper: being testable is a hallmark of science as opposed to pseudoscientific theories that can be reconciled with all kinds of empirical evidence. Popper's classical examples are psychoanalysis and Marxist economics. While pseudoscientific theories provide a lens to interpret the world rather than making statements about it (e.g., Marxists consider all economic developments as following the logic of class struggle), genuinely scientific theories make testable predictions and may be refuted empirically.

In Popper's characterization of corroboration testability is assigned a crucial role, too. Corroboration should be sensitive to the informativity and logical content of a theory, which is again related to its improbability. If one considers that degree of corroboration should guide our judgments of acceptance in NHST, this makes a lot of sense: good theories should agree with observed evidence and be informative (see the discussions in Hempel 1960; Levi 1963; Huber 2005b; Brössel 2013). Popper confirms that scientific theory assessment pursues both goals at once:

> Science does not aim, primarily, at high probabilities. It aims at a *high informative content*, well backed by experience. But a hypothesis may be very probable simply because it tells us nothing, or little. (Popper 1959/2002, 416, original emphasis)

Such a characterization of corroboration is attractive because it amalgamates two crucial cognitive values in theory assessment: high informative content and empirical confirmation. In NHST both values play a role, too, since a precise hypothesis (the null) is tested against a continuum of alternatives.

Let us now look at how Popper characterizes degree of corroboration. Transcribed into the framework of this book, hypothesis H and evidence E are instantiations of two variables H and E in a DAG G with a probability distribution p that is Markovian relative to G. Degree of corroboration $\zeta(\mathrm{H}, \mathrm{E})$ is then a function of the joint probability distribution of H and E. As before, we will omit reference to background assumptions and assume that they are implicit in the choice of the probability function p.

Note that such a probabilistic measure of corroboration does not capture all aspects of corroboration. Popper (1959/2002, 265–266, 402, 437) and also his modern followers (e.g., Rowbottom 2008, 2011) emphasize that corroborating evidence has to report the results of sincere and severe efforts to overturn the tested hypothesis. Obviously, such requirements cannot be formalized completely (see also Popper 1983, 154). We cannot infer reversely from high (probabilistic) degree of corroboration to a sound experimental design. The point of a probabilistic measure is rather to describe the degree of corroboration of a hypothesis if all important methodological requirements are met—as it is often the case in NHST.

Popper then specifies a set of adequacy criteria I–IX for degree of corroboration as a function of empirical performance.

I $$\left.\begin{array}{l}\zeta(H,E)>0\\ \zeta(H,E)=0\\ \zeta(H,E)<0\end{array}\right\}\text{ if and only if }\left\{\begin{array}{l}p(E|H)>p(E),\\ p(E|H)=p(E),\\ p(E|H)<p(E).\end{array}\right.$$

This is a classical statistical-relevance condition: E corroborates H just in case supposing H makes E more expected. This condition is also in line with Popper's remark that corroboration is, like preference, essentially contrastive (Popper 1979, 18).

II $-1 = \zeta(\neg H, H) \leq \zeta(H, E) \leq \zeta(H, H) \leq 1$.

III $\zeta(H, H) = 1 - p(H)$.

IV If $E \models H$ then $\zeta(H, E) = 1 - p(H)$.

V If $E \models \neg H$ then $\zeta(H, E) = -1$.

These conditions determine under which conditions corroboration measures take their extremal values. Minimal degree of corroboration is obtained if the evidence refutes the hypothesis (V). Conversely, the most corroborating piece of evidence E is a verification of H (II). In that case, degree of corroboration is equal to $1 - p(H)$, which expresses the informativity, testability and logical content of H (III, IV). This is especially plausible in Carnap's logical interpretation of probability, which Popper adopts for $p(H)$. But it also makes sense under a subjective Bayesian interpretation (see Popper 1959/2002, 268–269; Popper 1963, 385–387; Rowbottom 2012, 741–744).

VI $\zeta(H, E) \geq 0$ increases with the power of H to explain E.

VII If $p(H) = p(H') > 0$, then $\zeta(H, E) > \zeta(H', E')$ if and only if $p(H|E) > p(H'|E')$.

These conditions reiterate the statistical relevance rationale from Condition I, and make it more precise. Regarding Condition VI, Popper (1959/2002, 416) defines explanatory power according to the formula

$$\mathcal{E}(H, E) = (p(E|H) - p(E))/(p(E|H) + p(E)),$$

another measure of the statistical relevance between E and H. Condition VII states that corroboration essentially co-varies with posterior probability whenever two hypotheses are equiprobable at first. In that case, posterior probability is a good indicator of past performance.

VIII If $H \models E$, then

a) $\zeta(H, E) \geq 0$;

b) $\zeta(H, E)$ is an increasing function of $1 - p(E)$;

c) $\zeta(H, E)$ is an increasing function of $p(H)$.

IX If $\neg H$ is consistent and $\neg H \models E$, then

a) $\zeta(H, E) \leq 0$;

b) $\zeta(H, E)$ is an increasing function of $p(E)$;

c) $\zeta(H, E)$ is an increasing function of $p(H)$.

Condition VIII demands that corroboration gained from a successful deductive prediction co-vary with the informativity of the evidence and the prior probability of the hypothesis. Condition IX mirrors this requirement for the case $\neg H \models E$. These conditions can be motivated from the idea that if $H \models E$, then corroboration should not automatically transfer to hypotheses $H \wedge H'$ that contain an "irrelevant conjunct" H' which has not yet been tested. The same argument has been made for degree of confirmation in Variation 1 and explanatory power in Variation 7.

Popper (1954, 359) then proposes the corroboration measure $\zeta_P(H, E)$ which satisfies all of his constraints:

$$\zeta_P(H, E) = \frac{p(E|H) - p(E)}{p(E|H) + p(E) - p(E|H)\,p(H)}. \tag{9.3}$$

But we can easily see that one essential motivation behind a measure of degree of corroboration is not satisfied: $\zeta_P(H, E)$ is an increasing function of $p(H)$ for all values of $p(E|H)$ and $p(E)$. Hence the informativity and testability of a hypothesis, as measured by $1 - p(H)$, never contribute to its degree of corroboration. This violates Popper's informal characterization of the concept: bold and informative hypotheses should be preferred to less informative hypotheses, *ceteris paribus* (for more arguments against Popper's explication, see Díez 2011). We shall now show that this problem arises not only for Popper's measure $\zeta_P(H, E)$, but for all corroboration measures that are motivated from the same intuitions, that is, measures that aim at capturing statistical relevance and testability at the same time.

9.3 The Impossibility Results

Popper's nine adequacy conditions are specific requirements and too strong for the purpose of a general analysis of degree of corroboration. We will therefore weaken them and retain only those adequacy conditions that are indispensable for a conceptual analysis of corroboration. We then proceed to proving two impossibility results for corroboration measures that (i) build on statistical relevance between H and E and the predictive success of H for E; and (ii) preserve the intuition that corroboration should be responsive to the informativity and testability of the tested hypothesis (see also Sprenger 2018d, 145–150).

First, we impose a condition which is mainly representational in nature and is frequently used in Bayesian Confirmation Theory and formal epistemology more generally (see Variations 1, 6 and 7 of this book for details). Popper's own measure $\zeta_P(H,E)$ also conforms to it. Again, we assume that E and H are values of variables *E* and *H* in a causal Bayesian network with associated probability function *p*.

Probabilistic Dependence (Corroboration) There is a real-valued, continuous function $f\colon [0;1]^3 \to \mathbb{R}$ such that for any hypothesis H and piece of evidence E, the degree of corroboration that E confers on H, $\zeta(H,E)$, can be represented as

$$\zeta(H,E) = f(p(E|H), p(E), p(H)).$$

This condition relates degree of corroboration to the joint probability distribution of E and H. The three arguments of *f* determine that distribution in all non-degenerate cases, and they are the same quantities that figure in Popper's measure of corroboration $\zeta_P(H,E)$. This makes comparisons easier. Probabilistic Dependence means that two scientists who agree about all relevant probabilities will make the same corroboration judgments.

In a Popperian spirit, we now demand that corroboration track predictive success (e.g., Popper 1983, 241–243):

Weak Law of Likelihood (WLL) Suppose that for mutually exclusive hypotheses H_1 and H_2, conceptualized as disjunct ranges of a variable *H*,

$$p(E|H_1) \geq p(E|H_2) \qquad \text{and} \qquad p(E|\neg H_1) \leq p(E|\neg H_2),$$

with one inequality being strict. In that case, also $\zeta(H_1,E) > \zeta(H_2,E)$.

The WLL has been defended as capturing a "core message of Bayes' Theorem" (Joyce 2003): If H_1 predicts E better than H_2, and $\neg H_2$ predicts E better than $\neg H_1$, then E favors H_1 over H_2. Since WLL is phrased in terms of predictive performance, it is even more compelling for corroboration than for degree of confirmation: Hypotheses that perform better on the evidence are also better corroborated. After all, $p(E|\pm H_1)$ and $p(E|\pm H_2)$ measure how well H_1 and H_2 have stood up to a test with outcome E. The version given here is in one sense stronger and in one sense weaker than Joyce's original formulation: it is stronger insofar as only one inequality has to be strict (see also Brössel 2013, 395–396); and it is weaker insofar as the WLL has been restricted to mutually exclusive hypotheses, where our intuitions tend to be more reliable.

The next condition deals with the role of irrelevant evidence in corroboration judgments:

Screened-Off Evidence If E_2 is statistically independent of H, E_1 and $H \wedge E_1$, and $p(E_2) > 0$, then $\zeta(H, E_1) = \zeta(H, E_1 \wedge E_2)$.

Structurally identical versions of this condition prominently figure in explications of confirmation and explanatory power (e.g., Kemeny and Oppenheim 1952; Schupbach and Sprenger 2011). It is a weaker version of Condition (9.2), which demands, translated to corroboration, that $\zeta(H, E) = \zeta(H, E')$ if and only if $p(H|E) = p(H|E')$. To see this, just choose $E := E_1$, $E' := E_1 \wedge E_2$, and note that under the independence conditions of Screened-Off Evidence,

$$p(H \mid E_1 \wedge E_2) = \frac{p(H \wedge E_1 \mid E_2)}{p(E_1 | E_2)} = p(H|E_1).$$

Hence anybody who accepts something similar to Condition (9.2) for measures of corroboration will also endorse Screened-Off Evidence. However, Screened-Off Evidence is also very sensible on independent grounds: in an experiment where H has been tested and (relevant) evidence E_1 has been observed, irrelevant extra evidence ($E_2 \perp\!\!\!\perp E_1, H, E_1 \wedge H$) should not change the evaluation of the results. Imagine, for example, that a scientist tests the hypothesis that voices with high pitch are recognized more easily. As her university is interested in improving the planning of lab experiments, the scientist also collects data on when participants drop in, which days of the week are busy, which ones are quiet, etc. Plausibly, these data satisfy the independence conditions of Screened-Off Evidence. But equally plausibly, they do not influence the degree of corroboration of the hypothesis under investigation.

The next adequacy condition is motivated by the problem of irrelevant conjunctions for confirmation measures (e.g., Hawthorne and Fitelson 2004). Assume that hypothesis H asserts the wave nature of light. Taken together with a body of auxiliary assumptions, H implies the phenomenon E: the interference pattern in Young's double-slit experiment. Such an observation apparently corroborates the wave nature of light.

However, once we tack an utterly irrelevant proposition H', for example, that the chicken came before the egg, to the original hypothesis H, the evidence E corroborates $H \wedge H'$—the *conjunction* of the wave theory of light and the chicken–egg hypothesis—not more than H, if at all. After all, H' was in no way tested by the observations we made. It has no record of past performance to which we could appeal. This problem, familiar from Bayesian Confirmation Theory (see Variation 1), motivates the following constraint:

Irrelevant Conjunctions Assume the following conditions on H, H' and E are satisfied:

[9.1] H and H' are consistent and $p(H \wedge H') < p(H)$;

[9.2] $p(E) \in (0;1)$;

[9.3] $H \models E$;

[9.4] $p(E|H') = p(E)$.

Then it is always the case that $\zeta(H \wedge H', E) \leq \zeta(H, E)$.

This requirement states that for any non-trivial hypothesis H' that is consistent with H (Condition [9.1]) and irrelevant for E (Condition [9.4]), $H \wedge H'$ is corroborated no more than H whenever H non-trivially entails E (Conditions [9.2] and [9.3]). A similar requirement has been defended for measures of empirical justification (Atkinson 2012, 50–51). Indeed, it would be strange if corroboration (or justification) could be increased for free by attaching irrelevant conjunctions. That would also make it nearly impossible to reply persuasively to Duhem's problem, and to separate innocuous from blameworthy hypotheses after the failure of an experimental test. Degree of corroboration is supposed to guide our evaluation of hypotheses in the light of experimental results. But a measure which is invariant under logical conjunction of hypotheses cannot fulfil this function.

As a matter of fact, these adequacy conditions are closely related to Popper's original constraints:

Fact 9.1. *The above adequacy conditions can be derived from Popper's requirements as follows:*

- *Popper's Condition VII implies Weak Law of Likelihood for equiprobable hypotheses with probability greater than zero.*
- *Popper's Condition VII implies Screened-Off Evidence.*
- *Popper's Condition VIIIc implies Irrelevant Conjunctions.*

This shows that our adequacy conditions are motivated in the right way: they are either weaker versions of Popper's criteria, or closely related to them. We can thus be confident that our formal analysis of corroboration is on target and that our adequacy conditions do not track a different, incompatible concept.

However, unlike confirmation, corroboration contains an element of severe testing: the hypothesis should run a risk of being falsified. High informativity and testability contribute to this goal. As Popper states, "in many cases, the more improbable ... hypothesis is preferable" (Popper 1979, 18–19), and the purpose of a measure of degree of corroboration is "to show clearly in which cases this holds and in which it does not hold" (ibid.). This motivates the following desideratum:

Weak Informativity Degree of corroboration $\zeta(\mathrm{H},\mathrm{E})$ does not generally increase with the probability of H. In particular, there are DAG's with variables H and E, instantiations H and E and a (Markovian) probability function p such that

(1) $p(\mathrm{E}|\mathrm{H}) = p(\mathrm{E}|\mathrm{H}') > p(\mathrm{E})$;

(2) $1/2 \geq p(\mathrm{H}) > p(\mathrm{H}')$;

(3) $\zeta(\mathrm{H},\mathrm{E}) \leq \zeta(\mathrm{H}',\mathrm{E})$.

The intuition behind Weak Informativity can also be expressed as follows: Corroboration does not, in the first place, assess the probability of a hypothesis; therefore $\zeta(\mathrm{H},\mathrm{E})$ should not always increase with the probability of H. To this, the following condition—Strong Informativity—adds that low probability/high logical content can in principle be corroboration-conducive. Note that the requirement $1/2 \geq p(\mathrm{H}), p(\mathrm{H}')$ is purely technical and philosophically innocuous.

Strong Informativity The informativity/logical content of a proposition can increase degree of corroboration, *ceteris paribus*. In particular, there are DAG's with variables H and E, instantiations H and E and a (Markovian) probability function p such that

(1) $p(\mathrm{E}|\mathrm{H}) = p(\mathrm{E}|\mathrm{H}') > p(\mathrm{E})$;

(2) $1/2 \geq p(\mathrm{H}) > p(\mathrm{H}')$;

(3) $\zeta(\mathrm{H},\mathrm{E}) < \zeta(\mathrm{H}',\mathrm{E})$.

To our mind, any account of corroboration that denies Strong or Weak Informativity has stripped itself of its distinctive features with respect to degree of confirmation. At the very least, the Popperian characterization of corroboration as capturing both predictive success and testability would have to be abandoned, and links with NHST would loosen. The idea behind Strong/Weak Informativity has also recently been defended by Roberto Festa in his discussion of the "Reverse Matthew Effect": successful predictions reflect more favorably on powerful general theories than on restricted or weakened versions of them (Festa 2012, 95–100). Note that neither Strong nor Weak Informativity postulates that corroboration decreases in general with prior probability. They just deny the converse claim, namely the "Matthew Effect" that corroboration co-varies with prior probability (see also Roche 2014).

We will now demonstrate that the listed adequacy conditions are incompatible with each other. First, as a consequence of Weak Law of Likelihood, corroboration increases with the prior probability of a hypothesis. This clashes directly with Strong/Weak Informativity:

Theorem 9.1. *No measure of corroboration* $\zeta(H, E)$ *can satisfy Probabilistic Dependence, Weak Law of Likelihood and Strong/Weak Informativity at the same time.*

Since Probabilistic Dependence is a purely representational condition, this result means that Weak Law of Likelihood and Strong/Weak Informativity pull in different directions: the first condition emphasizes the predictive performance of the tested hypothesis, the second its logical strength. It is perhaps surprising that these two conditions are already incompatible, since it is a popular tenet of Critical Rationalism that informative hypotheses are also more valuable predictively.

Second, Strong Informativity clashes with Irrelevant Conjunctions and Screened-Off Evidence:

Theorem 9.2. *No measure of corroboration* $\zeta(H, E)$ *can satisfy Probabilistic Dependence, Screened-Off Evidence, Irrelevant Conjunctions and Strong Informativity at the same time.*

Thus the intuition behind Strong Informativity cannot be satisfied if other plausible adequacy constraints on degree of corroboration are accepted. In particular, if a measure of corroboration is insensitive to irrelevant evidence and does not reward adding irrelevant conjunctions, then it cannot give any bonus to informative hypotheses. The less informative and testable a hypothesis is, the higher its degree of corroboration, *ceteris paribus.*

Finally, the result of Theorem 9.2 can be extended to Weak Informativity if we make the assumption that irrelevant conjunctions dilute the degree of corroboration, rather than not increasing it (proof omitted). See also the corresponding remark in the motivation of Irrelevant Conjunctions on page 241.

Note that these results are meaningful even for those who are not interested in the project of explicating Popperian corroboration (e.g., because they are radical proponents of the Monist Thesis). Some of the above adequacy conditions have been proposed for measures of confirmation or explanatory power as well; others could be potentially interesting in this context. For instance, Brössel (2013) discusses the condition Continuity, which is similar to Strong/Weak Informativity: If the posterior probabilities of two hypotheses are almost indistinguishable from each other, we should prefer the hypothesis which was initially less probable. Hence the above results are also meaningful in the framework of Bayesian Confirmation Theory: they indicate the impossibility of statistical relevance measures that capture informativity and predictive success at the same time. See Variation 1 for more details.

All this does not imply that explicating degree of corroboration is a futile project. Rather, it reveals a fundamental and insoluble tension between the two main contributing factors of corroboration that Popper identifies: predictive success and testability/informativity. Weak Law of Likelihood, Screened-Off Evidence and Irrelevant Conjunctions all speak to the predictive-success intuition, whereas Strong/Weak Informativity rewards informative and testable hypotheses. In other words, the pre-theoretic concept of corroboration is overloaded with desiderata that point in different directions and create insoluble tensions. The point of Theorems 9.1 and 9.2 is to lay bare these tensions and to suggest ways out of the dilemma. Basically, we have four options: (i) to reject one of the (substantial) adequacy conditions; (ii) to split up degree of corroboration into different sub-concepts that preserve subsets of these intuitions; (iii) to conclude that the explication of degree of corroboration is hopeless and not worthy of further pursuit; and (iv) to reconcile the various desiderata in a different mathematical and conceptual framework.

Option (i) would come down to giving up either Weak Law of Likelihood, Screened-Off Evidence, Irrelevant Conjunctions or Strong/Weak Informativity. But each of these adequacy conditions for degree of corroboration has been carefully motivated. Such a step would therefore appear arbitrary and unsatisfactory.

For example, one could propose to endorse a statistical-relevance measure of degree of confirmation as a measure of corroboration, giving up the informativity intuition. This has the advantage of relating corroboration to a wealth of statistical and philosophical literature on degree of confirmation, but it comes at the price of stripping corroboration of its defining characteristics, and it runs into the objections against the Monist Thesis presented in Section 9.1.

Option (ii) amounts to endorsing pluralism for degree of corroboration. The model case for this option are probabilistic analyses of degre of confirmation: some measures, like $d(\mathrm{H},\mathrm{E}) = p(\mathrm{H}|\mathrm{E}) - p(\mathrm{H})$, capture the boost in degree of belief in H provided by E, while others, like $l(\mathrm{H},\mathrm{E}) = p(\mathrm{E}|\mathrm{H})/p(\mathrm{E}|\neg\mathrm{H})$, aim at the discriminatory power of E with respect to H and $\neg$H. However, it is not clear what similarly interesting subconcepts could look like for degree of corroboration.

Neither does the pessimistic option (iii)—to give up the explicative project—have much appeal, unless one proffers convincing reasons why scientists can dispense with the concept of corroboration, and hypothesis testing in general.

This leaves us with option (iv): changing the mathematical framework for explicating degree of corroboration. Perhaps it is neither necessary nor sufficient to base a corroboration judgment on the joint probability distribution of H and E? Instead of pitching H against $\neg$H with respect to their past performance, as implicit in Probabilistic Dependence, it may be appropriate to compare H to a multitude of distinct alternatives H_i with $i \in \mathbb{N}$. Perhaps corroboration judgments should be made with respect to the best-performing alternative in the hypothesis space, and not with respect to the aggregate of all these hypotheses. This is the option that we explore in the next section.

9.4 A New Explication of Corroboration

Suppose one of our friends—let's call him the coffee-tasting professor—claims to be able to taste whether a cup of *caffè* (known outside of Italy as espresso) has been prepared at water pressure of greater or smaller than 10 bar. Over a couple of weeks, we serve him a hundred coffee cups from two machines, one set to 10.5 bar and one to 9.5 bar, and count the number of cups he categorizes correctly. Simplifying a bit, we model the experiment as a Binomial trial with fixed sample size $N = 100$ and the null hypothesis H_0:

$\mu = 1/2$ that he is just guessing. The alternative hypothesis postulates that he is able to discriminate the two kinds of cups ($\mu \neq 1/2$). This example is modelled after R. A. Fisher's famous example of the tea-tasting lady (1935/74, 11–19).

How shall we evaluate the evidence we find? If the observed relative frequency of successes is close to 1/2 (e.g., E: $x = 53$), $\zeta(H_0, E)$ should not depend on the values of $p(x|\mu)$ for very large and very small values of μ. The performance of such alternatives is relevant for the posterior probability of H_0 (i.e., our degree of belief that the professor is just guessing), but not for testing whether this hypothesis is an adequate simplification of the complex statistical model where $\mu \in [0;1]$ is treated as a free parameter. However, for Bayesian statistical-relevance measures that compare $p(x \mid \mu = 1/2)$ to $p(x \mid \mu \neq 1/2)$, such a dependence is inevitable since $p(x \mid \mu \neq \mu_0) = \int_0^1 p(x|\mu)p(\mu)\,d\mu$. The probability of the data under the alternative is just the weighted average of the $p(x|\mu)$. This dependence of degree of corroboration on a weighting function (e.g., the prior probability function in Bayesian inference) is problematic in so far as the relative weight of alternative hypotheses should not really matter for assessing the tenability of the null hypothesis. A hypothesis test typically asks the question of whether *some* scientifically relevant alternative outperforms the null. This means that we should not conceptualize the alternative as a particular proposition H_1—for example, as a logical disjunction of precise alternatives—but as a *partition* of hypotheses $\mathcal{H}$ whose elements are scientifically meaningful alternatives. For example, they could be effect size ranges that correspond to different practical conclusions: Is the professor entitled to write a coffee tasting guide? How should we dose a drug in medical treatment? In other words, we replace the (Bayesian) question of whether the null hypothesis has survived a test against "the alternative" by the question of whether the null hypothesis has survived a test against a *set of alternatives*.

Thus, we explicate degree of corroboration of the null hypothesis H_0 with respect to **a partition of alternatives to the null.** As a consequence, Probabilistic Dependence has to be dropped and degree of corroboration becomes **partition-relative:** testing H_0 with alternative $\neg H_0$ leads to different corroboration judgments than testing H_0 with alternatives $\mathcal{H} = \{H_1, H_2, \ldots, H_n\}$ even if $\neg H_0 = \bigvee_{1 \leq i \leq n} H_i$ (cf. Good 1960, 1968a,b, 1975). Consider, for example, a test whether a medical drug is effective. The null corresponds to a particular parameter value H_0: $\mu = \mu_0$, indicating efficacy at placebo level, and the alternative to H_1: $\mu \neq \mu_0$. Dependent on the practical implications of certain effect sizes, we may divide the hypotheses in the following coarse-grained intervals: "worse than a placebo", "as good as a placebo", "slightly better than a placebo", "clearly better than a placebo", etc., whereas in other testing contexts (e.g., determining the value of a natural constant), a very fine-grained partition of the alternatives would be appropriate.

We now derive such a measure of corroboration on axiomatic grounds, focusing on measuring past performance while neglecting the testability intuition for the moment. It will resurface later, though. The first and most substantial adequacy condition states that corroboration judgments are made with respect to the best-performing alternative in the hypothesis space, and not with respect to *all* possible alternatives. This is because the corroboration of the null should not depend on whether $\mathcal{H}$ includes some implausible and poorly performing hypotheses. In other words, corroboration is defined as the minimal weight of evidence in favor of H_0. As in previous Variations, we suppress reference to a DAG and the associated probability distribution for the sake of convenience.

Corroboration = Minimal Weight of Evidence For a null hypothesis H_0 and a (possibly infinite) partition of alternative hypotheses $\mathcal{H} = \{H_1, H_2, \ldots\}$, the degree of corroboration that observation E provides for H_0 relative to $\mathcal{H}$ is defined as

$$\zeta_{\mathcal{H}}(H_0, E) = \min_{H_i \in \mathcal{H}} \omega(H_0, H_i, E), \tag{9.4}$$

where $\omega(H_0, H_i, E)$ quantifies the weight of evidence that E provides for H_0 and against the specific alternative H_i.

On this account, positive corroboration requires that no genuine alternative $H_i \in \mathcal{H}$ be evidentially favored over H_0. The tested hypothesis H_0 has to outperform each of them.

Next, we wonder what the weight-of-evidence function $\omega(H_0, H_1, E)$ for H_0, compared to an alternative H_1, should look like. In analogy to previous chapters, we posit the following condition, which makes weight of evidence a non-decreasing function of the past performance of H_0, and a non-increasing function of the past performance of H_1.

Difference-Making (Weight of Evidence) For hypotheses H_0 and H_1 and evidence E, there is a continuous function $g\colon [0;1]^2 \to \mathbb{R}$ such that

$$\omega(H_0, H_1, E) = g\big(p(E|H_0), p(E|H_1)\big)$$

and g is non-decreasing in the first argument and non-increasing in the second argument.

Making weight of evidence a function of the predictive performance of both hypotheses is in line with Popper's characterization of corroboration as indicating past performance. Similar requirements are also made in Good 1952, Bernardo 1999 and Williamson 2010.

Next, irrelevant evidence should not distort the relative weight of evidence that other observations provide. This argument has already appeared in the context of motivating Screened-Off Evidence, and also in earlier chapters: compare the Modularity condition in Variation 1, No Dilution in Variation 6 and Independent Background Theories in Variation 7.

Irrelevant Evidence (Weight of Evidence) Suppose that some observation E' does not discriminate between H_0 and H_1—that is, $p(E' \mid H_0) = p(E' \mid H_1)$—and moreover, for another observation E,

$$p(E \mid E', H_0) = p(E|H_0), \qquad p(E \mid E', H_1) = p(E|H_1), \tag{9.5}$$

In that case, also $\omega(H_0, H_1, E) = \omega(H_0, H_1, E \wedge E')$.

The argument in support of Irrelevant Evidence is that additional data E' without relevance for either H_0 or H_1 should not change the weight of evidence that E provides for either of them. Difference-Making and Irrelevant Evidence are already sufficient to narrow down the set of admissible weight-of-evidence measures to functions that are ordinally equivalent to $p(E|H_0)/p(E|H_1)$, in agreement with various proposals from the literature (Good 1960; Royall 1997; Taper and Lele 2004).

The final adequacy condition demands that weight of evidence be additive with respect to evidence from independent and identically distributed (i.i.d.) observations. This requirement allows us to aggregate evidence from various experiments with identical design in a convenient manner.

Additivity (Weight of Evidence) If for two observations E and E' and two hypotheses H_0 and H_1

$$\begin{aligned} p(E \wedge E' \mid H_0) &= p(E|H_0) \cdot p(E'|H_0), \\ p(E \wedge E' \mid H_1) &= p(E|H_1) \cdot p(E'|H_1), \end{aligned}$$

then the total weight of evidence provided by E and E' is equal to the sum of the individual weights:

$$\omega(H_0, H_1, E \wedge E') = \omega(H_0, H_1, E) + \omega(H_0, H_1, E'). \tag{9.6}$$

Accepting these four conditions as desiderata for a measure of corroboration allows us to state our main constructive result:

Theorem 9.3. *Corroboration = Minimal Weight of Evidence, Difference-Making, Irrelevant Evidence and Additivity jointly determine the weight-of-evidence function*

$$\omega(H_0, H_1, E) = k \log \frac{p(E|H_0)}{p(E|H_i)} \qquad \text{for } k > 0 \tag{9.7}$$

and the corroboration measure

$$\zeta_{\mathcal{H}}(H_0, E) = k \min_{H_i \in \mathcal{H}} \log \frac{p(E|H_0)}{p(E|H_i)} \qquad \text{for } k > 0. \tag{9.8}$$

The base of the logarithm and the scalar k may be chosen *ad libitum*; but in order to keep the scale consistent with logarithmic Bayes factors and the log-likelihood confirmation measure we suggest the natural logarithm and $k = 1$ (see also Variation 1, page 59).

Positive degree of corroboration entails that H_0 outperforms all relevant alternatives. If degree of corroboration is negative, one of the alternatives may become our new default hypothesis and be subjected to subsequent tests. The dependence of $\zeta_{\mathcal{H}}$ on the granularity of the partition $\mathcal{H}$ implies that degree of corroboration is a context-sensitive concept that depends on which alternatives have scientific relevance. Consider an unknown real-valued parameter μ and the null hypothesis H_0: $\mu = \mu_0$. For the maximal partition $\mathcal{H}_{\max}$, all precise hypotheses H_μ with $\mu \neq \mu_0$ are relevant alternatives to the null hypothesis. This is typical of pure estimation problems and in this case, degree of corroboration is given by the (negative) log-likelihood ratio with respect to the maximum likelihood estimate $\hat{\mu}$: $\zeta_{\mathcal{H}_{\max}}(H_0, E) = -\log(p(E|H_{\hat{\mu}})/p(E|H_0))$.

Conversely, consider the minimal partition $\mathcal{H}_{\min}$ consisting of only one element H_1: $\mu \neq \mu_0$, weighted by a prior distribution for μ. In that case, the testing problem is fully symmetric and degree of corroboration corresponds to the log-Bayes factor: $\zeta_{\mathcal{H}_{\min}}(H_0, E) = \log BF_{01}(E)$. There is thus no tension between degrees of corroboration and Bayesian hypothesis testing: they always agree in symmetric testing problems, but degree of corroboration extends evidential judgments to more general, asymmetric testing problems.

We illustrate this partition-dependence by returning to our example of the coffee-tasting professor and the experiment that tests his skills. Figure 9.2 plots the degree of corroboration of the null hypothesis that he is just guessing ($\mu = 1/2$), as a function of the number of successes (for $N = 100$ cups to taste) and three different partitions of the space of alternative hypotheses:

Minimal Partition: $\mathcal{H}_{\min} = \{[0;1]\}$—there is only one alternative hypothesis, which is a weighted mixture of all values of μ. (The weighting function is, as a matter of convention, based on the logistic function.)

Medium Partition: $\mathcal{H} = \{(0.5;0.6], (0.6;0.7], (0.7;0.8], \ldots\}$—the alternatives are intervals with width 0.1, corresponding to small, medium, large, etc. effect sizes.

Maximal Partition: $\mathcal{H}_{\max} = [0;1]$—there is an infinity of precise alternatives corresponding to the real numbers in the interval $[0;1]$.

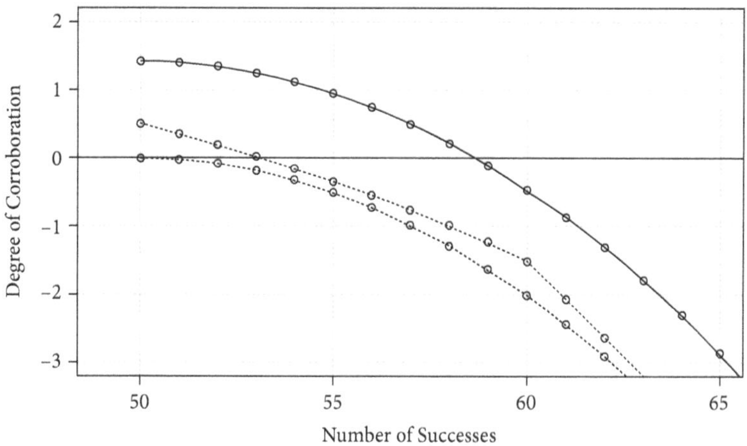

Figure 9.2: Degree of corroboration of the hypothesis H_0: $\mu = \mu_0$ plotted against number of observed successes in a Binomial experiment, with sample size $N = 100$. The solid line correspond to the unique alternative $\mathcal{H}_{\min} = \{[0;1]\}$, the dashed line in the middle corresponds to the interval partition $\mathcal{H} = \{[0;0.1), [0.1;0.2), \ldots\}$, and the dotted line at the bottom corresponds to the maximally fine-grained partition of alternatives $\mathcal{H}_{\max} = [0;1]$.

For the minimal partition $\mathcal{H}_{\min}$ which is standardly used in Bayesian hypothesis testing, degree of corroboration is positive until $x = 58$ (full line). This means that a success rate of roughly 60 % is required to find evidence against the null hypothesis that the professor does not have any coffee-tasting skills. For more fine-grained partitions such as effect size ranges of width 0.1, already $x = 54$ counts as evidence against the null (dashed line). In such a testing problem, already small departures from chance would be classified as coffee-tasting skills. Finally, for the maximally fine-grained partition $\mathcal{H}_{\max}$ where every point hypothesis is a potential alternative, degree of corroboration is always negative (dotted line). This is actually very natural: when each parameter value is a serious scientific option, how can a point null hypothesis be ever corroborated unless the sample mean agrees *exactly* with the hypothesized parameter value? Changing the weighting function for the intervals in $\mathcal{H}$ barely makes a difference to these results.

These findings suggest that more fine-grained partitions lead in general to a smaller degree of corroboration. Indeed we can verify this claim:

Theorem 9.4. *If $\mathcal{H}$ is a subpartition of $\mathcal{H}'$, then $\zeta_{\mathcal{H}}(\mathrm{H}_0, \mathrm{E}) < \zeta_{\mathcal{H}'}(\mathrm{H}_0, \mathrm{E})$, provided that the alternative hypotheses are weighted in the same way.*

This property implies that testability affects degree of corroboration. If the alternatives are highly specific and testable, the degree of corroboration of the null hypothesis is lower than if the alternatives are quite unspecific. In other words, what matters is not the (absolute) testability of H_0 as measured by its logical strength, but how testable H_0 is as opposed to the relevant alternatives. On Popper's original view, these two properties collapse into one, but as we see now, it is important to keep them apart. Hence, Popper's two crucial ingredients of corroboration—past performance and testability—can finally be reconciled.

9.5 Discussion

This Variation has motivated the need for a concept of corroboration in scientific inference, and in particular in NHST. Amending NHST by the concept of corroboration of the null hypothesis allows us to state evidential support (corroboration) for the null hypothesis and to provide an evidential interpretation of non-significant findings. By doing so, the corroboration measure acts as an antidote to the file drawer effect and publication bias (and thus, the replication crisis). In particular, if researchers are familiarized with the idea that support for null hypotheses can be expressed in a positive manner—rather than just stating "failure to reject the null hypothesis"—this will help them to overcome the pernicious and unfortunately still widespread idea that only a study with significant results *against* the null hypothesis counts as a successful experiment.

While corroboration judgments agree with logarithmic Bayes factors in symmetric testing problems, there are also several important features where our explication of corroboration departs from orthodox Bayesian inference:

1. The explication is asymmetric, respecting that the role of the null hypothesis and the alternative are not interchangeable (see Objection 3, page 234). This preserves an important feature of hypothesis testing in science without buying into the methodological shortcomings of NHST.

2. The explication of corroboration is sensitive to the partition of hypotheses against which the null is tested. This is our key conceptual move. The standard Bayesian framework for hypothesis testing conceptualizes the alternative hypothesis as the probabilistic mixture of all point values $\mu \neq \mu_0$. However, we argue that it is often fruitful to think about the alternative as a set of hypotheses, for example, intervals that correspond to certain scientific conclusions (e.g., small/sizeable/very large effect).

3. Our proposed explication is independent of one's (raise in level of) confidence in the tested hypothesis and therefore distinctive of corroboration as opposed to confirmation (see Objection 1, page 233). The measure of corroboration is constructed by comparing the null hypothesis to the best-performing alternative. This is in agreement with scientific reasoning, where we accept a theory only if it outperforms all relevant alternatives.

4. Hypotheses with prior probability zero can be corroborated (see Objection 2, page 234). That we are not prepared to bet on the truth of a precise point hypothesis, regardless of the betting odds, does not preclude that this hypothesis can perform well and be corroborated with respect to competitors.

What do these results mean for the project of a Bayesian philosophy of science? Isn't there a tension between our ambitions and the (non-Bayesian) explication we provide? Only partially so. Relative plausibility judgments are still present in corroboration-based hypothesis testing. First and foremost this happens in the partitioning of alternatives. The partition $\mathcal{H}$ against which H_0 is pitched depends on how we identify the relevant alternatives to H_0. What is a scientifically meaningful effect size? Which differences can be neglected? Answers to these questions require subjective plausibility judgments, and they will affect the degree of corroboration substantially, as Figure 9.2 demonstrates. Second, for interval alternatives (e.g., different effect size ranges like in the example on page 250), calculating the likelihoods will require a (subjective) weighting function for aggregating the likelihoods of the various precise hypothesis. Third, for the minimal partition $\mathcal{H}_{\min} = \{\neg H_0\}$, degree of corroboration corresponds to the logarithmic Bayes factor (and the log-likelihood ratio measure of confirmation). This means that Bayesian hypothesis testing can be represented as a special case of corroboration-based hypothesis testing: a classical Bayesian hypothesis test conceptualizes the alternative to H_0 as the negation of H_0. Corroboration judgments can be seen as a generalization of Bayesian inference to asymmetric hypothesis testing.

This relationship suggests the first question for future research: the reconciliation of Bayesian inference and NHST within a corroboration-centered perspective. In particular, one should examine the conjecture that our explication of corroboration unifies Bayesian and non-Bayesian hypothesis testing, and compare it to other proposals to reconcile Bayesian and frequentist hypothesis testing (Berger, Boukai and Wang 1997; Berger 2003; Nardini and Sprenger 2013). Second, the proposed corroboration measure needs to be applied to more complicated cases of statistical inference, including nuisance parameters, hierarchical models and model selection. In this context, it would also be challenging to see what kind of meaning the notorious p-value obtains within a corroboration-based framework. Third, it would be worthwhile to conduct case studies that reconstruct specific episodes of scientific reasoning as guided by corroboration judgments, and to see whether these episodes fit into a probabilistic explication of degree of corroboration.

The next chapter stays with the topic of statistical inference and focuses on the role of simplicity considerations in the process of model selection, that is, comparing parametrized classes of statistical hypotheses.

Appendix: Proofs of the Theorems

Proof of Fact 9.1: We begin with showing that Condition VII implies the Weak Law of Likelihood (WLL). Assume $p(H_1) = p(H_2)$. We distinguish two jointly exhaustive cases in which WLL may apply:

- Case 1: $p(E|H_1) > p(E|H_2)$.
- Case 2: $p(E|H_1) = p(E|H_2)$ and $p(E|\neg H_1) < p(E|\neg H_2)$.

For the first case, the proof is simple in virtue of the inequality

$$p(H_1|E) = p(H_1)\frac{p(E|H_1)}{p(E)} > p(H_2)\frac{p(E|H_2)}{p(E)} = p(H_2|E).$$

Then VII guarantees that $\zeta(H_1,E) > \zeta(H_2,E)$.

For the second case, let $x := p(E|H_1) = p(E|H_2)$ and $y := p(H_1) = p(H_2)$. We know that

$$\begin{aligned}
p(E|\neg H_1) &= \frac{1}{1-p(H_1)}\Big(p(E|H_2)\,p(H_2) + p(E|\neg H_1,\neg H_2)\,p(\neg H_1,\neg H_2)\Big) \\
&= \frac{1}{1-y}\Big(xy + p(E|\neg H_1,\neg H_2)\,p(\neg H_1,\neg H_2)\Big), \\
p(E|\neg H_2) &= \frac{1}{1-p(H_2)}\Big(p(E|H_1)p(H_1) + p(E|\neg H_1,\neg H_2)\,p(\neg H_1,\neg H_2)\Big) \\
&= \frac{1}{1-y}\Big(xy + p(E|\neg H_1,\neg H_2)\,p(\neg H_1,\neg H_2)\Big).
\end{aligned}$$

Hence, $p(E|\neg H_1) = p(E|\neg H_2)$. On the other hand, we have assumed that $p(E|\neg H_1) < p(E|\neg H_2)$. This shows that the second case can never occur and may be dismissed.

We now prove the second implication, that is, VII $\Rightarrow$ Screened-Off Evidence. To this end, remember that Condition VII reads

VII If $p(H) = p(H')$, then $\zeta(H,E) \leq \zeta(H',E')$ if and only if $p(H|E) \leq p(H'|E')$.

Assuming $H = H'$, it is easy to see that VII implies

VII′ If $p(H|E) = p(H|E')$, then $\zeta(H,E) = \zeta(H,E')$.

The reason is simple: If $p(H|E) = p(H|E')$ then also $p(H|E) \leq p(H|E')$, and the "$\Leftarrow$"-direction of VII implies $\zeta(H,E) \leq \zeta(H,E')$, where H has been substituted for H′. Now we repeat the same trick with the premise $p(H|E') \leq$

$p(\mathrm{H}|\mathrm{E})$ and we obtain $\zeta(\mathrm{H},\mathrm{E}') \leq \zeta(\mathrm{H},\mathrm{E})$. Taking both inequalities together yields the conclusion $\zeta(\mathrm{H},\mathrm{E}) = \zeta(\mathrm{H},\mathrm{E}')$ and thereby VII′.

Notice that under the conditions of Screened-Off Evidence, $p(H \mid \mathrm{E}_1 \wedge \mathrm{E}_2) = p(\mathrm{H}|\mathrm{E}_1)$. This is so because

$$\begin{aligned} p(h \mid \mathrm{E}_1 \wedge \mathrm{E}_2) &= p(\mathrm{H})\frac{p(\mathrm{E}_1 \wedge \mathrm{E}_2 \mid \mathrm{H})}{p(\mathrm{E}_1 \wedge \mathrm{E}_2)} \\ &= p(\mathrm{H})\frac{p(\mathrm{E}_1|\mathrm{H})\,p(\mathrm{E}_2)}{p(\mathrm{E}_1)\,p(\mathrm{E}_2)} = p(\mathrm{H})\frac{p(\mathrm{E}_1|\mathrm{H})}{p(\mathrm{E}_1)} = p(\mathrm{H}|\mathrm{E}_1). \end{aligned}$$

Hence we can apply VII′ to the case of Screened-Off Evidence, with $e := e_1$ and $e' := \mathrm{E}_1 \wedge \mathrm{E}_2$. This implies

$$\zeta(\mathrm{H}, \mathrm{E}_1 \wedge \mathrm{E}_2) = \zeta(\mathrm{H},\mathrm{E}_1),$$

completing the proof.

Finally, we have the implication VIIIc ⇒ Irrelevant Conjunctions. Let the conditions of Irrelevant Conjunctions ([9.1]–[9.4] on page 241) be satisfied for H, H′ and E. Since $\mathrm{H} \vDash \mathrm{E}$, VIIIc implies that $\zeta(\mathrm{H},\mathrm{E})$ and $\zeta(\mathrm{H}\wedge\mathrm{H}', \mathrm{E})$ are increasing functions of the probability of the tested hypotheses—$p(\mathrm{H})$ and $p(\mathrm{H}\wedge\mathrm{H}')$, respectively. But by assumption, we have $p(\mathrm{H}\wedge\mathrm{H}') < p(\mathrm{H})$. Hence, it follows that $\zeta(\mathrm{H}\wedge\mathrm{H}', \mathrm{E}) \leq \zeta(\mathrm{H},\mathrm{E})$. □

Proof of Theorem 9.1: First note that the f from Probabilistic Dependence needs to be considered on the set $\{(x,y,z) \mid 1+xz-z \geq y \geq xz\} \subset [0;1]$ only. This is evident from the equality

$$p(\mathrm{E}) = p(\mathrm{E}|\mathrm{H})\,p(\mathrm{H}) + p(\mathrm{E}|\neg\mathrm{H})\,(1-p(\mathrm{H})),$$

which implies, by setting $p(\mathrm{E}|\neg\mathrm{H})$ to its extremal values, the inequalities

$$p(\mathrm{E}) \geq p(\mathrm{E}|\mathrm{H})p(\mathrm{H}) \qquad \text{and} \qquad p(\mathrm{E}) < p(\mathrm{E}|\mathrm{H})\,p(\mathrm{H}) + 1 - p(\mathrm{H}).$$

Hence, outside the above set, there cannot be propositions E and H with probability distribution p such that $x = p(\mathrm{E}|\mathrm{H})$, $y = p(\mathrm{E})$ and $z = p(\mathrm{H})$.

After these preliminaries, we begin with the real proof. By Weak Informativity and Probabilistic Dependence, there are $x > y$ and $z > z'$ with $z+z' < 1$, $1+xz-z \geq y \geq xz$ and $1+xz'-z' \geq y \geq xz'$ such that

$$f(x,y,z) \leq f(x,y,z').$$

Choose a probability function p such that $p(\mathrm{H}_1) = z$, $p(\mathrm{H}_2) = z'$, $p(\mathrm{E}|\mathrm{H}_1) = p(\mathrm{E}|\mathrm{H}_2) = x$, $p(\mathrm{H}_1 \wedge \mathrm{H}_2) = 0$ and $p(\mathrm{E}) = y$. We now verify that this distribution satisfies the axioms of probability. Because of $xz > xz'$ and $1+xz-z < 1+xz'-z'$, it suffices to verify the inequalities $y \geq xz$ and $y \leq 1+xz-z$.

First note that

$$\begin{aligned} p(\mathrm{E}) &= \\ & p(\mathrm{E}|\mathrm{H}_1)\,p(\mathrm{H}_1) + p(\mathrm{E}|\mathrm{H}_2)\,p(\mathrm{H}_2) + p(\mathrm{E}|\neg\mathrm{H}_1,\neg\mathrm{H}_2)(1-p(\mathrm{H}_1)-p(\mathrm{H}_2)), \end{aligned}$$

which, setting $\omega := p(\mathrm{E}|\neg\mathrm{H}_1,\neg\mathrm{H}_2)$, translates as

$$y = xz + xz' + \omega(1-z-z').$$

This equation allows us to show the desired inequalities:

$$y - xz = xz + xz' + \omega(1-z-z') - xz = xz' + \omega(1-z-z') \geq 0,$$

$$\begin{aligned} 1+xz-z-y &= 1+xz-z-xz-xz'-\omega(1-z-z') \\ &= (1-z-xz') + \omega(1-z-z') \geq 0. \end{aligned}$$

In both cases, all summands are greater than or equal to zero because $z+z' < 1$ by assumption. This completes the proof that the above probability distribution is well-defined.

Now it is straightforward to show that

$$\begin{aligned} p(\mathrm{E}|\neg\mathrm{H}_1) &= \frac{1}{1-p(\mathrm{H}_1)}\Big(p(\mathrm{E}|\mathrm{H}_2)\,p(\mathrm{H}_2) + p(\mathrm{E}|\neg\mathrm{H}_1,\neg\mathrm{H}_2)\,p(\neg\mathrm{H}_1,\neg\mathrm{H}_2)\Big) \\ &= \frac{1}{1-p(\mathrm{H}_1)}\Big(p(\mathrm{E}|\mathrm{H}_1)\,p(\mathrm{H}_2) + p(\mathrm{E}|\neg\mathrm{H}_1,\neg\mathrm{H}_2)\,p(\neg\mathrm{H}_1,\neg\mathrm{H}_2)\Big), \\ p(\mathrm{E}|\neg\mathrm{H}_2) &= \frac{1}{1-p(\mathrm{H}_2)}\Big(p(\mathrm{E}|\mathrm{H}_1)\,p(\mathrm{H}_1) + p(\mathrm{E}|\neg\mathrm{H}_1,\neg\mathrm{H}_2)\,p(\neg\mathrm{H}_1,\neg\mathrm{H}_2)\Big). \end{aligned}$$

because by assumption, $p(\mathrm{E}|\mathrm{H}_1) = p(\mathrm{E}|\mathrm{H}_2)$. From this we can infer:

$$\begin{aligned} & p(\mathrm{E}|\neg\mathrm{H}_1) - p(\mathrm{E}|\neg\mathrm{H}_2) \\ &= \frac{p(\mathrm{E}|\mathrm{H}_1)\,p(\mathrm{H}_2)}{1-p(\mathrm{H}_1)} + \frac{p(\mathrm{E}|\neg\mathrm{H}_1,\neg\mathrm{H}_2)\,p(\neg\mathrm{H}_1,\neg\mathrm{H}_2)}{1-p(\mathrm{H}_1)} - \\ &\quad \frac{p(\mathrm{E}|\mathrm{H}_1)\,p(\mathrm{H}_1)}{1-p(\mathrm{H}_2)} - \frac{p(\mathrm{E}|\neg\mathrm{H}_1,\neg\mathrm{H}_2)\,p(\neg\mathrm{H}_1,\neg\mathrm{H}_2)}{1-p(\mathrm{H}_2)} \\ &= p(\mathrm{E}|\mathrm{H}_1)\left(\frac{p(\mathrm{H}_2)}{1-p(\mathrm{H}_1)} - \frac{p(\mathrm{H}_1)}{1-p(\mathrm{H}_2)}\right) + \end{aligned}$$

$$p(\mathrm{E} \mid \neg \mathrm{H}_1, \neg h_2)\,(1-p(\mathrm{H}_1)-p(\mathrm{H}_2))\left(\frac{1}{1-p(\mathrm{H}_1)}-\frac{1}{1-p(\mathrm{H}_2)}\right)$$

$$= p(\mathrm{E} \mid \mathrm{H}_1)\frac{p(\mathrm{H}_2)-p(\mathrm{H}_2)^2-p(\mathrm{H}_1)+p(\mathrm{H}_1)^2}{(1-p(\mathrm{H}_1))\,(1-p(\mathrm{H}_2))} +$$

$$p(\mathrm{E} \mid \neg \mathrm{H}_1, \neg h_2)\,(1-p(\mathrm{H}_1)-p(\mathrm{H}_2))\frac{p(\mathrm{H}_1)-p(\mathrm{H}_2)}{(1-p(\mathrm{H}_1))\,(1-p(\mathrm{H}_2))}$$

$$= p(\mathrm{E} \mid \mathrm{H}_1)\frac{(p(\mathrm{H}_1)-p(\mathrm{H}_2))\,(p(\mathrm{H}_1)+p(\mathrm{H}_2)-1)}{(1-p(\mathrm{H}_1))\,(1-p(\mathrm{H}_2))} +$$

$$p(\mathrm{E} \mid \neg \mathrm{H}_1, \neg h_2)\,(1-p(\mathrm{H}_1)-p(\mathrm{H}_2))\frac{p(\mathrm{H}_1)-p(\mathrm{H}_2)}{(1-p(\mathrm{H}_1))\,(1-p(\mathrm{H}_2))}$$

$$= \frac{(p(\mathrm{H}_1)-p(\mathrm{H}_2))\,(p(\mathrm{H}_1)+p(\mathrm{H}_2)-1)}{(1-p(\mathrm{H}_1))\,(1-p(\mathrm{H}_2))}\Big(p(\mathrm{E} \mid \mathrm{H}_1) - p(\mathrm{E} \mid \neg \mathrm{H}_1, \neg \mathrm{H}_2)\Big).$$

If we look at the signs of the involved factors, we notice first that $p(\mathrm{H}_1) - p(\mathrm{H}_2) = z - z' > 0$ and $p(\mathrm{H}_1) + p(\mathrm{H}_2) - 1 = z + z' - 1 < 0$. Then we observe that H_1 and H_2 were disjoint and that $p(\mathrm{E} \mid \mathrm{H}_1)$ and $p(\mathrm{E} \mid \mathrm{H}_2)$ are both greater than $p(\mathrm{E})$, implying $p(\mathrm{E} \mid \mathrm{H}_1) = p(\mathrm{E} \mid \mathrm{H}_2) > p(\mathrm{E} \mid \neg \mathrm{H}_1, \neg \mathrm{H}_2)$. Taken together, we can then conclude

$$p(\mathrm{E} \mid \neg \mathrm{H}_1) - p(\mathrm{E} \mid \neg \mathrm{H}_2) < 0.$$

Hence the conditions for applying the Weak Law of Likelihood are satisfied: H_1 and H_2 are two mutually exclusive hypotheses with $p(\mathrm{E} \mid \mathrm{H}_1) = p(\mathrm{E} \mid \mathrm{H}_2)$ and $p(\mathrm{E} \mid \neg \mathrm{H}_1) < p(\mathrm{E} \mid \neg \mathrm{H}_2)$. Thus we can conclude

$$f(x, y, z) = \zeta(\mathrm{H}_1, \mathrm{E}) > \zeta(\mathrm{H}_2, \mathrm{E}) = f(x, y, z'),$$

in contradiction with the inequality $f(x, y, z) \leq f(x, y, z')$ that we got from Weak Informativity. □

Lemma 9.1. *Any measure of corroboration* $\zeta(\mathrm{H}, \mathrm{E}) = f(p(\mathrm{E} \mid \mathrm{H}), p(\mathrm{E}), p(\mathrm{H}))$ *that satisfies Probabilistic Dependence and Screened-Off Evidence also satisfies the equality*

$$f(ax, ay, z) = f(x, y, z)$$

for $x > y > 0,\ z > 0$ *and* $0 < a \leq 1$ *with* $1 + xz - z \geq y \geq xz$.

Proof of Lemma 9.1: For any $0 < a \leq 1$, $x > y > 0$ and $z > 0$ with $1+xz-z \geq y \geq xz$, we can construct a causal model with variables H, E_1 and E_2 and instantiations H, E_1, E_2, and a probability function p such that

$$\begin{aligned} a &= p(\mathrm{E}_2), & p(\mathrm{E}_2,\mathrm{H}) &= p(\mathrm{E}_2)\,p(\mathrm{H}),\\ x &= p(\mathrm{E}_1|\mathrm{H}), & p(\mathrm{E}_1\wedge\mathrm{E}_2) &= p(\mathrm{E}_2)\,p(\mathrm{E}_1),\\ y &= p(\mathrm{E}_1), & p(\mathrm{E}_1\wedge\mathrm{E}_2\,|\,\mathrm{H}) &= p(\mathrm{E}_2)\,p(\mathrm{E}_1|\mathrm{H}),\\ z &= p(\mathrm{H}). \end{aligned}$$

Since our choice of p is not restricted, this is always possible. Now, the conditions of Screened-Off Evidence are satisfied, and it follows that $\zeta(\mathrm{H}, \mathrm{E}_1\wedge\mathrm{E}_2) = \zeta(\mathrm{H}, \mathrm{E}_1)$. By Probabilistic Dependence, we can also derive the equalities

$$\begin{aligned} \zeta(\mathrm{H}, \mathrm{E}_1\wedge\mathrm{E}_2) &= f\Big(p(\mathrm{E}_1\wedge\mathrm{E}_2\,|\,\mathrm{H}),\ p(\mathrm{E}_1\wedge\mathrm{E}_2),\ p(\mathrm{H})\Big)\\ &= f\Big(p(\mathrm{E}_2)\,p(\mathrm{E}_1|\mathrm{H}),\ p(\mathrm{E}_2)\,p(\mathrm{E}_1),\ p(\mathrm{H})\Big)\\ &= f(ax, ay, z),\\ \zeta(\mathrm{H}, \mathrm{E}_1) &= f(x, y, z). \end{aligned}$$

Taking all these equalities together delivers the desired result:

$$f(ax, ay, z) = \zeta(\mathrm{H}, \mathrm{E}_1\wedge\mathrm{E}_2) = \zeta(\mathrm{H}, \mathrm{E}_1) = f(x, y, z).$$

Finally we note that (ax, ay, z) is always in the domain of f when $a \leq 1$ and $1+xz-z \geq y \geq xz$:

$$\begin{aligned} ay &\geq ax/z,\\ ay &\leq a(1+xz-z) = axz + a(1-z) \leq 1+axz-z. \end{aligned}$$ □

Proof of Theorem 9.2: Choose a causal model with variables H_1, H_2 and E and instantiations H_1, H_2, E, and a probability function p such that the conditions of Strong Informativity are satisfied:

(1) $p(\mathrm{E}|\mathrm{H}_1) = p(\mathrm{E}|\mathrm{H}_2) > p(\mathrm{E})$;

(2) $1/2 \geq p(\mathrm{H}_1) > p(\mathrm{H}_2)$;

(3) $\zeta(\mathrm{H}_1, \mathrm{E}) < \zeta(\mathrm{H}_2, \mathrm{E})$.

With the definitions $x := p(\mathrm{E}|\mathrm{H}_1) = p(\mathrm{E}|\mathrm{H}_2)$, $y := p(\mathrm{E})$, $z_1 := p(\mathrm{H}_1)$ and $z_2 := p(\mathrm{H}_2)$, we then obtain

$$f(x, y, z_1) = \zeta(\mathrm{H}_1, \mathrm{E}) < \zeta(\mathrm{H}_2, \mathrm{E}) = f(x, y, z_2). \tag{9.9}$$

Since $\zeta(\mathrm{H}, \mathrm{E})$ satisfies Probabilistic Dependence and Screened-Off Evidence, by Lemma 9.1 it also satisfies the equality

$$f(ax, ay, z) = f(x, y, z)$$

for $x > y > 0,\ z > 0$ and $0 < a \leq 1$. It is easy to see that $(1, y/x, z)$ is in the domain of f if (x, y, z) is. Applying the above equality to the expressions in Equation (9.9) and choosing $a := x$, we then obtain

$$f(1, y/x, z_1) = f(x, y, z_1) \qquad \text{and} \qquad f(1, y/x, z_2) = f(x, y, z_2).$$

Then it follows from Inequality (9.9) and the above equalities that

$$f(1, y/x, z_1) < f(1, y/x, z_2) \tag{9.10}$$

for these specific values of x, y, z_1 and z_2.

We can now find sentences H, H′, E′ and a probability function $p'(\cdot)$ such that the conditions of Irrelevant Conjunctions are satisfied and at the same time, $p'(\mathrm{H}) = z_1$, $p'(\mathrm{H} \wedge \mathrm{H}') = z_2$ and $p'(\mathrm{E}') = y/x$. This implies $\zeta(\mathrm{H} \wedge \mathrm{H}', \mathrm{E}') \leq \zeta(\mathrm{H}, \mathrm{E}')$. By Probabilistic Dependence, we obtain

$$f(1, y/x, z_1) \geq f(1, y/x, z_2).$$

However, this inequality contradicts Equation (9.10), which we have shown before. Hence the theorem is proven. □

Proof of Theorem 9.3: First we show that by Difference-Making and Irrelevant Evidence, weight of evidence must be a monotonically increasing function of the ratio $p(\mathrm{E}|\mathrm{H}_0)/p(\mathrm{E}|\mathrm{H}_1)$. The proof proceeds analogously to the proof of the second part of Theorem 1.2. Assume that E′ is an irrelevant observation for evidence E and the pair $(\mathrm{H}_0, \mathrm{H}_1)$ in the sense of Equation (9.5). This implies for the function g describing weight of evidence:

$$\begin{aligned} g(p(\mathrm{E}|\mathrm{H}_0), p(\mathrm{E}|\mathrm{H}_1)) &= g(p(\mathrm{E} \wedge \mathrm{E}' \,|\, \mathrm{H}_0), p(\mathrm{E} \wedge \mathrm{E}' \,|\, \mathrm{H}_1)) \\ &= g(p(\mathrm{E}|\mathrm{H}_0)\, p(\mathrm{E}'|\mathrm{H}_0), p(\mathrm{E}|\mathrm{H}_1)\, p(\mathrm{E}'|\mathrm{H}_1)) \\ &= g(p(\mathrm{E}|\mathrm{H}_0)\, p(\mathrm{E}'|\mathrm{H}_0), p(\mathrm{E}|\mathrm{H}_1)\, p(\mathrm{E}'|\mathrm{H}_0)). \end{aligned}$$

Since we have not made any assumptions on the values of these probabilities, we can infer the general equality $g(x, x') = g(xy, x'y)$. In particular, for $a < b$ we can infer that

$$g(a, b) = g((a/b) \cdot b, b) = g(a/b, 1)$$

Thus, $\omega(H_0, H_1, E)$ depends on the ratio of $p(E|H_0)$ and $p(E|H_1)$ only; in the remainder of the proof we call this function f.

Now, assume that E and E′ satisfy the premises of Additivity. We can then infer

$$\begin{aligned}\omega(H_0, H_1, E \wedge E') &= f\left(\frac{p(E \wedge E' \mid H_0)}{p(E \wedge E' \mid H_1)}\right) \\ &= f\left(\frac{p(E|H_0)}{p(E|H_1)} \cdot \frac{p(E'|H_0)}{p(E'|H_1)}\right)\end{aligned}$$

Moreover, we know that

$$\omega(H_0, H_1, E) = f\left(\frac{p(E|H_0)}{p(E|H_1)}\right), \qquad \omega(H_0, H_1, E') = f\left(\frac{p(E'|H_0)}{p(E'|H_1)}\right)$$

and we restate Equation (9.6):

$$\omega(H_0, H_1, E \wedge E') = \omega(H_0, H_1, E) + \omega(H_0, H_1, E').$$

Combining all these equations yields the equality

$$f\left(\frac{p(E|H_0)}{p(E|H_1)} \cdot \frac{p(E'|H_0)}{p(E'|H_1)}\right) = f\left(\frac{p(E|H_0)}{p(E|H_1)}\right) + f\left(\frac{p(E'|H_0)}{p(E'|H_1)}\right).$$

Using the variables $x := p(E|H_0)/p(E|H_1)$ and $y := p(E'|H_0)/p(E'|H_1)$, we can then derive the general equality $f(x \cdot y) = f(x) + f(y)$, since we have not made any assumptions on the values of the above probabilities. This equality is only satisfied by functions of the form $f(x) = k \log_a x$, where a denotes the base of the logarithm. The rest of the proof is trivial; it follows by plugging in the resulting weight-of-evidence measure into the definition of the corroboration measure in Equation (9.4). □

Proof of Theorem 9.4: Without loss of generality, we can restrict ourselves to the case of two disjoint alternatives $\mathcal{H} = \{H_1 \vee H_2\}$ and $\mathcal{H}' = \{H_1, H_2\}$, with $p(E|H_1) > p(E|H_2)$. This is because the relative weighting of the elements of H_1 and H_2 stays the same. Now we observe

$$\begin{aligned}p(E \mid H_1 \vee H_2) &= \frac{1}{p(H_1 \vee H_2)}\Big(p(E|H_1)\, p(H_1) + p(E|H_2)\, p(H_2)\Big) > \\ &\frac{1}{p(H_1) + p(H_2)}\Big(p(E|H_2)\, p(H_1) + p(E|H_2)\, p(H_2)\Big) = p(E|H_2).\end{aligned}$$

Hence $\zeta_{\mathcal{H}}(H_0, E) > \zeta_{\mathcal{H}'}(H_0, E)$, irrespective of the value of $p(E|H_0)$. □

Variation 10: Simplicity and Model Selection

Numquam ponenda est pluralitas sine necessitate.
—*William of Occam*

Is simplicity a virtue of a good scientific theory? Are simpler theories more likely to be true or successful? Few questions in philosophy of science are older, and few have been debated more controversially. The thesis that simple demonstrations, scientific theories or ontological systems are more valuable than complex ones has already been defended by great philosophers such as Aristotle, Aquinas and Kant, for example,

> If a thing can be done adequately by means of one, it is superfluous to do it by means of several; for we observe that nature does not employ two instruments where one suffices. (Aquinas 1945, 129)

Aquinas backs the view of simplicity as a **cognitive value** by the ontological assumption that things are more likely to be simple than complex. Opponents reply that this belief is unjustified: we have no reason to assume that nature has a built-in preference for simple theories or states of affairs. On their view, simplicity is only a **pragmatic value** related to our cognitive limitations as human beings (e.g., van Fraassen 1980). Simple theories are easier to handle than complex ones, be it for purposes of prediction, explanation or further theoretical development. But there is no intrinsic connection between simplicity and truth or predictive accuracy.

This Variation explores whether Bayesians can make sense of the intuition that it is rational to prefer simpler theories to more complex ones for purely epistemic reasons. The first distinction concerns the **syntactic** and the **semantic dimension of simplicity.** The semantic dimension is concerned with the ontological implications of that theory. How many entities does a theory postulate? Are they all of the same kind? And so on. This dimension of simplicity is called **parsimony** (Nolan 1997; Baker 2003, 2016). The thesis that parsimonious theories are to be preferred to less parsimonious ones has again two aspects: one pertaining to the number of postulated entities, and one pertaining to plurality in the kinds of postulated entities. Both of them are fundamental questions in the metaphysics of science. We feel that there is little that the Bayesian framework—which is primarily a tool for uncertain reasoning—can contribute to deciding these ontological disputes.

For our purposes, the syntactic dimension of simplicity is more interesting. It deals with the way scientific theories are formulated: How many hypotheses are postulated? How complex are they? Can they be related to each other in a straightforward way? Discussions about the role of simplicity in curve-fitting, model selection and other omnipresent parts of scientific inference belong to this dimension, which the survey article by Baker (2016) calls **elegance.** In the remainder of the Variation, whenever we write "simplicity", we refer to the sense of simplicity as elegance.

The Variation proceeds as follows. In the first section, we motivate why statistical inference in general, and model selection in particular, is a good battleground for discussing theses about the epistemic value of simplicity (Section 10.1). We also explain a qualitative rationale for preferring simpler models to complex ones. Subsequently (Section 10.2), we take a closer look at Akaike's Information Criterion (AIC), since it has played a large role in discussions of simplicity in philosophy of science. In particular, AIC is often argued to measure an adequate trade-off rate between simplicity and goodness-of-fit (e.g., Forster and Sober 1994). Afterwards, we turn to Bayesian explications of the value of simplicity in model selection. Based on the discussion in Sprenger 2013c, we review three different model selection criteria which depend on the complexity of a model: the Bayesian Information Criterion (BIC, Section 10.3), the Minimum Message Length criterion (MML, Section 10.4), and the Deviance Information Criterion (DIC, Section 10.5). Readers without specific interest in model selection can omit the latter two case studies and jump directly to the final section. Finally, we summarize our findings (Section 10.6): Bayesian inference mainly plays an instrumental role in motivating model selection criteria. Rather than maximizing posterior probability, these criteria use the Bayesian calculus as a convenient mathematical tool for accomplishing diverse epistemological goals. Briefly, there are Bayesian ways of making simplicity matter, but there is no clear-cut Bayesian explication of the virtue of simplicity.

10.1 Simplicity in Model Selection

The debate about simplicity as elegance has, in recent decades, focused on the role of simplicity in statistical **model selection**—or perhaps more aptly, model comparison. This is a relatively young subfield of statistics, where various candidate models are compared and evaluated on the basis of their properties and their fit with the observed data. Model selection is a particularly apt area for discussing the epistemic value of simplicity since we can neatly quantify a model's elegance in terms of its number of adjustable parameters. This understanding of simplicity is manifest in many model selection criteria and allows for a rigorous treatment of the role of simplicity in statistical inference.

In modeling phenomena such as economic growth, social decision-making or the atmospheric system and climate change, statistical models are just idealizations of an excessively complex reality. Often it is unrealistic to assume that the "true model" (i.e., the data-generating process) is found among the candidate models: data sets are often huge and messy, the underlying processes are complex and hard to describe theoretically, and they contain lots of noise and confounding factors (for discussions of this problem in climate science, see Stainforth et al. 2007; Tebaldi and Knutti 2007). Furthermore, candidate models are often generated by automatic means (e.g., as linear combinations of potential predictor variables). This means that they usually do not provide the most striking mechanism for explaining a phenomenon. Rather, they are supposed to be a reliable device for future predictions. Therefore it has been argued that the real epistemological question surrounding simplicity is not whether simple models are more likely to be true but whether they are more likely to be **predictively accurate** (Forster 2002; Sober 2002). If so, simplicity has genuine cognitive, epistemic value because it contributes to attaining another cognitive value, namely predictive accuracy, whose epistemic significance stands undisputed (Kuhn 1977a; McMullin 1982, 2008; Douglas 2013).

We illustrate the basic intuitions behind valuing simplicity with the help of a curve-fitting problem—a special case of model selection. Consider fitting a scatterplot such as Figure 10.1 with a polynomial curve. Assume that we describe the relationship between an independent ("input") variable x and a dependent ("output") variable y either by a linear or by a quadratic polynomial, together with i.i.d. Gaussian (i.e., Normally distributed) noise terms ε_i (e.g., $\varepsilon_i \sim N(0,1)$) and coefficients α, β and γ. Then the data points

$E = \{(x_i, y_i) \colon 1 \leq i \leq N\}$ are described by the equations

$$y_i = \alpha + \beta x_i + \varepsilon_i, \quad \text{(LIN)}$$

$$y_i = \alpha + \beta x_i + \gamma x_i^2 + \varepsilon_i. \quad \text{(PAR)}$$

The linear model (LIN) now corresponds to the null hypothesis $H_0 \colon \gamma = 0$ and the quadratic model (PAR) to the more general alternative $H_1 \colon \gamma \in \mathbb{R}$. The classical method for fitting the linear model to the data is the method of ordinary least squares: the parameters α and β are estimated by the values $\hat{\alpha}$ and $\hat{\beta}$ such that the curve $y = \hat{\alpha} + \hat{\beta}x$ makes the data $E = (x_i, y_i)$ most likely among all pairs (α, β). If the error terms are i.i.d. and follow a Gaussian distribution, this is equivalent to minimizing the sum of the squares of the residuals, $\sum_i (y_i - \alpha - \beta x_i)^2$, that is, the variation in the data that cannot be explained by assuming the model (LIN). The same is done for the more complex model (PAR). In both cases, the estimator $(\hat{\alpha}, \hat{\beta})$ that minimizes this sum is called the **maximum likelihood estimate (MLE).**

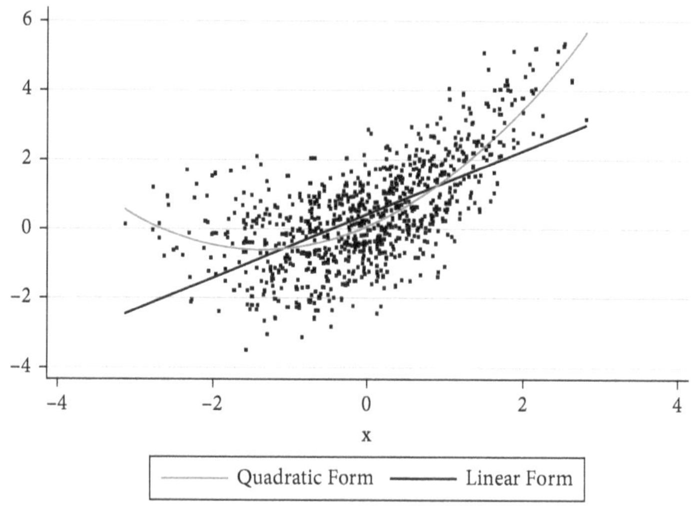

Figure 10.1: A linear model (LIN, straight line) and a quadratic model (PAR, curved line) are fitted to a scatterplot of data according to the ordinary least squares method.

Figure 10.1 shows how linear and quadratic curves are fitted to the data using the MLE. Because (PAR) contains (LIN) as a special case, it has more flexibility in the choice of parameter values and achieves a better fit with the data. However, there is also something strange about (PAR), namely that the values of y increase in both directions, although y seems to be, in general, increasing in x. For higher-order polynomials the effect would be even greater; the curve would be highly oscillatory and there would be many additional extremal and inflection points—features that we would struggle to explain.

Complex models can fit data better than simple models can. But these superior fitting resources can also be a vice: often we will fit a curve to the noise contained in the data, especially when the parameter set is large in comparison to the data set. This is known as the problem of **overfitting.** Besides, the more degrees of freedom a model has, the more difficult it is to estimate all model parameters reliably. Simultaneously estimating numerous parameters is more difficult and error-prone than only estimating one or two of them. So complex models have higher **estimation variance** than simple ones. For this reason, complex models can perform worse than simpler models even if they contain the data-generating process and the simple model doesn't.

This problem is aggravated by maximum likelihood estimation: MLE is overoptimistic with respect to the predictive performance of the chosen model, especially when the number of adjustable parameters is high. An MLE always selects the best-fitting hypothesis in a class. Projecting the current goodness-of-fit to future predictive accuracy just neglects the problem of high estimation variance and the danger of overfitting. While there is a strong epistemic intuition that we should try to find the correct model, it can be safer to work with a wrong, but simple and stable model: the joint estimation of the coefficients in the complex model will often lead to misleading parameter values, and thus to bad predictions. This explains why many philosophers of science have stressed the value of prediction as opposed to mere accommodation (Hitchcock and Sober 2004; Barnes 2008).

The problems of overfitting and estimation variance ground the epistemic value of simplicity in curve fitting and model selection. The issue occurs prominently in null hypothesis significance tests (NHST), too. When a null hypothesis H_0: $\theta = \theta_0$ is tested against the alternative H_1: $\theta \neq \theta_0$, the alternative is more complex than the null since it has one additional degree of freedom: the parameter θ takes no definite value under H_1. This allows H_1 to fit the data better than H_0. On the other hand, if the population mean is not exactly equal to θ_0, but quite close to it, the null hypothesis still does a good job. In that case, the null hypothesis makes more successful predictions than the alternative, and it is easier to use in theoretical developments. The choice of a threshold (e.g., $p < .05$) for calling observed data significant evidence against H_0 and in favor of the more complex H_1 reflects how much we take into account that complex hypotheses can more easily achieve a good fit with the data.

There is a striking resemblance between these considerations and Popper's idea that good scientific theories should trade off simplicity—which is associated with being informative—and predictive accuracy (Popper 1959/2002, 416). Our Variation 9 advances similar considerations: degree of corroboration is an explicit attempt to trade off the precision and simplicity of a null hypothesis with the more flexible fit of the alternative. Such a viewpoint can in turn be related to truthlikeness or verisimilitude as a primary goal of science, and it is possible to find a fruitful role for simplicity in that paradigm (e.g., Oddie 1986; Niiniluoto 1999).

One note of caution, though. The understanding of simplicity as number of free parameters works well for polynomial models and similar cases, but not across the board. Some models are deceptively simple. For example, with only two free parameters (α and β) we can construct a family of curves $f(x) = \alpha \sin(\beta x)$ such that *all* data points in a finite data set $E \subset \mathbb{R}^2$ are fitted up to an arbitrary amount of precision, notwithstanding the size of E (e.g., Romeijn 2017). But certainly such a model has many features that are hard to make sense of scientifically, such as the extreme oscillation of f as a function of x. Comparing different hypotheses in terms of simplicity is thus relative to a particular family of curves (e.g., polynomials) in which they are embedded.

We shall now review bolder claims, made by Forster and Sober (1994) and defended in a series of sequel papers (Forster 1995, 1999, 2000; Sober 2002, 2008; Forster and Sober 2010): the value of simplicity for predictive accuracy is established by means of the mathematical properties of one particular model comparison criterion, the Akaike Information Criterion (AIC). Forster and Sober connect their analysis of AIC to more general theses in philosophy of science, such as the replacement of truth by predictive accuracy as an *achievable* goal of science (Forster 2002), the value of successful prediction compared to mere accommodation (Hitchcock and Sober 2004), the realism–instrumentalism dichotomy (Mikkelson 2006; Sober 2008) and the aptness of Bayesian reasoning for statistical inference (Forster 1995; Bandyopadhyay, Boik and Basu 1996). Because of these far-reaching claims, and because it is useful to compare Bayesian to non-Bayesian model selection criteria such as the AIC, we investigate the rationale behind AIC in some detail. Against Forster and Sober, we shall argue that AIC does not establish an optimal trade-off rate between simplicity and goodness-of-fit: such links are tenuous and the value of simplicity in model selection is highly context-dependent.

10.2 The Akaike Information Criterion

The **Akaike Information Criterion (AIC)** was proposed by Akaike (1973) and Sakamoto, Ishiguro and Kitagawa (1986) as a way to make simplicity matter for predictive accuracy. To avoid equivocations, we first introduce some terminology, following Forster 2002, S127. A statistical (point) *hypothesis* is a specific probability distribution from which the data may have been generated, for example, the standard Normal distribution $N(0,1)$. A statistical *model*, by contrast, consists in a family of hypotheses, for example, all Normal distributions of the form $N(\theta, \sigma^2)$ with parameter values $\theta \in \mathbb{R}$ and $\sigma^2 \in \mathbb{R}^{\geq 0}$. A **model selection criterion** is a function of the observed data that assigns scores to point hypotheses or entire models. On the basis of that score, the different models or point hypotheses can be compared, ranked or averaged. Quite often we will identify point hypotheses with *fitted models*, namely when a particular hypothesis has been obtained by fitting parameters to the data. While some model selection procedures evaluate models in terms of expected predictive accuracy (e.g., Akaike 1973), others, typically classified as Bayesian, aim at the model with the highest posterior probability (e.g., Schwarz 1978).

AIC tries to estimate the discrepancy between a particular model (e.g., the best representative of a class of hypotheses indexed by θ) and the data-generating process. A popular metric for this discrepancy is the **Kullback–Leibler divergence,** here taken with respect to the Lebesgue measure between the model's probability density function g_θ and the density function f of the (unknown) true model:

$$\begin{aligned} \mathrm{D_{KL}}(f, g_\theta) &:= \int f(x) \log \frac{f(x)}{g_\theta(x)}\, dx \\ &= \int f(x) \log f(x)\, dx - \int f(x) \log g_\theta(x)\, dx. \end{aligned} \tag{10.1}$$

This divergence function is used to measure the loss of content when estimating an unknown distribution by an approximating density function g_θ (see also the introductory chapter, page 24, and Variations 1 and 4).

Of course we cannot compute $\mathrm{D_{KL}}(f, g_\theta)$ directly. First, we do not know the true probability density f. This implies that we can only *estimate* $\mathrm{D_{KL}}(f, g_\theta)$. Second, g_θ stands for an entire class of hypotheses with parameter θ, and we have to opt for a particular representative. The maximum likelihood estimate (MLE) $g_{\hat{\theta}}$ is a particularly natural candidate: it is the hypothesis $\theta = \hat{\theta}$ that maximizes the probability of the data given the model. However, if one used the maximum likelihood estimate to estimate KL-divergence without any corrective terms, one would overestimate the closeness to the true model. Third, we are not interested in KL-divergence per se, but in predictive success. So we should in some way relate Equation (10.1) to the predictive performance of a model. Akaike's (1973) famous mathematical result addresses all these worries:

Akaike's Theorem. *For observed data y and a candidate model class g_θ with K adjustable parameters (or an adjustable parameter θ of dimension K), the model comparison criterion*

$$\textsc{aic}(g_\theta, N) \; := \; -2 \log g_{\hat{\theta}(y)}(y) + 2K$$

is an asymptotically unbiased estimator of $\mathbb{E}_x \mathbb{E}_y \left[\log(f(x)/g_{\hat{\theta}(y)}(x))\right]$—the "expected predictive success of $g_{\hat{\theta}}$".

In the above equation, $g_{\hat{\theta}(y)}$ denotes the probability density of the maximum likelihood estimate $\hat{\theta}(y)$. To better understand the double expectation in the last term, note that the maximum likelihood estimate $g_{\hat{\theta}}$ is determined with the help of the data set y. Then, the KL-divergence between $g_{\hat{\theta}}$ and f is evaluated with respect to another dataset x. This justifies the name "predictive success", and taking the expectation two times—over training data y and test data x—justifies the name **expected predictive success.**

In other words, AIC estimates expected predictive success by subtracting the number K of parameters from the log-likelihood of the data under the maximum likelihood estimate $g_{\hat{\theta}}$. It gives an *asymptotically unbiased estimate* of predictive success—an estimate that will, in the long run, center around the true value. The more parameters a model has, the more do we have to correct the MLE in order to obtain an unbiased estimate. Among all hypotheses, we will favor the one which minimizes AIC. What matters is not the absolute AIC score but the relative score compared to the other models. According to Forster and Sober,

> Akaike's theorem shows the relevance of goodness-of-fit *and* simplicity to our estimate of what is true. But of equal importance, it states a precise rate-of-exchange between these two conflicting considerations; it shows how the one quantity should be traded off against the other. (Forster and Sober 1994, 11)

Moreover, they use Akaike's Theorem to counter the (empiricist) idea that simplicity is a merely pragmatic virtue and that "hypothesis evaluation should be driven by data, not by *a priori* assumptions about what a 'good' hypothesis should look like" (Forster and Sober 1994, 27). By means of Akaike's Theorem, simplicity is assigned a specific weight in model selection and established as a cognitive value.

However, the argument does not stand on firm grounds. First, not all unbiased estimators are reliable. The goodness of an estimator $\hat{\theta}$ relative to the true value θ is usually measured by the mean square error, which can be written as the square of the bias plus its variance:

$$\text{MSE}[\hat{\theta}] = \left(\mathbb{E}[\hat{\theta} - \theta]\right)^2 + \mathbb{E}\left[(\hat{\theta} - \theta)^2\right].$$

If $\hat{\theta}$ is unbiased, the first term will disappear, but this does not ensure low overall error—an unbiased estimator may have high variance, dissipate far from the true value and be awfully bad in practice. In particular, it may be outperformed by an estimator with low variance that is only slightly biased. That AIC provides (asymptotically) unbiased estimators is, in itself, no sufficient reason to prefer it over competing model selection criteria.

This objection may be countered by noting that unbiasedness is an advantage, *ceteris paribus*. Forster and Sober (2010) note that AIC and one of its rivals, the Bayesian Information Criterion (BIC), just differ by a constant. If estimators differ by a constant, they have the same variance. Since mean square error = square of bias + variance, the unbiased estimator will have the lower mean square error. Hence BIC seems to be a worse estimator of predictive accuracy than the unbiased AIC.

Apart from the fact that BIC pursues a different estimation target (see the next section), this argument is based on an oversight which many authors in the debate commit (Forster and Sober 1994, 2010; Kieseppä 1997): AIC is *not* an unbiased estimator—it is just *asymptotically* unbiased; in other words, the property of unbiasedness is only realized for very large samples. To the excuse of these authors, it should be added that also the standard textbook on AIC uses this formulation in passing (Sakamoto, Ishiguro and Kitagawa 1986, 69—though see also pages 65, 77 and 81).

We invite the reader to have a look at the mathematical details in the appendix of this chapter. There, the dependence of Akaike's Theorem on the asymptotical, and not the actual, normality of the maximum likelihood estimate becomes clear. This has substantial consequences and speaks against an overly normative interpretation of Akaike's findings. For medium-sized and small data sets, there is no general reason why we should base model selection on AIC. As long as we don't know the speed of convergence, the asymptotic properties do not tell us much about the performance on an actual, finite data set. There are, of course, usually conditions where AIC outperforms its competitors (Burnham and Anderson 2002; Yang 2005), but they have to be defined specifically and do not hold across the board. In the light of these observations, claims that AIC establishes a trade-off rate between simplicity and goodness-of-fit shine in a dim light. But do Bayesian model selection criteria fare any better?

10.3 The Bayesian Information Criterion

The transition from a prior to a posterior probability function is a cornerstone of Bayesian inference. Consequently, a model selection procedure is called Bayesian if it is based on the posterior distribution of degrees of belief, or on the divergence between prior and posterior probabilities. This was also the rationale behind the manifold measures of confirmation presented in Variation 1.

An example for such a procedure is model selection based on **Bayes factors.** They compare the performance of the rivalling models H_1 and H_0 by means of the ratio

$$B_{10}(E) := \frac{p(H_1|E)}{p(H_0|E)} \cdot \frac{p(H_0)}{p(H_1)} = \frac{p(E|H_1)}{p(E|H_0)}.$$

Citing work by Spiegelhalter and Smith (1980), Kass and Raftery (1995, 790) argue that Bayes factors act as a "fully automatic Occam's razor" for nested models: when the Bayes factor favors a simple model (e.g., H_0: $\theta = 0$), the complex model will be penalized for hosting lots of poor-fitting hypotheses (e.g., H_1: $\theta \neq 0$). In that case, the loss in predictive accuracy incurred by accepting the simpler model will be negligible. The search for an *explicit* trade-off rate between simplicity and goodness-of-fit is replaced by the rate that is implicit in Bayesian inference and Bayesian measures of evidence.

However, this orthodox Bayesian method of model selection is not frequently put into practice. First of all, there is a plethora of practical and methodological problems, such as the computational costs of calculating posterior distributions or handling nested models in a Bayesian framework. Second, when prior probabilities are assigned, reliable expert opinion is usually hard to elicit, so that the choice of the prior is often dominated by mathematical convenience. Furthermore, results may be highly sensitive to the prior distribution. This has triggered a search for computationally feasible approximations of Bayes factors and/or posterior probabilities. Schwarz's

Bayesian Information Criterion (BIC) is the most prominent representative of that group; hence we will investigate its foundations in some detail.

The BIC aims at the posterior probability of a parametric model M_θ, that is, at the weighted sum of the posterior probabilities of the hypotheses in M_θ that correspond to different values of θ. Thus it has a different target than AIC, which compares the best-performing representatives of various models. We will now reconstruct and analyze the motivation of BIC, following Schwarz 1978.

Assume that M_θ is one of our candidate models, indexed by vectors θ with dimension K. We would like to approximate the posterior probability of M_θ. Assume further that all probability densities for data x (with respect to the Lebesgue measure) belong to the exponential family and that they can be written as

$$p(x\,|\,\theta) \;=\; \exp\Big(N\big(A(x) \;-\; \lambda|\theta - \hat{\theta}(x)|^2\big)\Big). \qquad (10.2)$$

Here, $\hat{\theta}(x)$ denotes the maximum likelihood estimate of the unknown θ, and N the sample size, assuming independent sampling. This specific form of the likelihood function seems to make a substantial presumption, but in fact the densities in Equation (10.2) comprise the most familiar distributions, such as the Normal, Uniform, Fisher, Poisson and Student's t-distribution. For that reason, from a practical point of view the assumption is plausible.

Then we take a standard Bayesian approach and write the posterior probability of M_θ as proportional to the prior probability $p(M_\theta)$ and the averaged likelihood of the data x under M_θ:

$$\begin{aligned} p(M_\theta\,|\,x) \;&\sim\; p(M_\theta)\int_{\theta\in\Theta}\exp\Big(N\big(A(x) \;-\; \lambda|\theta - \hat{\theta}(x)|^2\big)\Big)d\theta \\ &=\; p(M_\theta)\,\exp(NA(x))\int_{\theta\in\Theta}\exp\big(-N\lambda|\theta - \hat{\theta}(x)|^2\big)d\theta. \end{aligned}$$

After two linear transformations of the integration variable θ (adding $\hat{\theta}(x)$ and multiplying with $1/\sqrt{N\lambda}$), and realizing that for the maximum likelihood estimate $\hat{\theta}(x)$, $p(x\,|\,\hat{\theta}(x)) = \exp(NA(x))$, we obtain

$$\begin{aligned} &\log p(M_\theta\,|\,x) \\ &\sim\; \log p(M_\theta) + NA(x) + \log(1/N\lambda)^{K/2} \\ &\quad + \log\int_{\theta\in\Theta}\exp(-|\theta|^2)\,d\theta \qquad (10.3) \\ &=\; \log p(M_\theta) + NA(x) + 1/2\,K\log(1/N\lambda) + \log\sqrt{\pi}^K \\ &=\; \log p(M_\theta) + \log p(x\,|\,\hat{\theta}(x)) - 1/2\,K\log(N\lambda/\pi). \end{aligned}$$

Let us take stock. On the left-hand side, we have the log-posterior probability, which can be interpreted as a subjective Bayesian's natural model selection criterion. As we see from Equation (10.3), this term is proportional to the sum of three terms: log-prior probability, the log-likelihood of the data under the maximum likelihood estimate, and a penalty proportional to the number of model parameters. This derivation, whose assumptions are relaxed subsequently in order to yield more general results, forms the mathematical core of BIC. The number K of parameters enters the calculations because the expected likelihood of the data depends on the dimension of the model, via the skewness of the likelihood function.

In practice, it is difficult to elicit sensible subjective prior probabilities of the candidate models (i.e., classes of hypotheses), and the computation of posterior probabilities is computationally demanding. Therefore Schwarz suggests to estimate log-posterior probability in Equation (10.3) by a large-sample approximation. We neglect the terms that make only constant contributions and focus on the terms that increase in N. So $\log p(M_\theta)$ drops out of the picture. In the long run, the model with the highest posterior probability will be the model that minimizes

$$\text{BIC}(M_\theta, x) \;=\; -2\log p(x \mid \hat{\theta}(x)) \;+\; K \log {}^{N}\!/\!_{\pi}. \tag{10.4}$$

BIC is intended to select the model that accumulates, in the long run, the most posterior mass. However, it neglects the contribution of the priors when comparing the models to each other. Keeping in mind the identity

$$\log p(\mathrm{H}|\mathrm{E}) \;=\; \log p(\mathrm{H}) + \log\left(\frac{p(\mathrm{E}|\mathrm{H})}{p(\mathrm{E})}\right) \tag{10.5}$$

and comparing this expression to Equation 10.3, we see that BIC could as well be described as an approximation to the log-ratio measure of confirmation $\log p(\mathrm{H}|\mathrm{E}) - \log p(\mathrm{H}) \;=\; \log(p(\mathrm{H}|\mathrm{E})/p(\mathrm{H}))$, up to addition of a constant (see Variation 2).

We conclude that BIC does not possess a strict Bayesian justification: while (log-ratio) confirmation may be suitable for comparing models on the basis of past performance, it does not conform to classical subjective Bayesian inference: the priors drop out of the picture, as witnessed by the transition from (10.3) to (10.4). Instead of conforming to the subjective Bayesian rationale, the BIC uses the Bayesian calculus as a convenient mathematical tool for meeting goals such as selecting models with strong performance on past data. There is nothing specifically Bayesian about the estimation target of

BIC. This finding is, by the way, in agreement with Schwarz's note that BIC extends "beyond the Bayesian context" (Schwarz 1978, 461; see also Forster and Sober 1994, 23–24). What is more, frequentist properties are sometimes invoked in an attempt to justify the practical use of BIC (e.g., Burnham and Anderson 2002). To strengthen this conclusion further, note that BIC is quite different from a numerical large-sample approximation for posterior degrees of belief: the posterior approximated by BIC is detached from subjective prior probability.

Neither does the statistical consistency of BIC provide a genuinely Bayesian justification. An estimator is called *consistent* if and only if, as sample size increases, it converges in probability to the true model. Both Bayesians and frequentists regard consistency as a necessary constraint on good estimators, and BIC is consistent as long as the class of all models contains the data-generating process. But even when this optimistic assumption holds, consistency alone has no implications for speed of convergence to the true value. A familiar problem from the analysis of AIC resurfaces: for medium-sized data sets, desirable statistical properties of the model do not guarantee good performance.

All in all, BIC lacks, in spite of the extensive use of Bayesian formalism, a fully Bayesian rationale. This diagnosis is supported by the variety of purposes to which BIC is put. Sometimes it is regarded as an approximation to the Bayes factor (Kass and Raftery 1995). Raftery (1995) proposes an interpretation of BIC as an approximation to marginal likelihood, which is easily derived on the basis of the above calculations. Romeijn, van de Schoot and Hoijtink (2012) see different worries with a Bayesian understanding of BIC and propose to anchor it more securely in Bayesian reasoning by taking into account the size of the parameter space. This would allow for a more informative assignment of prior probabilities. Hence, what the asymptotic analysis of BIC approximates is not determined by the mathematics only: it depends on the general perspective one adopts. All this shows that for BIC, Bayesian inference constitutes no consistent philosophical underpinning (e.g., as a logic of belief revision), but only a flexible framework for motivating a specific estimator of posterior probability.

10.4 The Minimum Message Length Principle

Another attempt to construct a properly Bayesian model selection criterion is the **Minimum Message Length (MML) principle** (Wallace 2005; Dowe 2011).

MML is a statistical inference procedure aiming at inferring the hypothesis ("theory")

> that allows the data to be stated in the shortest two-part message, where the first part of the message asserts the theory, and the second part of the message asserts the data under the assumption that the asserted theory is true. (Dowe, Gardner and Oppy 2007, 717)

The basic idea is to infer the best-explaining hypothesis, which is explicated as the explanation with the shortest expected message length in a probabilistic code. That is, the explanation has to trade off the plausibility of the hypothesis with the likelihood of the data under the hypothesis.

We illustrate this idea by means of an example (cf. Dowe, Gardner and Oppy 2007, 721–722). Assume we want to estimate the parameter θ in a repeated Bernoulli (i.e., success vs. failure) experiment, where X quantifies the number of successes in N independent and identically distributed trials. Then MML partitions the sample space $\mathcal{X} = \{0, \ldots, N\}$ into K interval sets $I_k = \{c_{k-1}, \ldots, c_k - 1\}$ with $c_0 = 0$ and $c_K = N+1$. Let k_j be a weakly monotonic sequence such that $j \in I_{k_j}$. Then, we define for each I_{k_j} a corresponding point estimate $\hat{\theta}_{k_j}$ of θ such that any $0 \leq j \leq N$ is mapped to $\hat{\theta}_{k_j}$.

Assuming a uniform prior over θ, the expected message length of estimator $\hat{\theta}$ is measured by the term

$$L := -\sum_{j=0}^{N} p(X=j)\Big(\log p(\hat{\theta}_{k_j}) + \log p(X=j \mid \hat{\theta}_{k_j})\Big). \qquad (10.6)$$

In the case of $N = 100$, the optimal partition works with 10 different point estimates (see Table 10.1). Notably, the "natural" unbiased estimator X/N does not perform well on this count: the low prior probability of the associated intervals, which only consist of a singleton set each, diminishes the overall score of X/N.

From a Bayesian point of view: The two components of L correspond to the two core components of Bayesian inference: the (log-)prior of the hypothesis (here: $\hat{\theta}_{k_j}$) and the (log-)likelihood of the data, given that hypothesis (here: $p(X = j \mid \hat{\theta}_{k_j})$). MML proponents then argue that an inference to the theory that allows for the shortest two-part message will also be an inference to the most probable theory (or model).

However, since we measure *expected* total message length, the optimal trade-off depends on the design of the experiment and in particular the sample size, cf. Equation (10.6). This is actually admitted by the inventors of MML:

I_{k_j}	0	1–6	7–17	18–32	33–49	50–66	67–81	82–93	94–99	100
$\hat{\theta}_{k_j}$	0	0.035	0.12	0.25	0.41	0.58	0.74	0.875	0.965	1

Table 10.1: The optimal MML partitioning of the sample space (i.e., the possible numbers of successes) into intervals I_{k_j} and the corresponding parameter estimates $\hat{\theta}_{k_j}$, for the case of the Binomial distribution $B(100, \theta)$ with a uniform prior. See Wallace 2005, 157–160; Dowe, Gardner and Oppy 2007, 721–722.

> The receiver of an explanation message is assumed to have prior knowledge on the set X of possible data, and the message is coded on that assumption. ... The optimum explanation code requires that one assertion or estimate value serve for a range of distinct but similar possible data values. Hence, it seems inevitable that the assertion [i.e., the hypothesis] used to explain the given data will *depend to some extent on what distinct but similar possible data values might have occurred but did not*. (Wallace 2005, 254, our emphasis)

What is more, for the entire idea of the "shortest explanation", we have to choose between different conceptualizations of the hypothesis space, dependent on the chosen experimental design. This situation is in itself remarkable: while classical Bayesian reasoning considers the set of candidate models as fixed, MML aims at finding the partition of the hypothesis space that allows for the most efficient encoding of hypothesis and data.

These dependencies conflict, however, with one of the core principles of Bayesian inference when applied to statistical inference: the **Likelihood Principle** (Birnbaum 1962). Here is one of its classical formulations:

> [A]ll the information about θ obtainable from an experiment is contained in the likelihood function $L_{\mathrm{E}}(\theta) = p(\mathrm{E}\,|\,\theta)$ for θ given E .... Two likelihood functions for θ (from the same or different experiments) contain the same information about θ if they are proportional to one another. (Berger and Wolpert 1984, 19, notation adapted)

To see how closely the Likelihood Principle aligns with Bayesian inference, imagine that H and ¬H each make a specific claim about θ (e.g., $\theta = 0$ and $\theta = 1$, respectively) and recall the identity

$$p(\mathrm{H}\,|\,\mathrm{E}) = \left(1 + \frac{p(\neg\mathrm{H})}{p(\mathrm{H})}\,\frac{p(\mathrm{E}\,|\,\neg\mathrm{H})}{p(\mathrm{E}\,|\,\mathrm{H})}\right)^{-1},$$

which is just another way of expressing Bayes' Theorem. From a Bayesian point of view, the likelihood function encompasses all relevant experimental information that is not already contained in the priors.

The Likelihood Principle demands in particular that one's inference not depend on the space of possible outcomes, or on the sampling protocol, whereas in an MML inference, the same data will lead to different best estimates of θ when obtained from a Binomial design or a Negative Binomial design, respectively.

At this point, one may doubt that the Likelihood Principle is compelling for a Bayesian statistician, so much the more as the wording chosen by Berger and Wolpert is admittedly vague. Therefore it is important to realize that it is actually equivalent to the conjunction of the following two principles:

Sufficiency Let $\mathcal{E}$ be an experiment with a statistical model parametrized by $\theta \in \Theta$ and random variable X. If $T(X)$ is a sufficient statistic for θ, that is, if it satisfies $p(X = x \mid T = t, \theta) = p(X = x \mid T = t)$, and $\mathcal{E}_T$ is the experiment where any outcome x of $\mathcal{E}$ is represented by reporting the value $T(x)$, then $\mathcal{E}$ and $\mathcal{E}_T$ yield the same evidence about θ.

(Strong) Conditionality If $\mathcal{E}$ is any experiment having the form of a mixture of component experiments $\mathcal{E}_i$, then for each outcome $(\mathcal{E}_i, x_i)$ of $\mathcal{E}$, ... the evidential meaning of any outcome x of any mixture experiment $\mathcal{E}$ is the same as that of the corresponding outcome x_i of the experiment $\mathcal{E}_i$ which has actually been performed, ignoring the overall structure of the mixed experiment. (Birnbaum 1962, 270–271, notation adapted)

Since the Sufficiency Principle is unanimously endorsed by both Bayesian and frequentist statisticians, we can focus our inquiry on the (Strong) Conditionality Principle. Informally, Conditionality can be described as asserting the evidential irrelevance of experiments that were actually not performed. From a Bayesian point of view, this is eminently sensible; after all, the entire idea of Bayesian Conditionalization is based on taking into account (only) evidence that has actually been observed. Indeed, a lot of Bayesian arguments in statistical methodology and experimental design (e.g., with respect to optional stopping) are based on the soundness of that principle and explicitly on the irrelevance of which results *might* have been observed (cf. Royall 1997; Sprenger 2009a). So Conditionality and MML are directly at odds with each other.

The conflict is, by the way, known from other foundational debates in statistical inference. For instance, reference Bayesians such as Bernardo (2012) determine reference priors for Bayesian inference as a function of the sample space. Seen in that light, MML is somewhat typical for modern Bayesian statistics and its departure from Bayesian orthodoxy. It exemplifies a hybrid approach, where the Bayesian machinery is primarily a mathematical and conceptual toolbox for solving a specific problem whose definition does not depend on the Bayesian framework itself: determining the shortest explanation, the most efficient coding of theory and evidence. This is not meant to doubt that MML shares more with Bayesian inference than related model selection criteria do (e.g., Minimum Description Length—MDL; Grünwald 2005). But it is illuminating to see that extensive use of the Bayesian formalism and an explicit appeal to Bayesian reasoning can still hide disagreements with classical principles of Bayesian inference.

10.5 The Deviance Information Criterion

The analysis of BIC and MML shows that it is not straightforward to delimit Bayesian from non-Bayesian model selection criteria. Sometimes it is suggested to draw the division along the lines of inferential targets (Burnham and Anderson 2002, 2004). According to that proposal, even if the inferential strategies employed are not properly Bayesian at every step, as we have seen for MML and BIC, the target of inference—the posterior probability of a model or a fitted model—can only be formulated within a Bayesian framework. In support of this view, it is sometimes asserted that "Bayesians assess an estimator by determining whether the values it generates are probably true or probably close to truth" (Forster and Sober 2010, 535) or that "the model selection literature often errs that AIC and BIC selection are directly comparable, as if they had the same objective target model" (Burnham and Anderson 2004, 299). That is, where frequentist methods, such as AIC, estimate the predictive performance of fitted models, Bayesian methods, such as BIC, estimate the posterior probability of a given model, or construct estimators that minimize mean error with respect to the posterior distribution. To show that such a dichotomy is untenable, we conduct a further case study, namely on the Deviance Information Criterion (DIC).

The DIC is another model selection criterion that is commonly placed in the Bayesian family. Many model selection criteria, such as AIC and BIC, can be written and interpreted as an explicit trade-off of goodness-of-fit and complexity. This is difficult in a specific context that we often encounter in practice: complex, hierarchical models (e.g., Henderson et al. 2010). That is, when we represent the marginal distribution of the data x in a probability model as

$$p(x) = \int_{\theta \in \Theta} p(x \mid \theta)\, p(\theta)\, d\theta \qquad (10.7)$$

with parameter θ and prior density $p(\theta)$, we may sometimes choose to represent that prior as being governed by a hyperparameter ψ:

$$p(\theta) = \int_{\psi \in \Psi} p(\theta \mid \psi)\, p(\psi)\, d\psi. \qquad (10.8)$$

However, it is now unclear what should be considered the likelihood function of the data: $p(x \mid \theta, \psi)$, $p(x \mid \theta)$ or $p(x \mid \psi)$ (Bayarri, De Groot and Kadane 1988)? Consequently, it is unclear how the complexity of the model should be measured: should we base our understanding of complexity on the dimension of θ, the dimension of ψ or an aggregate of both? Apart from this ambiguity, the complexity of a model also depends on the amount of available prior information on the parameter values. The more information we have, the less complex a model is. Straightforwardly measuring complexity as the number of free parameters, as in the case of BIC, is therefore inappropriate as a general procedure.

Therefore Spiegelhalter et al. (2002) propose to measure complexity by comparing the *expected deviance* in the data (under the posterior distribution) to the deviance in the estimate $\tilde{\theta}(x)$ that we would like to use. In other words, complexity manifests itself in terms of "difficulty in estimation". The authors propose to measure surprise or deviance in the data x relative to a point hypothesis parametrized by $\theta \in \Theta$ by means of the function $f(\theta, x) = -\log p(x \mid \theta)$. This logarithmic scoring rule has a variety of theoretically appealing properties that distinguish it vis-à-vis common alternatives, including the quadratic scoring rule discussed on page 15 (Good 1952; Williamson 2010). The Bayesian twist of DIC, as opposed to frequentist approaches, consists in incorporating prior information on the parameters: "it seems reasonable that a measure of complexity may depend on both the prior information concerning the parameters in focus and the specific data that are observed" (Spiegelhalter et al. 2002, 585).

In particular, $\tilde{\theta}(x)$ denotes the Bayes estimator of the quantity of interest θ, usually the posterior mean of θ. Then we can compare the expected deviance in the data (conditional on the posterior distribution of θ) to the deviance we observe under our estimate of $\tilde{\theta}(x)$. This quantity p_D indicates how difficult it is to efficiently fit the parameters of a model M_θ:

$$p_D(M_\theta, x) = \mathbb{E}_{\theta|x}\left[-2\log p(x \mid \theta)\right] - 2\Big(-\log p(x \mid \tilde{\theta}(x))\Big)$$

$$= \quad 2\log p(x\,|\,\tilde{\theta}(x)) - 2\int_{\theta\in\Theta} \log p(x\,|\,\theta)\, p(\theta\,|\,x)\, d\theta, \qquad (10.9)$$

where $\mathbb{E}_{\theta|x}$ refers to the posterior expectation with respect to $p(\theta\,|\,x)$. Reading Equation (10.9) in yet another way, p_{D} measures the extent to which our estimate $\tilde{\theta}(x)$ is expected to overfit the data and how much deviance we can expect to observe in the future. This interpretation connects p_{D} to the predictive performance of our estimate.

Indeed, p_{D} has been used regularly for assigning scores to candidate models, and it serves as the basis of the **Deviance Information Criterion (DIC),** a model selection criterion defined as

$$\mathrm{DIC}(M_\theta, x) \;=\; \mathbb{E}\left[\mathrm{D}_f(\theta, x)\right] + p_{\mathrm{D}}(M_\theta, x), \qquad (10.10)$$

where the function $\mathrm{D}_f(\,\cdot\,,\,\cdot\,)$ is defined as

$$\mathrm{D}_f(\theta, x) \;=\; -2\log p(x\,|\,\theta) + 2\log f(x) \qquad (10.11)$$

for some standardized function $f(x)$ of the data. Taking into account that Equation (10.11) is mainly a function of the deviance between model M_θ and data x, we can regard the overall DIC score in Equation (10.10) as a trade-off between goodness of fit (the D_f-term) and the expected overfit (p_{D}).

The form of DIC illustrates that its target of inference is not particularly Bayesian. The difficulty of accurately fitting a model is relevant for the practitioner (e.g., for checking the adequacy of a model), but not of intrinsic interest for the orthodox Bayesian reasoner. On the other hand, there are many Bayesian elements in DIC: the estimator $\tilde{\theta}(x)$, whose deviance is estimated in Equation (10.9), is nothing but the posterior mean of θ, and it is evaluated with respect to the posterior distribution of θ. Also, Spiegelhalter et al. (2002) show how DIC can be understood as an approximate estimator of posterior expected loss.

The inventors of p_{D} and DIC are actually aware of that tension and clarify that they believe a rigorous Bayesian justification to be neither available nor necessary:

> Our approach here can be considered to be semiformal. Although we believe that it is useful to have measures of fit and complexity, and to combine them into overall criteria that have some theoretical justification, we also feel that an overly formal approach to model 'selection' is inappropriate since so many other features of a model should be taken into account before using it as a basis for reporting inferences .... (Spiegelhalter et al. 2002, 602)

DIC is thus a formidable example of a hybrid, eclectic approach to inference in model selection: it is inspired by Bayesian inference, frequentism and statistical decision theory. Notably, this eclecticism can go either way. For instance, if the amount of prior information is substantial compared to the data set, then the classical, frequentist AIC can be calibrated to asymptotically approximate the Bayes factor of different models (Kass and Raftery 1995), or it can be represented as a more general Bayesian criterion (Forster and Sober 2010).

From this analysis, we see that the idea of identifying Bayesian model selection by means of its inferential targets is not convincing. In particular, DIC clearly demonstrates that the targets of Bayesian model selection procedures are much more nuanced and varied than just posterior probabilities or Bayes estimates. Second, and more generally, target and justification of a model selection procedure are usually intertwined and hard to separate from each other. This implies that also the role and the weight of simplicity in Bayesian model selection are hard to characterize in general.

10.6 Discussion

Simplicity is a complex and ambiguous concept. This ambiguity may explain why people are so divided over whether or not it does have cognitive value in science. From a Bayesian point of view, it is most fruitful to investigate the concept of simplicity as elegance, that is, as referring to the number and the complexity of hypotheses in a scientific theory. The ontological dimension of simplicity ("parsimony") is left out of the picture.

Typically, simplicity as elegance has been investigated in the framework of model selection and curve-fitting, that is, in fitting a scientific hypothesis (i.e., a particular curve) to a set of data points. In this context, simplicity has epistemic value as an antidote to estimation error and overfitting. However, this qualitative finding does not answer the question of whether there is an optimal trade-off rate between simplicity and other cognitive values in model selection, most notably goodness-of-fit, and whether there is a specific Bayesian way to reason about simplicity.

To answer these questions, we have reviewed a non-Bayesian model selection criterion (AIC) and three Bayesian criteria (BIC, MML and DIC). First, we have taken issue with Forster and Sober's claim that an optimal trade-off rate between simplicity and goodness-of-fit is established by the Akaike Information Criterion (AIC). Then we moved on to the Bayesian model selection criteria. We observed that BIC, MML and DIC conform only partially to Bayesian reasoning, even if they are firmly anchored within the Bayesian formalism. Rather, they should be described as hybrid procedures: the Bayesian calculus may serve a different goal (MML: efficient coding), some crucial elements of Bayesian reasoning may be dropped (BIC: subjective priors) and ideas and techniques from different philosophies (DIC: Bayesian inference, decision theory, frequentism) may be mixed. This diagnosis need not conflict with their classification as Bayesian model selection procedures, but it highlights differences in target, justification and intended application context.

Accordingly, the question of what *justifies* these procedures cannot be answered in full generality. Neither of them has a general frequentist or Bayesian justification—it is even possible to find a non-Bayesian justification for BIC, and a Bayesian justification for AIC (Romeijn, van de Schoot and Hoijtink 2012). Consequently, the adequacy of a chosen model selection criterion depends on whether the implicit assumptions in the derivations of the criteria are satisfied. For example, BIC discounts the priors and focuses on asymptotic behavior, whereas DIC is particularly apt in hierarchical models. MML, on the other hand, does not work with a fixed set of candidate models: efficiently partitioning the model space is already an essential part of the inference problem! This is a crucial difference to BIC and DIC. The practitioner faces the non-trivial task of ensuring that a model selection criterion is adequate for a given application context.

So it is misleading to attach model selection criteria, and the implied weights of simplicity, to particular philosophical schools. The judgment on when these criteria do and do not work is highly context-sensitive. The Bayesian calculus involved in the derivation of BIC might be characterized as an **instrumental Bayesianism**—an approach to statistical inference which is happy to use Bayes' Theorem as a scientific modeling tool, without however taking the Bayesian elements too literally, as expressions of subjective uncertainty. See Sprenger 2009b, Sections 3 and 4, and Sprenger 2013c for more discussion of these topics.

Thus BIC, MML, or DIC do not exemplify a "Bayesian philosophy of model selection" in a substantive philosophical sense. This has repercussions on attempts at using the behavior of model selection criteria for a general assessment of Bayesian inference. For instance, Dowe, Gardner and Oppy (2007) defend MML on grounds of its generality, efficiency and invariance under transformations of the parameter space. They conclude:

> Since MML is a Bayesian technique we should conclude that the best philosophy of science is Bayesian. (Dowe, Gardner and Oppy 2007, 712)

However, we have seen that an implicit premise of such arguments—namely that Bayesian model selection (here instantiated by MML) is firmly anchored in Bayesian philosophy—is not satisfied. Therefore it is hard to use Bayesian model selection for promoting Bayesian approaches to simplicity, and Bayesian inference as a general philosophical framework.

On the other hand, as the late Dutch football legend Johan Cruyff used to say, each disadvantage has its advantage ("Elk nadeel heeft zijn voordeel"). Analogous arguments *against* Bayesian inference are suspicious on the same grounds. For example, Forster and Sober write with respect to the AIC:

> Bayesianism is unable to capture the proper significance of considering *families* of curves ... Akaike's reconceptualization of statistics does recommend that the foundations of Bayesian statistics require rethinking. (Forster and Sober 1994, 26, original emphasis)

As we have seen in the previous sections, this conclusion is overblown. First, Bayesian and non-Bayesian model selection criteria stand on equal footing. Second, the link between particular model selection criteria and philosophical schools is not particularly tight. Third, in calculating Bayes factors and approximations to them (such as the BIC), Bayesian inference considers the significance of families of curves (models) as opposed to single curves (fitted models). It may also be noted that Forster and Sober do not repeat this claim in later publications, probably realizing that their conclusion went too far.

This observation also suggests several avenues for further research. First, we can search for a philosophically appealing way of systematizing the context-sensitive considerations for particular model selection criteria (cf. Burnham and Anderson 2002, 2004). In other words, can we extract general guidelines how criteria should be matched with types of datasets and scientific fields? Second, we could extend our analysis to other model selection criteria (e.g., Grünwald's MDL) in order to see whether our conclusions regarding the philosophically eclectic foundations of model selection criteria remain valid. More specifically, it would be exciting to see whether our thesis of instrumental Bayesianism transfers to an instrumental frequentism for the case of AIC and other non-Bayesian model selection criteria. For example, can we find cases of scientific reasoning where schools of uncertain reasoning are treated as a mathematical quarry for practical solutions rather than as a philosophical basis that justifies specific inferences? Third, one may compare the role of simplicity in model selection to its role in other domains of scientific reasoning, such as the problem of trading off simplicity and goodness-of-fit in choosing between various causal Bayesian networks.

In the following chapter, Variation 11, we respond to a foundational objection to (subjective) Bayesian inference: It is not objective enough to ground evidential judgments and public policy decisions. In doing so, we combine a conceptual analysis of scientific objectivity with arguments from the practice of Bayesian statistics.

Appendix: Sketch of the Derivation of the Akaike Information Criterion

At the end of this Variation, we summarize the main steps of the derivation of AIC below, with a focus on the philosophical rationale that motivates this model selection criterion. Detailed treatments can be found in Chapter 4.3 of Sakamoto, Ishiguro and Kitagawa 1986 and Chapter 7.2 in Burnham and Anderson 2002.

The AIC aims at estimating the "expected predictive success" of a model, identified with its maximum likelihood estimate (MLE) $g_{\hat{\theta}(y)}$:

$$\mathbb{E}_x\mathbb{E}_y\left[\log \frac{f(x)}{g_{\hat{\theta}(y)}(x)}\right] = \mathbb{E}_x\mathbb{E}_y\left[\log f(x)\right] - \mathbb{E}_x\mathbb{E}_y\left[\log g_{\hat{\theta}(y)}(x)\right]. \qquad (10.12)$$

The first term on the right-hand side of (10.12) is equal for all candidate models. When comparing them, it drops out as a constant. Hence we can neglect it in the remainder and focus on the second term in (10.12).

The AIC is usually derived by a double Taylor expansion of the log-likelihood function. The general formula of Taylor expansion for an analytic, real-valued function f is

$$f(x) = \sum_{k=0}^{\infty} f^{(k)}(x_0)(x-x_0)^k.$$

In our case, we expand the term $\log g_{\hat{\theta}(x)}(y)$—our MLE—around θ_0, the value of θ that minimizes Kullback–Leibler divergence from the true model. The expansion is truncated at $k=2$, yielding

$$\begin{aligned}\log g_{\hat{\theta}(y)}(x) \approx\ & \log g_{\theta_0}(x) + N\left(\left(\frac{\partial}{\partial\theta}\log g_\theta(x)\right)(\theta_0)\right)(\hat{\theta}(y)-\theta_0) \\ & + 1/2 N(\hat{\theta}(y)-\theta_0)^T\left(\left(\frac{\partial^2}{\partial\theta^2}\log g_\theta(x)\right)(\theta_0)\right)(\hat{\theta}(y)-\theta_0).\end{aligned} \qquad (10.13)$$

The matrix

$$J := \left(-\frac{\partial^2}{\partial\theta^2}\log g_\theta(x)\right)(\theta_0),$$

that also occurs in (10.13), is called the Fisher information matrix of the data. It plays a crucial role in an asymptotic approximation of the maximum likelihood estimate that holds under plausible regularity conditions:

$$\sqrt{N}(\hat{\theta}(y)-\theta_0) \rightarrow \mathcal{N}(0, J^{-1}).$$

This asymptotic normality of the maximum likelihood estimate can be used to simplify (10.13). The term

$$\sqrt{N}(\hat{\theta}(y) - \theta_0)^T(-J)\sqrt{N}(\hat{\theta}(y) - \theta_0) \tag{10.14}$$

is asymptotically χ^2-distributed with K degrees of freedom. Hence, the expectation of (10.14) is K. By taking a double expectation over x and y, we thus obtain that

$$\mathbb{E}_x\mathbb{E}_y\left[1/2N(\hat{\theta}(y) - \theta_0)^T\left(\left(\frac{\partial^2}{\partial\theta^2}\log g_\theta(x)\right)(\theta_0)\right)(\hat{\theta}(y) - \theta_0)\right] \approx K/2. \tag{10.15}$$

Moreover, the linear term in (10.13) vanishes because the maximum likelihood estimate is an extremal point of the log-likelihood function. Thus, the mean of the first derivative is also zero:

$$\mathbb{E}_x\mathbb{E}_y\left[N\left(\left(\frac{\partial}{\partial\theta}\log g_\theta(x)\right)(\theta_0)\right)(\hat{\theta}(y) - \theta_0)\right] = 0. \tag{10.16}$$

Combining (10.13) with (10.15) and (10.16), we obtain for large samples that

$$\mathbb{E}_x\mathbb{E}_y\left[\log g_{\hat{\theta}(y)}(x)\right] \approx \mathbb{E}_x\left[\log g_{\theta_0}(x)\right] - K/2. \tag{10.17}$$

Repeating the Taylor expansion around the maximum-likelihood estimate and applying the same arguments once more gives us

$$\mathbb{E}_x\mathbb{E}_y\left[\log g_{\theta_0}(x)\right] \approx \mathbb{E}_y\left[\log g_{\hat{\theta}(y)}(y)\right] - K/2. \tag{10.18}$$

Finally, by combining (10.17) and (10.18), we obtain AIC as an estimate of "expected predictive accuracy":

$$\mathbb{E}_x\mathbb{E}_y\left[\log g_{\hat{\theta}(y)}(x)\right] \approx \mathbb{E}_y\left[\log g_{\hat{\theta}(y)}(y)\right] - K.$$

The conventional multiplication with -2 concludes the derivation.

Variation 11: Scientific Objectivity

The authority of science relies, to a large extent, on the objectivity of scientific method: it should not be affected by value commitments, personal bias or personal interests such as financial or career incentives (Reiss and Sprenger 2014). Objectivity in this sense contributes to the reliability of scientific research and conveys an image of epistemic authority.

By contrast, perceived lack of objectivity often undermines our trust in science. In the 2009 "Climategate" affair about leaked email communication in the Climatic Research Unit of the University of East Anglia, climate scientists were charged with presenting data in a way that was supposed to reflect alarmist tendencies and environmentalist values rather than scientifically established facts about global warming. Although later investigations cleared the scientists of the charges, finding "no evidence of any deliberate scientific malpractice" and recommending to include more professional statisticians in climate research (Oxburgh et al. 2010), the ensuing controversy harmed the public image of climate science substantially.

Similar qualms about scientific objectivity occur in disciplines that have trouble replicating experimental results from earlier studies. The most prominent example is experimental psychology, where a large-scale, multi-site replication project managed to obtain statistically significant results for only 36 % of all studies that had originally reported a significant finding (Open Science Collaboration 2015). Effect size estimates dropped dramatically, too. Assessments of replications in medical science and experimental economics support similar conclusions (e.g., Ioannidis 2005a; Prinz, Schlange and Asadullah 2011; Begley and Ellis 2012; Camerer et al. 2016; Nosek and Errington 2017). Lack of objectivity in experimentation and analysis, and the presence of questionable research practices ("QRPs"), are often cited as reasons for the low replication rates, which in turn lead to dwindling trust in published research results (e.g., Simmons, Nelson and Simonsohn 2011;

Bakker, Wicherts and van Dijk 2012; Francis 2012; Simonsohn, Nelson and Simmons 2014a,b).

Of course there is a large philosophy of science literature that questions the role of objectivity as a guiding ideal for scientific reasoning and its definition in terms of value freedom or adoption of an impersonal perspective (e.g., Feyerabend 1975; Kuhn 1977b; Harding 1991; Megill 1994). In this chapter, we shall touch this debate only peripherally. Instead, we focus on a different question: Given that some degree of objectivity is vital for scientific reasoning, especially when science enters the public arena (such as in the Climategate affair or the replication crisis), is objectivity compatible with Bayesian reasoning? Or do the subjective elements in Bayesian reasoning preclude any claims to objectivity?

There is a prima facie tension between subjective Bayesian inference and the pursuit of scientific objectivity: the subjective elements in Bayesian reasoning, above all the choice of a prior distribution, barely match widely endorsed senses of scientific objectivity, such as intersubjectivity, value freedom and conformity to standardized inference protocols. Since the epistemic authority of science in guiding public policy leans on the objectivity of scientific inference, it is sometimes claimed that "a notion of probability as personalistic degree of belief ..., by its very nature, is not focused on the extraction and presentation of evidence of a public and objective kind" (Cox and Mayo 2010, 298). It is noteworthy that this objection is put forward in a joint effort by two well-known methodologists: a statistician (David Cox) and a philosopher of science (Deborah Mayo). Their view is echoed in classical treatises on scientific method such as Fisher 1956, Popper 1959/2002 and Mayo 1996—see also Senn 2011 for a contemporary practitioner's perspective.

This debate about the merits of subjective Bayesian inference is, however, rarely buttressed by a conceptual analysis of scientific objectivity. By transferring insights from the objectivity debate (e.g., Longino 1990; Megill 1994; Douglas 2009b) to the context of statistical inference, we try to assess the probative value of the above objections, and the degree to which subjective Bayesian inference can be objective. To this end, we reproduce the arguments made in Sprenger 2018c and embed them into a broader perspective on statistical inference and Bayesian reasoning. First, we present the criticisms of Bayesian inference in somewhat greater detail (Section 11.1) and we show that two tempting responses are insufficient (Section 11.2). Then we argue that the criticisms apply equally to the main competitor of Bayesian inference in scientific reasoning—frequentist inference with null hypothesis significance tests (NHST, Section 11.3). What is more, the criticisms are based on restricted and outdated (yet popular) readings of scientific objectivity (Section 11.4). Then we explain why Bayesian inference promotes relevant senses of scientific objectivity (e.g., robustness, transparency, facilitating discussion and criticism) that are not captured by traditional accounts (Section 11.5). We support our argument with a case study from social psychology. Finally, we wrap up our main insights and discuss implications for future work (Section 11.6).

Thus, this chapter not only defends the objectivity of Bayesian inference; it also shows how conceptual work on scientific objectivity bears on questions in the methodology of statistical inference. A comparison of subjective and objective Bayesian inference would also be of great interest, but beyond the scope of this chapter—especially because objective Bayesian inference ("Objective Bayesianism") is no monolithic block, but contains different varieties and approaches (e.g., Jeffreys 1961; Jaynes 1968; Williamson 2010; Bernardo 2012). Some glosses on this topic can be found in the concluding section.

11.1 The Objections

Objectivity is a label that can be attached to different aspects of science: to the claims of a theory in relation to the world, to the process of gathering data, to individual reasoning about scientific theories and to the social dimension of producing scientific knowledge (Longino 1990; Douglas 2004, 2009b; Reiss and Sprenger 2014). Depending on what is supposed to be objective, there are different senses of the word. We follow Heather Douglas's taxonomy, which distinguishes eight senses of objectivity, and we adapt them (where appropriate) to the context of statistical inference. Five of them will be discussed later on: **manipulable objectivity** (i.e., reliability and reproducibility of real-world interventions), **value-neutral objectivity** (i.e., balancing various values in evidence judgments), **detached objectivity** (i.e., values should not replace evidence), **convergent objectivity** (i.e., validation of a finding from different perspectives) and **interactive objectivity** (i.e., enabling transformative criticism of scientific research). Now, we will focus on those senses of objectivity that are frequently invoked in statistical reasoning (e.g., Efron 1986; Cox and Mayo 2010) and that pose an obvious challenge for subjective Bayesian inference:

Concordant Objectivity (Intersubjectivity) Different speakers or community members agree on the reality of an observation, an evidence claim or a judgment on a theory. This sense of objectivity is purely factual; it does not concern the way agreement is reached.

Value-Free Objectivity Values and subjective judgments are banned from the process of scientific reasoning (e.g., in assessing theories on the basis of observed evidence). This sense of objectivity sees (noncognitive) values as detrimental to the unbiasedness and impartiality of scientific research.

Procedural Objectivity Experimentation and reasoning processes are standardized according to specific protocols. This sense of objectivity tries to eliminate individual idiosyncrasies: regardless of who performs an experiment or data analysis, the result should be the same (see also Porter 1996). It is particularly influential in the life sciences, where strict standards regulate the design, conduct and interpretation of medical trials.

Let us begin with concordant objectivity, or equivalently, intersubjectivity. This notion has a long philosophical tradition. For instance, Quine (1992, 5) states that "the requirement of intersubjectivity is what makes science objective". Subjective Bayesian inference violates concordant objectivity because different scientists may use different priors for analyzing one and the same data set, leading to different posterior distributions and different conclusions. This prompts the question of which (and whose) probability assessments should inform judgments on theories and evidence-based public policy.

The failure of subjective Bayesian inference with respect to value-free objectivity may be even more worrying. In sensitive areas such as climate science and the biomedical sciences, financial and ethical stakes are high, and consequences of wrong decisions are severe. These fields strive for inference methods that are as impartial and evidence-based as possible, and the pronounced role of personal degrees of belief in subjective Bayesian inference seems to jeopardize that aim. In the words of the medical methodologist Lemuel Moyé:

> Without specific safeguards, use of Bayesian procedures will set the stage for the entry of non-fact-based information that, unable to make it through the "evidence-based" front door, will sneak in through the back door of "prior distributions". There, it will wield its influence, perhaps wreaking havoc on the research's interpretation. (Moyé 2008, 476)

The objection can be rephrased as saying that the choice of a prior, which cannot always be based on hard information, will bias the final result in a particular direction. It is clear that such a liberal procedure cannot be objective in the sense of being value-free. Similarly, the discretion to choose a prior distribution at will is at odds with the goal of attaining procedural objectivity by means of standardized, uniform statistical analysis procedures.

All these tensions between subjective Bayesian inference and various senses of scientific objectivity support Cox and Mayo's intuition that (subjective) Bayesian methods fail to quantify objective evidence for use in science and public policy. Indeed one may even conclude that Subjective Bayesians commit a category mistake. Their formalism answers the question of what we may reasonably believe, but it does not quantify the (objective) evidence for a scientific claim (Royall 1997, 4).

The following sections respond to these worries. Before that, however, we would like to explain why two popular defenses of subjective Bayesian inference fail to counter the objections.

11.2 Convergence Theorems and Bayes Factors

A standard reply to the above worries contends that concordant objectivity may not hold at the beginning of a research process, but will be attained in the long run. For example, Adrian Smith, a well-known statistician and coauthor of a seminal textbook on Bayesian statistics, writes, "If a fairly sharp consensus of views emerges from a rather wide spread of initial opinions, then, and only then, might it be meaningful to refer to 'objectivity' " (Smith 1986, 10). And indeed, the famous **merging-of-opinions** or **washing-out theorems** (Blackwell and Dubins 1962; Gaifman and Snir 1982) show that Bayesians eventually reach such a consensus (see also page 26 in the introductory chapter). The theorems study the limiting behavior of two agents' degrees of belief when they are informed by the same body of evidence. In a nutshell, their posterior degrees of belief p^N and q^N will converge when collecting more and more information ($N \to \infty$), as long as the prior probability distributions p and q are *absolutely continuous* with respect to each other: they assign probability zero to the same propositions (i.e., for any X: $p(X) = 0$ if and only if $q(X) = 0$). So differences in prior probability will eventually wash out. Notably, the convergence is uniform, that it, it holds simultaneously for all propositions in the algebra.

Unfortunately, this observation fails to alleviate the worries from the previous section. The merging-of-opinion theorems make purely asymptotic claims and do not apply to the small and medium-sized data sets that we encounter in practice. Neither do they bound the speed of convergence (see also Earman 1992, 148–149). The theorems do not state sufficient but only *necessary* conditions for a calculus of degrees of belief that pursues the goal of intersubjectivity—just as statistical consistency (i.e., convergence to the true value as sample size increases) is a necessary, but not a sufficient, property for reliable statistical estimators. In particular, the merging theorems do not justify the claim that subjective Bayesian inference achieves concordant objectivity.

Another reply proposes to shift the objectivity discourse away from posterior distributions. This proposal acknowledges that posterior distributions are "contaminated" by subjective prior distributions and therefore not suitable for expressing evidence in an objective way. Instead, we should search for a measure of evidence that is free of the influence of subjective opinion and that allows the decision-maker to combine it with her own subjective probability and utility functions. In particular, while decision-makers may often disagree about their prior plausibility judgments, they should at least be able to agree on the strength of observed evidence. **Bayes factors** (Kass and Raftery 1995; Rouder et al. 2009) provide judgments about the weight of a particular body of evidence: they describe how strongly data E favors hypothesis H_1 over its competitor H_0:

$$\mathrm{BF}_{10}(E) := \frac{p(\mathrm{H}_1 | \mathrm{E})}{p(\mathrm{H}_0 | \mathrm{E})} \cdot \frac{p(\mathrm{H}_0)}{p(\mathrm{H}_1)} = \frac{p(\mathrm{E} | \mathrm{H}_1)}{p(\mathrm{E} | \mathrm{H}_0)}. \tag{11.1}$$

In other words, the Bayes factor measures the discriminative power of E with respect to H_1 and H_0 by comparing the (average) probability of E under H_1 to the (average) probability of E under H_0. The higher $\mathrm{BF}_{10}(\mathrm{E})$ is, the more E speaks for H_1, and vice versa. Table 11.1 provides a standardized interpretation scheme for Bayes factors.

Notably, the Bayes factor is independent of how strongly one is convinced of H_1 as opposed to H_0 a priori, because $p(\mathrm{H}_0)$ and $p(\mathrm{H}_1)$ cancel out when applying Bayes' Theorem to $p(\mathrm{H}_1|\mathrm{E})$ and $p(\mathrm{H}_0|\mathrm{E})$. However, there *is* a crucial dependence on prior probabilities. Assume that μ is a real-valued parameter and that we test the point hypothesis H_0: $\mu = 0$ against the alternative H_1: $\mu \neq 0$. This is a very common setting in science. In such a case, the Bayes factor will typically depend on how spread out the prior distribution for H_1 is: the extreme values of μ will badly fit the observed data, driving down $p(\mathrm{E}|\mathrm{H}_1) = \int_{\mu \in \mathbb{R}} p(\mathrm{E}|\mu)\, p(\mu)\, d\mu$ and thus also $\mathrm{BF}_{10}(\mathrm{E})$. The phenomenon is reversed for priors that are concentrated around $\mu = 0$

Bayes Factor BF_{10}	Interpretation
> 100	Extreme evidence for H_1
$30-100$	Very strong evidence for H_1
$10-30$	Strong evidence for H_1
$3-10$	Moderate evidence for H_1
$1-3$	Anecdotal evidence for H_1
1	No evidence for either hypothesis
$1/3-1$	Anecdotal evidence for H_0
$1/3-1/10$	Moderate evidence for H_0
$1/10-1/30$	Strong evidence for H_0
$1/30-1/100$	Very strong evidence for H_0
$< 1/100$	Extreme evidence for H_0

Table 11.1: Classification of Bayes Factors according to Lee and Wagenmakers 2014, adjusted from Jeffreys 1961. Reprinted from M.D. Lee and E.J. Wagenmakers, "Bayesian Cognitive Modeling: A Practical Course", Cambridge. Reproduced with permission of Cambridge University Press. © 2014, doi: 10.1017/CBO9781139087759.

(compare Section 9.4). Hence, individual differences in shaping the prior distribution influence—and possibly bias—the final evidential claims. While we agree that Subjective Bayesians should use Bayes factors for quantifying statistical evidence, this move alone is not enough to rebut the objections concerning lack of value freedom, procedural uniformity and intersubjective agreement.

11.3 Frequentism and Scientific Objectivity

In this section, we argue that the concerns about the objectivity of subjective Bayesian inference apply equally to its main competitor: frequentist inference. In that school of statistics, hypotheses are either true or false, and not objects of subjective uncertainty. Inferences are justified on the basis of their long-run properties, not on the basis of posterior probabilities. Among the many varieties of frequentist inference (Fisher 1956; Neyman and Pearson 1967; Mayo 1996), we focus on the most widespread one: Null Hypothesis Significance Tests (NHST) and the use of p-values for quantifying statistical evidence. Notably, none of the arguments below depend on the frequent misuse and misinterpretation of NHST in statistical practice. We focus on the classical problem of testing a point null hypothesis H_0 against an unspecific alternative H_1.

For the sake of simplicity, suppose that we analyze Normally distributed data $D = (X_1, \ldots, X_N)$ with unknown mean μ—the parameter of interest—and unknown variance σ^2. The data are assumed to be independent and identically distributed (i.i.d.). We calculate the sample mean $\overline{X}$ and the (corrected) standard deviation S by

$$\overline{X} = \frac{1}{N}\sum_{i=1}^{N} X_i \qquad \text{and} \qquad S = \sqrt{\frac{1}{N-1}\sum_{i=1}^{N}(X_i - \overline{X})^2}.$$

Now, the statistic

$$t = \frac{\overline{X} - \mu_0}{S/\sqrt{N}}$$

measures the divergence of the data from the null hypothesis H_0: $\mu = \mu_0$. Under H_0, it follows Student's t-distribution with $N-1$ degrees of freedom. This allows us to calculate the p-value: the probability that, if H_0 were true, t would display an even higher divergence from μ_0 than the actually observed discrepancy:

$$p := p_{H_0}\Big(\big|t(X)\big| \geq |t|\Big).$$

In this case, we conduct a two-tailed test and consider divergences in both directions from $\mu = \mu_0$. The inequality $p < .05$ is commonly interpreted as significant evidence against the null hypothesis. Still smaller p-values denote strong ($p < .01$) and very strong ($p < .001$) evidence against the null hypothesis. Standardly, researchers "reject the null" and infer the alternative on the basis of such p-values (see also Variation 9).

On the face of it, p-values are way more objective measures of evidence than Bayes factors: they do not involve subjective value judgments and they follow straightforwardly from the statistical model and the observed data. Each researcher will obtain the same p-value for a given data set. For many statisticians, a p-value "reassures oneself and others that the data have been interpreted fairly" (Efron 1986, 4), without influence of personal values and thereby contributing to scientific objectivity. However, these impressions dissolve upon closer inspection.

First, NHST and p-values take an asymmetrical stance on the hypothesis testing problem. As we have argued in Variation 9, there is presently no systematic guidance on how to interpret a nonsignificant result ($p \geq .05$). Statistics textbooks (e.g., Chase and Brown 2000; Wasserman 2004) basically restrict themselves to a purely negative interpretation: $p \geq .05$ means failure to demonstrate a statistically significant phenomenon. The founding father of NHST, R. A. Fisher (1935/74), even stressed that the only purpose of an experiment is to *disprove* the null hypothesis, and that we cannot say whether the results confirm the null hypothesis. Obviously, this leads to problems whenever the null hypothesis is of substantial scientific interest: independence of two variables in a causal model, the safety of a medical drug, or absence of parapsychological forces (Gallistel 2009; Wetzels et al. 2009; Morey et al. 2014; Sprenger 2018d). It is possible to remedy this problem by introducing a measure of corroboration, as proposed in Section 9.4, but this would amount to abandoning classical NHST and moving to a hybrid of Bayesian and frequentist inference.

In other words, phrasing a scientific inference problem in terms of NHST introduces a value judgment by ruling out the possibility of evidence in favor of the null hypothesis. Especially when the topic under investigation is delicate and politically contentious (e.g., the null hypothesis claims that a particular factor does not contribute to global warming), the asymmetry of NHST undermines their value freedom: there is no way how we could ever quantify the evidence for that claim and inform policy-makers accordingly. NHST are tied to a falsificationist methodology, which fails to be impartial when comparably important hypotheses are pitched against each other. Sometimes meta-analysis is supposed to fill this gap and failure to find significant evidence against the null in a series of experiments counts as evidence *for* the null. But neither does this move provide a systematic, principled theory of statistical evidence, nor does it answer the important question of how data support the null hypothesis in a single experiment.

Second, frequentist inferences are justified by the long-run properties of the procedures that generate them. This makes good experimental design vital for the justification of the entire inference. In particular, experiments are considered to be reliable when the type I error—the probability of observing significant evidence against a null hypothesis that is actually true—is bounded at a low level (e.g., $\alpha = .05$) and the power of the experiment to appraise a true alternative is reasonably high (e.g., $1-\beta = .8$). Power is always relative to a particular, representative effect size. It needs to be decided beforehand which effect sizes are plausible alternatives to the null, so that one does not end up with an underpowered experiment and an unreliable inference (Cohen 1988; Ioannidis 2005b). There is no surrogate for sound individual scientific judgment in this task. Subjective plausibility judgments are not only essential to Bayesian inference: they are part and parcel of NHST, too, and in fact, any method of scientific inference (for a practitioner's perspective, see Gelman and Hennig 2017). This observation undermines the alleged superiority of NHST vis-à-vis subjective Bayesian inference in terms of concordant, value-free and procedural objectivity.

Third, p-values are computed relative to a statistical test, whose direction is a matter of subjective judgment. In a recent experiment, the well-known social psychologist Daryl J. Bem (2011) published a study which presented evidence for various precognitive skills ("extrasensory powers", "*psi*"), including retroactive influence of a future event on an individual's behavior. In a Bernoulli (success/failure) experiment, Bem tested the null hypothesis H_0: $\mu = .5$ that success and failure are equally likely against the (one-sided) alternative H_1: $\mu > .5$ that precognition would lead to higher success than failure rates. In most experiments, the null was rejected at the $\alpha = .05$ level ($p < .05$). However, this finding was sensitive to whether the test was conducted as a two-sided test (H_1: $\mu \neq .5$) or as a one-sided test where only departures in a specific direction are considered (H_1: $\mu > .5$). The authors who critically discussed Bem's findings (Rouder and Morey 2011; Wagenmakers et al. 2011a) insisted that a more stringent two-tailed test be performed, where deviations in *both* directions would count as evidence against the null. Ultimately, the evaluation of the statistical evidence depends on whether success rates smaller than $\mu = .5$ are serious alternatives to the null hypothesis. This case shows that p-values are not as intersubjectively agreed, mechanically reproducible and value-free as a look at statistics textbooks may suggest. Of course, the general problem of choosing a suitable statistical test arises equally for frequentists and Bayesians.

Fourth and last, supporting scientific judgments with p-values is not straightforward. Critics of Bem's experiment insist that given the extraordinary nature of Bem's theoretical claims, p-values against the null of no precognition have to be much more convincing than the conventional $p < .05$ (Wagenmakers et al. 2011a). Indeed, what counts as substantial evidence against the null seems to be highly context-sensitive. While the psychological community usually conforms to the $p < .05$ criterion, standards are much more demanding in disciplines such as particle physics, where a divergence of five standard deviations from the null hypothesis ($p \approx \mathcal{O}(10^{-6})$) is required for announcing a finding of major importance, such as the recent discovery of the Higgs Boson. It may be argued that this is just a contingent sociological problem, but if so, it is highly persistent and unlikely to disappear soon.

The more general issue hiding here is the well-known problem of inductive risk in the assessment of scientific theories. Since the works of Rudner (1953) and Hempel (1965b) it is known that weighing uncertainties in statistical reasoning involves value judgments—be it by the individual scientist or by the scientific community in a field (Levi 1960; Douglas 2000). Both Bayesians and frequentists face this problem: translating p-values into a scientifically meaningful conclusion requires no less subjective judgment and consideration of context than do Bayes factors.

We have named four factors that impair the objectivity of frequentist inference with NHST: (1) the asymmetrical design of NHST and the impossibility to confirm the null hypothesis, (2) the need for plausibility judgments in power analysis and experimental design, (3) + (4) the contentious calculation and interpretation of p-values, including the general problem of inductive risk. All three senses of objectivity from Section 11.4—value-free, procedural and concordant objectivity—are affected by these factors.

It could be argued that our arguments make a good case against the classical NHST method that is still pervasive in statistical practice, but fail to hold for more sophisticated forms of frequentism. For instance, Cohen's (1988) power-centered perspective constructs frequentist inference as a decision procedure where it is also possible to (pragmatically) accept the null hypothesis when it fits the data better than the alternative hypothesis of a scientifically meaningful effect. But this is essentially a *decision procedure* for statistical inference and gives up on attempts to quantify *evidence* for the null—which is even conceded by proponents of that paradigm (Machery 2012, 816–818). The same holds true for the estimation-centered paradigm that proposes to replace NHST by estimation with confidence intervals (Cumming 2014). Since confidence intervals have a valid pre-experimental, but no valid post-experimental interpretation, the estimation paradigm—whatever its other merits may be—does not answer the important question of how to quantify statistical evidence for and against the tested null hypothesis (see also Gallistel 2009; Morey et al. 2014).

All in all, subjective Bayesian inference and frequentist inference with NHST face similar problems when they are evaluated in terms of concordant, value-free and procedural objectivity. Dismissing subjective Bayesian inference on the grounds of not meeting these criteria is therefore premature. What is more, we will now argue that the above senses of objectivity have limited epistemic value. For assessing the objectivity of a statistical inference method, they need to be replaced, or at least complemented, by other relevant senses of scientific objectivity.

11.4 Beyond Concordant, Value-Free and Procedural Objectivity

In this section, we question the epistemic value of concordant, value-free and procedural objectivity. Arguing these claims in detail would require another chapter, but we would like to cast some doubt on these senses of scientific objectivity and to motivate that we have to consider other senses, too.

First, the pursuit of procedural objectivity by means of banning subjective judgment and promoting standardized protocols has often contributed to a mindless use of statistical techniques (Cohen 1994; Gigerenzer 2004; Ziliak and McCloskey 2008): significance levels expressed by p-values replace proper scientific thinking and lead to the suppression of scientifically valuable, but statistically nonsignificant, research (e.g., Rosenthal 1979; Ioannidis 2005b). Here is a particularly illuminating quote:

> All psychologists know that statistically significant does not mean plain-English significant, but if one reads the literature, one often discovers that a finding reported in the Results sections studded with asterisks becomes in the Discussion section highly significant or very highly significant, important, big! (Cohen 1994, 1001)

In other words, procedural objectivity may have its merits in highly politicized or bias-prone areas of research, but at the same time it tends to promote mechanical, mindless and possibly misleading use of statistical inference procedures. It is therefore highly context-dependent whether procedural objectivity is epistemically beneficial. Satisfaction of procedural objectivity should therefore not be a *conditio sine qua non* for a school of statistical inference.

This diagnosis has implications for concordant objectivity, too. This sense of objectivity as intersubjectivity is purely factual; it does not make claims to intrinsic epistemic value. After all, scientists often differ in their disciplinary training, experience or methodological approach. These differences will lead them to different assumptions (e.g., different prior distributions) and different conclusions. It is completely unclear, however, whether there is any epistemic problem: the diversity of rational approaches to a data analysis problem is nothing that should be swept under the carpet. Meta-analysis and evidence aggregation are a more promising place for concordant objectivity: under ideal circumstances, a free exchange of information and argument may lead to individually rational belief states and intersubjective agreement at the same time (e.g., Lehrer and Wagner 1981). If we want to defend the epistemic value of concordant objectivity, it must not act as a constraint on data analysis, but at the level of amalgamating research findings.

Finally, there is value-free objectivity. We have already mentioned the problem of inductive risk. By now it is commonly accepted that complete value freedom cannot be achieved in scientific reasoning. Even matters of experimental design, such as choosing the significance level and power of a statistical test, reveal an implicit value judgment on the severity of various possible errors (e.g., Rudner 1953; Douglas 2000). As a consequence, Douglas (2004, 2009b) proposes to replace value-free objectivity by **detached objectivity:** values may have a place in scientific reasoning as long as they do not *replace* the evidence. This proposal implements the idea that objectivity implies impartiality: we do not want scientific evidence to be put aside, and to be trumped by political or religious ideologies. At the same time, it is a more modest characterization of objectivity than complete value freedom. With respect to detached objectivity, however, the criticism of subjective Bayesian inference loses much of its sting: the shape of the prior distribution affects the outcomes of a Bayesian analysis, but it is less clear why this is problematic. As long as the prior distribution can be justified by reference to past experience or theoretical considerations, the particular choice of the data analyst does not replace evidence by values and violate the ideal of detached objectivity.

Similarly, subjective Bayesian inference fits the bill with respect to Douglas's **value-neutral objectivity,** which means "taking a position that is balanced or neutral with respect to a spectrum of values" and avoiding positions that "are more extreme than they are supportable" (Douglas 2004, 460). For sure, prior probabilities can express extreme positions, but criticizing and varying them is a routine part of Bayesian inference (more on this will be said in Section 11.5). If value-neutral objectivity fails in practice, it is not because subjective Bayesian inference is methodologically flawed, but because its methods are abused—just as NHST and *p*-values are often abused, too.

Of course, all these judgments are gradual rather than categorical. Moreover, this brief overview cannot replace a thorough discussion of scientific objectivity (e.g., McMullin 1982; Longino 1990; Megill 1994; Lacey 1999; Douglas 2009b; Reiss and Sprenger 2014). Here, we only wanted to motivate why value freedom, concordance and procedural standardization need not make science more reliable and truth-conducive. These senses of objectivity do not contribute to the epistemic authority of scientific research, or at least not as much as it may appear at first sight. Other conceptions, such as detached and value-neutral objectivity, appear more reasonable, but then it is less clear why subjective Bayesian inference struggles to meet these ideals.

We will now introduce two other senses of scientific objectivity and show that on these counts, subjective Bayesian inference outperforms the rivalling frequentist framework. A case study—Bem's experiment on extrasensory powers—shall buttress our claims.

11.5 Interactive and Convergent Objectivity

This section moves our discussion of scientific objectivity from the level of individual reasoning to the social aspects of knowledge production. A prominent sense of objectivity in that domain is Helen Longino's

Interactive Objectivity "A method of inquiry is objective to the degree that it permits transformative criticism" (Longino 1990, 76). This includes, among others, the existence of (1) avenues for criticism of the obtained results, (2) shared standards for assessing theories and (3) equality of intellectual authority among qualified practitioners (see also Harding 1991; Douglas 2004).

In the first place, the concept of interactive objectivity aims at the social structures that regulate and facilitate scientific communication. That is, science must be structured in such a way that arguments can flow freely, that constructive criticism is possible and that discussions are not stifled by the power and authority of a particular subgroup in the scientific community.

Interactive objectivity can be applied at the level of statistical inference, too. Subjective Bayesian inference promotes interactive objectivity because it makes crucial assumptions behind a statistical inference transparent, such as structural assumptions on the unknown parameter, or expectations on the observed effect size. This move opens exactly those avenues for mutual criticism that are demanded from objective science (cf. Gelman and Hennig 2017). In frequentist inference, however, such assumptions are often hidden behind the curtain (see Section 11.3).

The debate about the Bem 2011 study on precognition, mentioned in Section 11.3, illustrates these abstract considerations. Bem conducted a series of nine similar experiments that tested the null hypothesis of no precognition. In one experiment, participants were asked which of two pictures on a computer screen they liked better. Later, the computer would randomly select one of the pictures as the "target" that would be displayed subliminally during the rest of the experiment. More often than chance would allow for, participants preferred the target which would later be selected by the computer. The other experiments, too, tested for the existence of retroactive influence patterns with binary response variables (success/failure).

Bem's study stirred a great deal of controversy. There were doubts about the interpretability of some of Bem's experiments (e.g., Rouder and Morey 2011), and indications for questionable research practices, such as selective reporting and selling exploratory as confirmatory research (Wagenmakers et al. 2011a; Francis 2012). Moreover, subsequent experiments failed to replicate Bem's effects (Galak et al. 2012). For the sake of the argument, we will assume that the experiments were conducted and reported in an orderly fashion, and focus on the statistical analysis of the data. Bem pitched the null hypothesis H_0: $\mu = .5$ (i.e., no precognition, participants are guessing) against the alternative H_1: $\mu > .5$ (i.e., participants do systematically better than guessing). Using a one-sided t-test as a large-sample approximation of testing the mean of a Binomial distribution, Bem observed statistically significant evidence against the null hypothesis ($p < .05$) in eight out of nine experiments. Effect size was calculated by Cohen's d—the difference between the observed value of μ and the hypothesized value μ_0, divided by the observed standard deviation σ:

$$d = \frac{\mu - \mu_0}{\sigma}. \qquad \text{(Cohen's } d\text{)}$$

The mean effect size over all experiments was $d = 0.22$, a small-to-moderate value. Overall, the results were supposed to indicate strong evidence against the null hypothesis of no precognition.

Wagenmakers et al. (2011a) and Wagenmakers et al. (2011b) conducted a Bayesian re-analysis of Bem's original data. In their critique of Bem's original data analysis, Wagenmakers et al. used a hierarchical Bayesian model where uncertainty about the true effect size δ is described by a Normal distribution $N(\mu, \sigma^2)$, centered around zero ($\mu = 0$) and with unknown variance $\sigma_\delta^2 \sim 1/\chi^2(1)$. When integrating out the influence of the variance σ^2, the prior for the effect size δ follows a Cauchy distribution with probability density f_r given below. The slope of the distribution is described by a scale parameter r (Rouder et al. 2009):

$$f_r(\delta) = \frac{1}{\pi r} \cdot \frac{r^2}{r^2 + \delta^2}.$$

Using the default choice $r = 1$, whose theoretical motivation goes back to Harold Jeffreys (1961), Wagenmakers and colleagues obtained a specific

Experiment No.	1	2	3	4	5	6(a+b)	7	8	9
BF_{10}, Wagenmakers et al.	1.64	1.05	1.82	0.58	0.88	1.10	0.13	0.47	5.88
BF_{10}, Bem et al.	4.94	3.45	5.35	1.76	2.74	3.78	0.50	1.62	10.12

Table 11.2: Calculation of the Bayes factor BF_{10} for the experimental data in Bem 2011. The null hypothesis H_0: $\mu = .5$ is tested against the alternative hypothesis H_1: $\mu \neq .5$ in a two-tailed t-test, according to the Cauchy prior of Wagenmakers et al. (2011a) and the knowledge-based prior of Bem, Utts and Johnson (2011).

prior distribution for δ: $f_1(\delta) = 1/(\pi \cdot (1+\delta^2))$. With that distribution, they performed a two-tailed Bayesian t-test and concluded that a majority of the experiments report evidence in favor of the null hypothesis H_0: $\mu = .5$. Only one experiment provides moderate evidence for the alternative hypothesis H_1: $\mu > .5$ ($10 > BF_{10} > 3$—see Tables 11.1 and 11.2), a further four experiments report anecdotal evidence in favor of H_0 ($3 > BF_{10} > 1$) or no evidence at all ($BF_{10} \approx 1$). The other experiments support H_0 to various degrees, leading to the overall conclusion that the evidence did not favor H_1.

In their response to the paper by Wagenmakers and colleagues, Bem, Utts and Johnson (2011) argue that this Bayesian analysis was based on prior distributions which are not suitable for analyzing data in parapsychological research. They note that effect sizes in psychology are typically small to moderate ($d \approx 0.25$, Bornstein 1989; Richard, Bond and Stokes-Zoota 2003), and even smaller for parapsychological effects. This does not square well with the prior distributions of Wagenmakers et al., where given H_1, we believe with probability .43 that the absolute value of the effect will be greater than .8. If we could rationally expect to observe such large effect sizes, there would be no debate about the reality of parapsychological phenomena.

The same argument is applied to argue against the use of Cauchy priors with their relatively thick, diffuse tails (see Figure 11.1). Placing any substantial probability on really large effects seems to be plainly inconsistent with the inconclusive and disputed evidence from parapsychological experiments (Hyman and Honorton 1986; Utts 1991; Storm, Tressoldi and Di Risio 2010). Yet, the prior distribution which is chosen by Wagenmakers and colleagues places a 6% chance on effects greater than $\delta = 10$. This, too, is clearly an unrealistic assumption. Choosing wide priors tends to favor the null hypothesis since the alternative hypothesis contains lots of extreme hypotheses which have, under spread-out priors, a substantial weight in the calculation of the Bayes factor.

Instead, Bem, Utts and Johnson (2011, 717–718) advocate a "knowledge-based prior" on δ, which differs in two crucial respects: (1) it is based on a Normal instead of a Cauchy distribution, leading to flat tails and highly improbable observations of large effects; (2) the variance σ^2 is chosen such that one's degree of belief in $\delta \in [-0.5, 0.5]$ is equal to 90 %, provided there is an effect at all. Figure 11.1 plots this prior distribution against the prior used by Wagenmakers and colleagues. Using their "knowledge-based prior", Bem, Utts and Johnson obtain moderate evidence in favor of the alternative hypothesis in most experiments (see Table 11.2). Multiplying the Bayes factors from the individual experiments, they argue that the total picture provides overwhelming evidence against the null hypothesis.

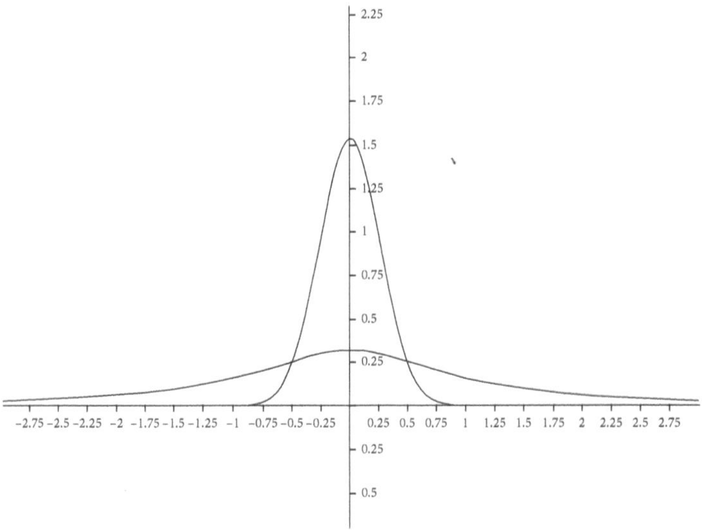

Figure 11.1: The default Cauchy prior ($r = 1$, flat curve) advocated by Rouder et al. (2009) and used by Wagenmakers et al. (2011a) versus the "knowledge-based prior" (steep curve) based on the Normal distribution by Bem, Utts and Johnson (2011). Reproduced from "The Objectivity of Subjective Bayesianism" by Jan Sprenger, *European Journal for Philosophy of Science*, Volume 8, No. 3, pp. 539–558, with permission of Springer Nature. © 2018, doi: 10.1007/s13194-018-0200-1.

Unsurprisingly, Wagenmakers and colleagues fail to be convinced. They object to the multiplication of Bayes factors across experiments. Moreover, they dispute the assumption that effect sizes should be that small, citing a survey of published articles in *Psychonomic Bulletin & Review* and *Journal of Experimental Psychology: Learning, Memory and Cognition* (Wetzels and

Wagenmakers 2012). These data suggest that substantial effects are as likely as small effects.

The point of reporting this debate is not to argue that either party is right or wrong. Rather, we would like to illuminate how the subjective Bayesian framework enhances transparency in statistical reasoning, and how it facilitates reasoning and argumentation about the assumptions involved. Phrasing their assumptions in terms of a prior distribution enables the two parties to trace the roots of their disagreement in a principled way, to identify the scientific propositions on which they disagree, and to correct potential errors. In the Bem case, the disagreement is primarily fed by three factors: (1) the technical question whether Bayes factors can be multiplied across experiments; (2) the mixed evidence about the distribution of effect sizes in psychology; (3) the methodological question of whether the specific nature of parapsychological experiments should be taken into account when shaping the alternative hypothesis.

It is not easy to imagine a similarly productive discourse in frequentist statistics, where a mechanistic interpretation of p-values prevails, and no (probabilistic) judgments about the plausibility of specific effect sizes are allowed. Sure, there are discussions about the right test statistic or significance level, such as the recent proposal to replace the common $p < .05$ criterion by $p < .005$ (Benjamin et al. 2018). However, for avoiding adhocness, such arguments must ultimately make reference to the relative plausibility of competing hypotheses and effect sizes. In fact, the authors of the cited paper compare the significance thresholds $p < .05$ and $p < .005$ by rephrasing their implications in Bayesian terms. Subjective Bayesian inference is much more intuitive and transparent than NHST: by making prior distributions explicit, readers, stakeholders and policy makers can decide to what extent they agree with the conclusions of a research study. Both sides in the Bem debate agree, incidentally, that this transparency is a great advantage of Bayesian statistics, both epistemically and socially, and they affirm that subjective assumptions are inevitable in scientific modeling (Bem, Utts and Johnson 2011, 718–719; Wagenmakers et al. 2011b, 11–12).

Moreover, we would like to point out a particular technique used by Wagenmakers and colleagues: **robustness analysis,** that is, assessing the sensitivity of the conclusions with respect to prior assumptions. Since any Bayesian inference relies on a prior probability distribution, Subjective Bayesians usually try to secure their evidential claims by showing that they are invariant under a variety of prior distributions. This technique is supposed to dispel worries that the final result is the product of idiosyncratic or extreme assumptions. Regulatory bodies even regard robustness analysis as an essential part of Bayesian reasoning:

> We recommend you be prepared to clinically and statistically justify your choices of prior information. In addition, we recommend that you perform sensitivity analysis to check the robustness of your models to different choices of prior distributions. (US Food and Drug Administration 2010)

Checking the robustness of a Bayesian inference with respect to the priors is built into the epistemic framework of subjective Bayesian inference, up to the point that it is a standard option in Bayesian statistics packages (e.g., JASP). Given the contentious nature of expectations about effect sizes in parapsychology, Wagenmakers and colleagues conducted a robustness analysis that varied the value of the scale parameter *r*. The results are taken from the online appendix of Wagenmakers et al. 2011a and reproduced in Figure 11.2.

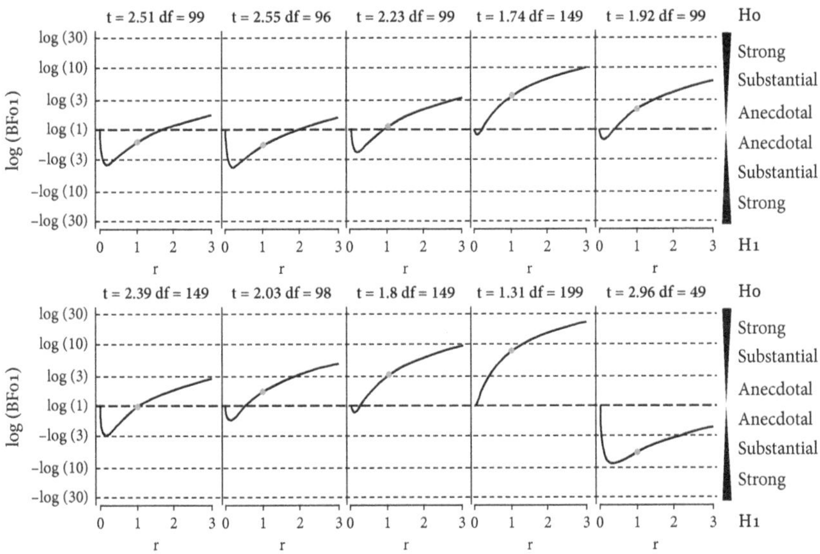

Figure 11.2: The robustness analysis of Wagenmakers et al. 2011a for the Bayes factor BF_{10} in Bem's (2011) experiments. Experiment 1–5 in the upper row, Experiment 6–9 in the lower row. Experiment 6 was split into two data sets. Reproduced from E.J. Wagenmakers, R. Wetzels, D. Borsboom, and H. van der Maas, Online Appendix for *"Why Psychologists Must Change the Way They Analyze Their Data: The Case of Psi"*: *A Robustness Analysis*, by permission of the authors: http://www.ejwagenmakers.com.

From these figures it becomes clear that only for a single experiment—Experiment #9, reproduced in the figure's bottom right corner—is there stable evidence in favor of H_1. By contrast, most experiments show that there is stable evidence in favor of H_0, independent of the choice of r. Only for very specific choices of r does the overall picture favor H_1. Wagenmakers et al. also point out that the choice of the scale parameter in Bem et al.'s knowledge-based prior clearly favors the alternative: almost all other values of that parameter would, within the Gaussian model chosen by Bem, Utts and Johnson (2011), lead to a more favorable assessment of the null hypothesis. These analyses suggest that Wagenmakers et al.'s evidence claim in favor of H_0 is more robust, and therefore more stable, than Bem's claim in favor of H_1.

Robustness analysis contributes to the assessment of the resilience of an evidential claim, and it is a good check against violations of value-neutral objectivity (Section 11.4). Evidence claims that require very specific, and possibly extreme, prior assumptions, such as in the case of Bem 2011 and Bem, Utts and Johnson 2011, hardly qualify for the "balanced judgment" that distinguishes value-free objectivity. That said, robustness analysis also matches another important sense of scientific objectivity (Douglas 2004, 2009b):

Convergent Objectivity A scientific result is objective to the extent that it is validated from independent assumptions and perspectives. Stability of a result under those variations increases confidence in its reliability.

Originally, this definition was meant to apply in a wide sense—for example, to invariance of a phenomenon or inference under different experimental designs and theoretical models. Famously, Scheele, Priestley and Lavoisier independently conducted groundbreaking work that led to the recognition of oxygen as a chemical element, and the rejection of the phlogiston theory. But there is no reason why this concept should not encompass sensitivity analysis with respect to Bayesian priors too. We have seen how robustness considerations in subjective Bayesian inference, and their entrenchment in regulatory constraints, foster the pursuit of convergent objectivity.

Of course, robustness analysis in statistics is more general than this particular application (e.g., Huber 2009; Staley 2012), but frequentist theory mostly deals with *distributional robustness*: deviations from the assumed sampling distribution, such as violations of Normality or homoscedasticity. It does not involve robustness with respect to expectations on the size of the observed effect (though see Mayo and Spanos 2006). Yes, this is not necessarily the fault of the NHST framework as opposed to the current practice of NHST, but robustness analysis with respect to prior expectations is easier, and more intuitive to implement, in Bayesian statistics.

All in all, subjective Bayesian inference promotes objectivity in various ways, especially in the social dimension of scientific inquiry: By adopting the ideal of robustness with regard to variations in the priors, subjective Bayesian inference secures the stability of evidential claims and contributes to convergent as well as value-neutral objectivity. By means of transparent prior distributions, it opens up avenues for criticism by scientific peers and contributes to interactive objectivity. Arguments for or against specific priors have to be justified by theoretical considerations or past data, not by authority or fiat. Choices are only as good and reasonable as the arguments that support them. These aspects of subjective Bayesian inference facilitate the "transformative criticism" (Longino 1990, 76) that distinguishes interactive objectivity in the production of scientific knowledge.

11.6 Discussion

In this chapter we have investigated to what degree (subjective) Bayesian inference can provide "objective evidence" for use in science and policy. To this end, we have split up the different senses that scientific objectivity can take in the context of statistical inference, and we have compared Bayesian inference to its fiercest competitor: frequentist inference with NHST. First, we have shown that frequentist inference suffers from the very same objectivity-related problems as Bayesian inference, albeit in a way that is less conspicuous to the naked eye. But the problems are equally severe (Section 11.3). Second, the senses of objectivity that support the claim that Bayesian inference falls short of objective science (i.e., concordance, value freedom, procedural uniformity) have dubious epistemic value. This observation makes us doubt that subjective Bayesian inference undermines the epistemic authority of science (Section 11.4). Finally, Bayesian inference promotes two epistemically and socially relevant senses of scientific objectivity, namely interactive and convergent objectivity (Section 11.5). Bayesian inference makes robustness analysis an integral part of assessing evidential claims, it increases the transparency of statistical reasoning and it facilitates a critical discussion of crucial modeling assumptions, such as the shape of the prior distribution. The fact that subjective judgment is inevitable in scientific reasoning does not imply that a school of reasoning that makes these elements explicit is less objective than its competitors, quite the contrary.

Obviously our investigations should be extended to other schools of statistical inference, such as objective Bayesian inference. Objective Bayesians claim that there is a privileged or "default" function that describes an agent's rational degree of belief given the evidence. Since the available evidence usually does not determine uniquely which prior (or posterior) probability functions one should adopt, supplementary principles are required. Objective Bayesians do not agree on the choice of those principles. As stated in the introductory chapter, some suggest maximizing the entropy of the probability function (MaxEnt priors; Jaynes 1968, 2003) and justify this choice by a decision-theoretic argument (Williamson 2010), while others prefer prior distributions that are invariant under bijective mappings of the parameter space (Jeffreys 1961; Bernardo 1979b, 2012). These schools of inference appear to be more objective than subjective Bayesian inference in the sense that they have less degrees of freedom in inference, leading to higher procedural uniformity and intersubjective agreement about posterior distribution.

Two replies suggest themselves. First, the choice among the different objective Bayesian approaches is itself a matter of judgment or philosophical preference, given that they are all based on reasonable and appealing principles. Second, a similar standardization is possible within the subjective Bayesian paradigm, too. There are various proposals for default priors that should be used in the absence of more precise information. The Cauchy prior that we encountered in the Bayesian re-analysis of Bem's study is a popular and well-studied default prior which can be used as an auxiliary device for comparing levels of evidence across different data sets and research teams. In other words, should the situation require it, the Subjective Bayesian can go as objective as the Objective Bayesian. We do think that default priors can play a useful role in inference as long as one handles them with care and does not treat them as an automatic route to inference that perpetuates the tendency toward "mindless statistics" that is widespread among NHST practitioners (Gigerenzer 2004; Gigerenzer and Marewski 2015).

That said, it remains a challenging project for future research to compare the objectivity of subjective and objective Bayesian inference in more detail. It would also be interesting to see which role Bayesian inference can play in overcoming the replication crisis, and how it interacts with proposals for social and methodological reform of the scientific enterprise. Would scientific inference be more reliable if we moved from frequentist inference with NHST to Bayesian inference (e.g., O'Hagan 2012; Lee and Wagenmakers 2014)? Would broader methodological reform, such as compulsory pre-registration of experiments and sharing data regardless of the outcome, decrease the rate of failed replications (Assen et al. 2014; Quintana 2015; Munafò et al. 2017)? Or is the replication crisis rather the product of inefficient social mechanisms and misguided incentives in the academic world? In that case, we might want to change the structure of the publication process and reform the credit reward scheme for scientific achievements (e.g., Koole and Lakens 2012; Fanelli 2016; Romero 2017, 2018).

Other important questions for future research are related to the use of subjective degrees of belief in public policy and group decision-making. For example, the periodic assessment reports by the Intergovernmental Panel on Climate Change (IPCC) amalgamate expert judgment and model outcomes when judging the likelihood of a particular outcome (e.g., a particular temperature increase). That is, the probabilities which they use to qualify a possible outcome as "virtually certain" or "likely" are partially based on subjective degrees of belief. To what extent should policy-makers rely on such probabilities, which may, or may not, be compromised by subjective or idiosyncratic value judgments?

Finally, how should disagreement on subjective probabilities be handled in reaching an "objective" outcome or consensus? There is a large literature on the aggregation of subjective probabilities in a group of rational agents, both from a philosophical and a mathematical-statistical perspective (e.g., Lehrer and Wagner 1981; Lindley 1983; Genest and Zidek 1986). But which of these procedures, or the adequacy conditions that characterize them, promote the achievement of more objective outcomes, in the senses specified above? In a nutshell, there is a fascinating variety of open research combining Bayesian probability and scientific objectivity.

We have almost reached the end of the book. There remains, however, one open question that haunts Bayesian theoreticians and practitioners alike: How should we interpret complex and idealized Bayesian models in statistical inference? Is it really plausible to assume that one of these models is a true description of reality, and to assign them positive degrees of belief? Isn't it rather the case that "all models are wrong, but some are useful" (Box 1976)? How does this insight square with the Bayesian interpretation of probability? The challenge of making sense of Bayesian modeling under those circumstances is addressed in the twelfth and final Variation.

Variation 12: Models, Idealizations and Objective Chance

The final chapter of our book returns to a foundational epistemological issue with Bayesian inference. For a scientific hypothesis H, the expression $p(\mathrm{H})$ is usually interpreted as an agent's subjective prior degree of belief that H is true. However, many statistical models are strong idealizations of reality. They abstract substantially from the features of reality that they are supposed to model, leaving out causally relevant variables or simplifying the functional dependencies between them. This is a universal feature of scientific modeling and occurs in disciplines as diverse as social psychology, medicine and engineering, to give just a few examples.

In those disciplines, it is often irrational to have a strictly positive degree of belief in the (approximate) truth of a particular hypothesis. It would be even less rational to accept a bet on the truth of that hypothesis, as operational interpretations of subjective probability demand. While the hypothesis may make reliable predictions, it does not aim at a faithful picture of reality; it is no candidate for (approximate) truth.

Thus we face the trilemma of having to reject one of the following three, jointly inconsistent, propositions:

(1) In Bayesian inference, $p(\mathrm{H})$ denotes an agent's degree of belief that H is true.

(2) Such a hypothesis H is part of a general statistical model $\mathcal{M}$ with a partition of hypotheses $\mathcal{H} = \{\mathrm{H}, \mathrm{H}_1, \mathrm{H}_2, \ldots\}$.

(3) Many of these models $\mathcal{M}$ are strong idealizations of reality—and likely, or known, to be false.

Propositions (2) and (3) are uncontentious aspects of statistical modeling. To the extent that this book would like to give a rational account of actual scientific reasoning, we cannot give them up. So we must reject (1) and get used to the idea that probabilities in Bayesian inference need not always denote subjective degrees of belief—at least not if the involved models are sufficiently idealized.

Note that it will not help to add a "catch-all hypothesis" $\widetilde{\mathrm{H}} = \neg \bigvee_{\mathrm{H}_i \in \mathcal{H}} \mathrm{H}_i$ to $\mathcal{H}$. By the above arguments, all probability mass would be concentrated on $\widetilde{\mathrm{H}}$ and we could not distinguish between the other hypotheses in $\mathcal{H}$.

There are various ways of escaping the dilemma. In an ingenious contribution, Vassend (forthcoming) defends the view that the subjective probability of H expresses the degree of belief that H is the most similar to the truth among all (false) hypotheses in $\mathcal{M}$. This approach connects Bayesian inference to the verisimilitude paradigm, where the goal of scientific inference consists in gradually approaching the truth and theories are judged according to their estimated closeness to truth (Niiniluoto 2011; Cevolani and Tambolo 2013; see also page 27 in the introductory chapter).

In this chapter, we propose to resolve the trilemma on page 311 by means of a suppositional or hypothetical analysis of probability in Bayesian inference. The probability $p(\mathrm{H})$ expresses in reality not an unconditional probability, but a conditional and model-relative probability: the degree of belief that we would have in H if we supposed that the target system is fully and correctly described by one of the hypotheses in $\mathcal{M}$. On this reading, the entire Bayesian inference is relative to $\mathcal{M}$. The degree of belief that H is true is in reality a **conditional degree of belief that H is true on the supposition that $\mathcal{M}$.** Motivating and working out such an account is our project in this final Variation.

First, we focus on conditional degrees of belief of the type $p(\mathrm{E}|\mathrm{H})$ given a statistical hypothesis H and evidence E. These degrees of belief are in practice constrained by the corresponding probability densities. But is this alignment rationally required? We argue that chance–credence coordination principles fail to justify it (Section 12.1). Starting from there we motivate a suppositional account of probability in Bayesian inference: conditional degrees of the type $p(\mathrm{E}|\mathrm{H})$ are the degrees of belief that we have in the occurrence of E upon supposing that the target system is fully described by H. This proposal solves the problem of chance–credence coordination by transferring coordination principles such as the Principal Principle from the actual world to counterfactual scenarios (Section 12.2). Then we explore the wider implications of the suppositional analysis, arguing that all probabilities in Bayesian inference should be understood suppositionally and model-relative. Consequently we explain why one can base predictions, theory evaluations and decisions on Bayesian models, even if they are highly idealized and not faithful to reality (Section 12.3). Notably, on the proposed account Bayes' Theorem should be interpreted as a coherence constraint between different probability functions rather than as a mathematical theorem (Section 12.4). Finally, we summarize how conditional degrees of belief should be understood, and how this understanding changes our view of Bayesian inference in science. We also mention projects for future research (Section 12.5).

12.1 The Equality and Chance–Credence Coordination

Prior probabilities in (subjective) Bayesian inference represent the subjective opinion of the agent who entertains them. Why are posterior probabilities—the basis for Bayesian theory assessment and decision-making—not equally subjective? Presumably because they are influenced by evidence. By Bayes' Theorem,

$$p(\mathrm{H}\,|\,\mathrm{E}) = p(\mathrm{H})\,\frac{p(\mathrm{E}\,|\,\mathrm{H})}{p(\mathrm{E})} = \left(\sum_{\mathrm{H}_i \in \mathcal{H}} \frac{p(\mathrm{E}\,|\,\mathrm{H}_i)}{p(\mathrm{E}\,|\,\mathrm{H})} \cdot \frac{p(\mathrm{H}_i)}{p(\mathrm{H})}\right)^{-1},$$

posterior probabilities of a hypothesis H depend on the conditional degree of belief in observed evidence E given H (and its competitors $\mathrm{H}_1, \mathrm{H}_2, \ldots$). Since prior probabilities are not constrained by evidence, any force that drives rational agreement must stem from agreement on the value of the conditional probabilities $p(\mathrm{E}\,|\,\mathrm{H})$, $p(\mathrm{E}\,|\,\mathrm{H}_1)$, $p(\mathrm{E}\,|\,\mathrm{H}_2)$, etc. If agreement on these values breaks down, we cannot justify that posterior probabilities are something else than arbitrarily chosen numbers, or that they are reliable guides for decisions and policy-making.

The same argument can be made for Bayes factors, the Bayesian's preferred measure of evidence (Jeffreys 1961; Kass and Raftery 1995; Lee and Wagenmakers 2014). Bayes factors are the (averaged) likelihood ratios of evidence E under two competing hypotheses H_0 and H_1:

$$\mathrm{BF}_{10}(\mathrm{E}) := \frac{p(\mathrm{H}_1\,|\,\mathrm{E})}{p(\mathrm{H}_0\,|\,\mathrm{E})} \cdot \frac{p(\mathrm{H}_0)}{p(\mathrm{H}_1)} = \frac{p(\mathrm{E}\,|\,\mathrm{H}_1)}{p(\mathrm{E}\,|\,\mathrm{H}_0)}.$$

In the subjective interpretation of probability that undergirds Bayesian inference, the probabilities $p(\mathrm{E}\,|\,\mathrm{H}_0)$ and $p(\mathrm{E}\,|\,\mathrm{H}_1)$ are read as a scientist's personal degree of belief that E is the case, given H_0 or H_1. Yet we often observe universal consensus on the values of those probabilities: any Bayesian reasoner in science plugs in the corresponding values of the probability densities in the relevant statistical model. To give a very simple example, if H_0: $\mu = .5$ or H_1: $\mu = .6$ correspond to different hypotheses about the fairness of a coin and E = "the coin comes up heads two times", then all of us will presumably report the degrees of belief $p(E|H_0) = .25$ and $p(E|H_1) = .36$—as the corresponding statistical model predicts. Although this agreement sounds trivial and in any case seems covered by a chance–credence coordination principle, we will soon see that it is far from easy to justify.

Let's look at this example in more detail. Suppose that we toss a coin in order to learn about its bias—a simple and transparent case of inductive inference. The hypotheses H_μ (where $\mu \in [0;1]$) describe the possible chances of the coin's coming up heads on any individual toss. When we toss the coin N times, our sample space is $\mathcal{S} = \{0,1\}^N$. Under the assumption that the tosses are independent and identically distributed (henceforth, "i.i.d."), we can describe the probability of observation E_k (i.e., k heads and $N-k$ tails) by the Binomial probability distribution and the corresponding **probability density function** $\rho_{H_\mu}(E_k) = \binom{N}{k}\mu^k(1-\mu)^{N-k}$. More generally, the sample space $\mathcal{S}$ together with the set of probability distributions over $\mathcal{S}$ induced by the ρ_{H_μ} constitutes a **statistical model** (e.g., Cox and Hinkley 1974; Bernardo and Smith 1994; McCullagh 2002).

Suppose $N = 2$ and let H be the hypothesis that the coin is fair, H: $\mu = 1/2$. Then the probability of observing two heads in two tosses (E: $k = 2$) is described by $\rho_H(E) = \binom{2}{2}(1/2)^2(1-1/2)^{2-2} = 1/4$. Bayesian reasoners in the sciences take these probabilities as plug-ins for their conditional degrees of belief $p(E|H_\mu)$ (e.g., Bernardo and Smith 1994; Howson and Urbach 2006). And since the former are uniquely determined, so are the latter. Agents with divergent priors will approach each other's views in the posterior distribution. Uniquely rational values for $p(E|H_\mu)$ have an important epistemic function: they work toward establishing (approximate) consensus on posterior distributions, and they lead to unanimous assessments of the strength of observed evidence. For instance, the Bayes factor between two hypotheses H_1 and H_0 is just the ratio of $p(E|H_1)$ and $p(E|H_0)$. If these degrees of belief are objectively determined, so is the Bayes factor. In other words, we need a satisfactory answer to the following question:

The Equality What justifies the equality

$$p(E|H) = \rho_H(E) \qquad \text{(THE EQUALITY)}$$

between conditional degrees of belief and the corresponding probability densities?

The most traditional analysis of conditional degrees of belief, contained in various textbooks on Bayesian inference and decision-making (e.g., Earman 1992; Skyrms 2000; Howson and Urbach 2006), is the so-called **Ratio Analysis**:

$$p(\mathrm{E}|\mathrm{H}) = \frac{p(\mathrm{E} \wedge \mathrm{H})}{p(\mathrm{H})} \qquad \text{whenever } p(\mathrm{H}) > 0.$$

As a mathematical constraint on conditional probability (and hence, on conditional degree of belief), Ratio Analysis is generally accepted. However, we doubt that Ratio Analysis provides a convincing philosophical analysis of conditional degree of belief: it does not explain our inference practices, in particular how we form and evaluate conditional degrees of belief. The reasons are explained in detail on pages 17–21 in the introductory chapter.

A more promising attempt to justify THE EQUALITY consists in deriving it from a general epistemic norm: a **chance–credence coordination principle** that calibrates subjective degree of belief with objective chance (e.g., Lewis 1980; Strevens 1999; Williamson 2007, 2010). For instance, according to the **Principle of Direct Inference (PDI)** (e.g., Reichenbach 1949; Kyburg 1974; Levi 1977), if you know that a coin is fair, you should assign degree of belief 1/2 that heads will come up. David Lewis (1980) formalized a related intuition in his **Principal Principle (PP):** the initial credence function of a rational agent, conditional on the proposition that the physical chance of E takes value x, should also be equal to x.

Do these principles apply to THE EQUALITY? At first sight, yes. Presumably, the Principle of Direct Inference (or its close cousin, the Principal Principle) would recommend to carry over the value of $\rho_{\mathrm{H}}(\mathrm{E})$ to the conditional degree of belief $p(\mathrm{E}|\mathrm{H})$. After all, ρ_{H} does not depend on subjective epistemic attitudes, just on the properties of the statistical model. In this sense, it seems to qualify as an objective chance.

Note, however, that objective chances are commonly taken to make empirical statements: their values are constituted by patterns of events and processes in the *actual* world (Hoefer 2007, 549), such as the setup of an experiment or the physical composition of the coin we toss. This is true of frequencies, propensities and best-system chances alike. Our case is different: the truth values of sentences such as "$\rho_{\mathrm{H}}(\mathrm{E}) = 1/4$" are fully determined by the properties of the statistical model (see also Sprenger 2010b). If H denotes the hypothesis that the coin is fair and E denotes the observation of two heads in two tosses, then it is part of the *meaning* of H that the probability of E given H is 1/4. That is, the sentence

> When we perform two i.i.d. tosses of a fair coin, the chance of observing two heads is 1/4 (+)

has no empirical content—it has a distinctly analytical flavor. It does not refer to properties or events in the actual world; in fact, no fair coins need to exist for this sentence to be true. In other words, $\rho_H(E)$ describes an objective, but not an ontic probability; its value is subject-independent, but not a *physical chance* (Rosenthal 2004).

Hence, neither the Principle of Direct Inference nor the Principal Principle applies to Bayesian statistical inference straightforwardly. These principles coordinate our degrees of belief with properties of the actual world, such as observed relative frequencies or known propensities. But the density functions ρ_H in a statistical model do not make any statements about the actual world. Therefore, the usual chance–credence coordination principles cannot justify THE EQUALITY. The rest of the chapter is devoted to finding a semantics for conditional degrees of belief that explains THE EQUALITY and solves the trilemma from the introduction.

12.2 The Suppositional Analysis

In this section, we elaborate a **suppositional analysis of conditional degrees of belief.** According to this view, we determine our degrees of belief in E given H by *supposing* that H is true. We call our account "suppositional" instead of "counterfactual" since in some cases, no counterfactual assumptions are necessary. On the other hand, many applications of statistical models in scientific inference imply straightforward counterfactual assumptions; and whenever we want to emphasize this point, we use the adjective "counterfactual" (analysis, interpretation, ...) instead of the more general "suppositional".

The suppositional analysis goes back to Bruno de Finetti (1972, 2008) and Frank P. Ramsey and is based on Ramsey's (1926) famous analysis of conditional degrees of belief:

> If two people are arguing 'if H will E?' and both are in doubt as to H, they are adding H hypothetically to their stock of knowledge and arguing on that basis about E. We can say that they are *fixing their degrees of belief in* E *given* H. (Ramsey 1926, our emphasis)

In other words, $p(\text{E}|\text{H})$ is the rational degree of belief in E if we suppose that H is true and assess the plausibility of E in the light of this assumption. Implications of this analysis for the epistemology of conditionals and evaluating subjective probabilities of the form $p(\text{H} \rightarrow \text{E})$ (e.g., Stalnaker 1968, 1975; Levi 1996) are beyond the scope of this book.

To make Ramsey's proposal precise, consider a statistical hypothesis H which describes the behavior of a target system S (e.g., "the coin is fair"). Supposing that H is true means conceiving of the target system S as governed by physical chance, with the probability of an observation E given by the probability density $\rho_{\text{H}}(\text{E})$. Since several possible worlds satisfy this description, we have to explain which possible world is relevant for determining our conditional degrees of belief.

Define a relation $\omega \sim_S \omega'$ on the space of possible worlds $\mathcal{W}$ that holds if and only if the same probability law applies to S in the possible worlds ω and ω'. It is easy to check that this relation is reflexive, symmetric and transitive and thus an equivalence relation. Thus, for the space of possible worlds $\mathcal{W}$, we can define the quotient space $\mathcal{W}/{\sim_S}$ that divides $\mathcal{W}$ into equivalence classes of possible worlds where the same probability law is assigned to S. The set $[\omega_{\text{H},S}] \in \mathcal{W}/{\sim_S}$ is the class of worlds where S is governed by H, the set $[\omega_{\text{H}',S}]$ is the class of worlds where S is governed by H′, and so on.

Differences between possible worlds in a single equivalence class are irrelevant. We want to learn something about S, and to this purpose we have set up a Bayesian model which includes H as well as rivaling statistical hypotheses H′, H″ and so on. Such models typically abstract from features of the world beyond S. In other words, supposing H means restricting the space of relevant possible worlds to the equivalence class of worlds where $\rho_{\text{H}}(\text{E})$ describes the objective chance of an event E in target system S. Sometimes these suppositions are counterfactual: a given coin may not be perfectly fair, the tosses are not i.i.d., and so on. How these counterfactual degrees of belief relate to conclusions about the real world is a question for the next section.

Objective chances within such a hypothetical scenario then determine conditional degrees of belief—an argument that is expounded at greater length in Sprenger 2018a and that we summarize as follows. Let H denote the hypothesis that the coin is fair. To simplify the argument, we replace the equivalence class $[\omega_{\text{H},S}]$ by one of its (arbitrarily chosen) representatives, ω_{H}. Supposing H then means that the objective, physical chance of two heads in two tosses is 1/4, the chance of two heads in three tosses is 3/8, and so on. The

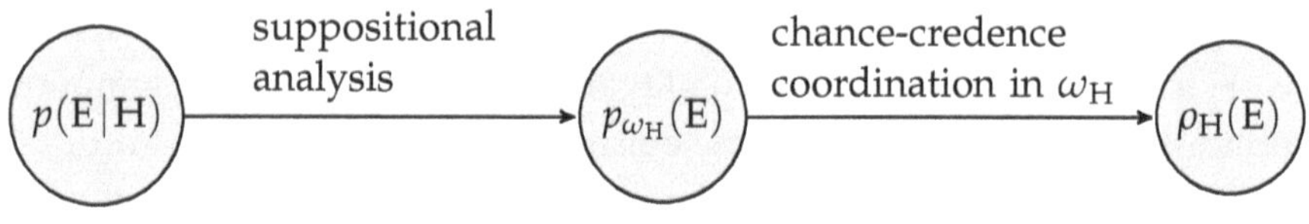

Figure 12.1: A graphical representation of how the suppositional analysis of conditional degree of belief and chance-credence coordination work together to justify THE EQUALITY.

existence of such an objective chance within ω_H allows us to apply a chance–credence coordination principle such as the Principal Principle (PP) or the Principle of Direct Inference (PDI). Upon supposing that we are in the possible world ω_H (i.e., we suppose the truth of H), we transfer the probabilistic predictions of H to our degrees of belief in E: $p_{\omega_H}(E) = \rho_H(E)$. Moreover, the suppositional analysis yields that our ("actual") conditional degrees of belief in E given H, equal our degree of belief in E on the supposition that H. See Figure 12.1. Combining the suppositional analysis with chance-credence coordination, we can then derive THE EQUALITY: $p(E|H) = \rho_H(E)$. Probability density functions determine conditional degrees of belief. The suppositional analysis of conditional degrees of belief explains THE EQUALITY, and in particular the analytic character of sentences such as (+). The agreement also transfers to statistical measures of evidence that are derived from these conditional degrees of belief and play a pivotal role in statistical inference, such as Bayes factors.

There is an ongoing discussion about whether objective chances exist outside some subfields of physics and biology (and even there, the claim may be contested; compare Frigg and Hoefer 2015). For applying chance–credence coordination in the actual world with an eye on THE EQUALITY, the probability densities ρ_H would have to correspond to some form of ontic, physically realized chance. As argued in the previous section, such a claim is difficult to establish, if not outright wrong. It is much more attractive to read these probabilities as physical chances in hypothetical scenarios and to apply chance–credence coordination upon the supposition of such a scenario.

Let us wrap up the essence of our proposal. We interpret $p(E|H)$, the conditional degree of belief in observation E given statistical hypothesis H, by supposing that H is the true model of the target system S. In this hypothetical or counterfactual scenario $[\omega_{H,S}]$, we coordinate our degree of belief in E with the known objective chance of E, given by $\rho_H(E)$ (Lewis 1980). What the probability density functions inform are not our actual degrees of belief but our hypothetical degrees of belief given a particular statistical hypothesis.

We will now demonstrate that the suppositional account is also a fruitful framework for interpreting prior and posterior degrees of belief and for explaining how scientists reason with idealized Bayesian models.

12.3 Suppositional Prior Probabilities and the Trilemma Resolution

So far, we have been silent on the demarcation of the statistical hypothesis H vis-à-vis the background assumptions that are part of the statistical model $\mathcal{M}$. Consider the coin-tossing case. When we evaluate $p(\mathrm{E}|\mathrm{H})$ with H = "the coin is fair", we assume that the individual tosses of the coin are independent and identically distributed. However, this assumption is not part of H itself: H just describes the tendency of the coin on any particular toss. If we contrast H to some alternative H′, we notice that differences between them are typically expressed in terms of parameter values, such as H: $\mu = 1/2$ versus H′: $\mu = 2/3$, H″: $\mu > 1/2$, etc. Crucial assumptions on the experimental setup, such as independence and identical distribution of the coin tosses, do not enter the particular hypothesis we are testing. In a sense, $p(\mathrm{E}|\mathrm{H})$ is underdetermined because so much supplementary information about the experiment is missing.

Thus, we have to make the role of our background assumptions explicit. There are two layers of statistical modeling. First, there is the general statistical model $\mathcal{M} = (\mathcal{S}; \mathcal{P})$ where $\mathcal{S}$ denotes the sample space and $\mathcal{P}$ denotes the set of probability distributions over $\mathcal{S}$. What counts as an element of $\mathcal{P}$ depends on several factors: if we describe the repeated coin toss by a set of Binomial distributions $B(N, \mu)$, then we also assume that the tosses are i.i.d. and that we toss the coin exactly N times. The choice of specific hypotheses such as H: $\mu = 1/2$, H′: $\mu = 2/3$, and so on, is only possible after choosing a family of distributions for $\mathcal{P}$.

This implies that the conditional degree of belief $p(\mathrm{E}|\mathrm{H})$ is not only conditional on H, but also conditional on the general model $\mathcal{M}$. Indeed, a Bayesian inference about the probability of heads in the coin-tossing example takes $\mathcal{M}$ as given from the very start. Strictly speaking, any occurrence of $p(\mathrm{E}|\mathrm{H})$ so far should have read $p_{\mathcal{M}}(\mathrm{E}|\mathrm{H})$.

The same diagnosis applies to the assignment of prior probabilities $p(\mathrm{H})$: Bayesian inference regarding particular parameter values is relative to a model $\mathcal{M}$ into which all hypotheses are embedded, such as the Binomial model $\mathcal{M} = \left(\mathcal{S} = \{0,1\}^N;\ \mathcal{P} = B(N, \mu),\ \mu \in [0,1]\right)$ for a repeated coin toss with fixed sample size. Degrees of belief are distributed only over elements of $\mathcal{P}$. In particular, also the prior and posterior degrees of belief, $p(\mathrm{H})$ and $p(\mathrm{H}|\mathrm{E})$, are relative to a model $\mathcal{M}$. For instance, in a Binomial model, we distribute prior probabilities over the different values of μ, and over no other statistical hypotheses. Of course there are also hierarchical models which involve submodels of different types (e.g., the Binomial model vs. a model with dependent outcomes), but this does not change the general diagnosis: all probabilities in Bayesian inference are relative to a statistical model (see also Wenmackers and Romeijn 2016). In fact, this observation was the starting point of this chapter.

This perspective on prior probabilities, which flows logically from our take on conditional degrees of belief, resolves the trilemma from page 311: we deny Premise (1), that the probability $p(\mathrm{H})$ of a statistical hypothesis H denotes an actual degree of belief. Any such probability is, like the probabilities of evidence given a hypothesis, essentially suppositional and should be read as $p_{\mathcal{M}}(\mathrm{H})$.

We have to address one potential objection: Supposing $\mathcal{M}$ may not yield a uniquely rational value for $p_{\mathcal{M}}(\mathrm{H})$. There is no objective chance of H in the hypothetical world $\omega_{\mathcal{M}}$. So the analogy with the case of $p(\mathrm{E}|\mathrm{H})$ breaks down. Does this mean that the suppositional analysis is not applicable to prior degrees of belief and fails to say anything meaningful?

The answer is No. It is the very essence of subjective Bayesian inference that different rational choices of the prior distribution are possible, and that supposing a model does not force us to adopt a specific prior distribution. The suppositional analysis provides a general framework for interpreting degrees of belief and understanding how we reason with Bayesian models; it does not aim at determining uniquely rational (conditional or unconditional) degrees of belief.

What is more, having a prior probability distribution over an unknown parameter μ within a statistical model $\mathcal{M}$ does not imply that we are ready to bet real money on the true value of μ, with betting odds corresponding to our degrees of belief. The prior distribution just describes how our degrees of belief and betting odds would be if $\mathcal{M}$ were indeed a faithful model of reality.

The suppositional interpretation naturally transfers to posterior probabilities $p_{\mathcal{M}}(\mathrm{H}|\mathrm{E})$. According to Bayes' Theorem, $p_{\mathcal{M}}(\mathrm{H}|\mathrm{E})$ depends on $p_{\mathcal{M}}(\mathrm{H})$, $p_{\mathcal{M}}(\mathrm{E}|\mathrm{H})$, $p_{\mathcal{M}}(\mathrm{H}_i)$ and $p_{\mathcal{M}}(\mathrm{E}|\mathrm{H}_i)$, the conditional degrees of belief in E given the alternatives H_i:

$$p_{\mathcal{M}}(H|E) = \left(\sum_{H_i \in \mathcal{H}} \frac{p_{\mathcal{M}}(E|H_i)}{p_{\mathcal{M}}(E|H)} \cdot \frac{p_{\mathcal{M}}(H_i)}{p_{\mathcal{M}}(H)} \right)^{-1}. \qquad (12.1)$$

We see that the impact of $p_{\mathcal{M}}(H)$ on the posterior probability diminishes when there is sufficiently strong evidence in favor of or against H, as expressed by the ratio $p_{\mathcal{M}}(E|H_i)/p_{\mathcal{M}}(E|H)$ of conditional probabilities. The value of the posterior probability $p_{\mathcal{M}}(H|E)$ is rationally constrained because (i) Bayes' Theorem binds it to the value of other conditional degrees of belief and (ii) the conditional degrees of belief $p_{\mathcal{M}}(E|H_i)$ in E given the competing statistical hypotheses H_i obey THE EQUALITY and track the corresponding probability densities.

Finally we address the question of how such counterfactual degrees of belief affect our epistemic attitudes about real-world events E. It will not surprise the reader that we propose to interpret the marginal likelihood $p(E) = p_{\mathcal{M}}(E)$ in a model-relative, suppositional way too. In fact, this follows from the Law of Total Probability: the model-relative probability $p_{\mathcal{M}}(E)$ is the weighted average of the conditional probabilities of E,

$$p_{\mathcal{M}}(E) = \sum_{H_i \in \mathcal{H}} p_{\mathcal{M}}(E|H_i) \cdot p_{\mathcal{M}}(H_i),$$

and is thus our subjective expectation that E occurs if $\mathcal{M}$ is the case. Suppose that $\mathcal{M}$ is an ensemble of hypotheses (e.g., global climate models) that make statistical predictions for specific events such as E: "On emission scheme S, global mean temperature in 2100 will be at least 2°C higher than in 1990." Suppose that $p_{\mathcal{M}}(E) \approx 1$ because most hypotheses in $\mathcal{M}$ assign a high probability to E. Do these predictions affect our actual epistemic attitudes regarding the occurrence of E? After all, on the suppositional interpretation, all probabilities are relative to a statistical model $\mathcal{M}$. But we care for whether E is *actually* likely to occur, not whether it is likely to occur given a certain model. In other words: How does the suppositional analysis link model predictions to the real world? This question is especially urgent in contexts such climate science, where doubts about the adequacy of large-scale models for making probabilistic predictions are widespread:

> It is ... inappropriate to apply any of the currently available generic techniques which utilize observations to calibrate or weight models to produce forecast probabilities for the real world. (Stainforth et al. 2007, 2145)

In general, confidence in the predictions of a statistical model $\mathcal{M}$ depends on whether its constituents capture the relevant aspects of the target system.

When the target system is highly complex and hard to predict, we will have to preserve a healthy dose of skepticism toward the predictions of $\mathcal{M}$. On the other hand, the better the constituents of $\mathcal{M}$ describe relevant aspects of the target system, the more justification do we have for an inference by analogy, and for transferring their predictions to our actual epistemic attitudes.

In this respect, Bayesian inference is just another form of model-based reasoning in science: its soundness depends on whether the overall model $\mathcal{M}$ is well chosen or inadequate (Weisberg 2007; Frigg and Hartmann 2012; Suárez 2016). Converting knowledge about the model into knowledge about the target system is equally difficult for Bayesian and non-Bayesian model builders. The strategies for dealing with the model–reality mismatch are also similar. A non-probabilistic modeler may correct an estimate because she knows that the model neglects a particular feature of the target system. Similarly, a Bayesian modeler may not fully align his actual degree of belief in E with the probability predicted by $\mathcal{M}$. Correcting upwards or downwards may be necessary to balance limitations of the model. If you build, for example, a statistical model of financial markets where you neglect the effects of psychological chain reactions on asset prices (e.g., because the model would get too complicated), you will underestimate the probability of a stock exchange crash. Our actual degrees of belief should, to some extent, compensate for such modeling decisions. Put differently, we should not read off our actual degrees of belief from a Bayesian model; instead, the model informs our degrees of belief and our predictions by showing what they would be under reasonable idealizing assumptions.

12.4 Bayes' Theorem Revisited

Before concluding, we invite the reader to reflect on the meaning of Bayes' Theorem under the new, suppositional interpretation of degrees of belief in scientific inference. Remember that the model-relative version of Bayes' Theorem looks as follows:

$$p_{\mathcal{M}}(\mathrm{H}\,|\,\mathrm{E}) = \left(\sum_{\mathrm{H}_i \in \mathcal{H}} \frac{p_{\mathcal{M}}(\mathrm{E}\,|\,\mathrm{H}_i)}{p_{\mathcal{M}}(\mathrm{E}\,|\,\mathrm{H})} \cdot \frac{p_{\mathcal{M}}(\mathrm{H}_i)}{p_{\mathcal{M}}(\mathrm{H})} \right)^{-1}. \tag{12.1}$$

Operationalized in terms of bets, this means the following: If we understand the above probabilities as indicating fair odds for conditional bets that are called off if the proposition behind the dash is false, then the only way to avoid Dutch books is to make sure that one's betting odds (and the implied probabilities) respect Equation (12.1)—the model-relative version of Bayes' Theorem.

Note, however, that these conditional degrees of belief are derived from supposing different (representatives of sets of) possible worlds (e.g., $\omega_{\mathcal{M}}$, $\omega_{\mathcal{M},\mathrm{H}}$ and $\omega_{\mathcal{M},\mathrm{E}}$), corresponding to different probability functions. Seen this way, Bayes' Theorem states how the different functions that represent our conditional degrees of belief ($p_{\mathcal{M}}$, $p_{\mathrm{H},\mathcal{M}}$ and $p_{\mathrm{E},\mathcal{M}}$) need to coordinate in order to avoid a Dutch book. This is in fact very similar to the standard justification of Bayes' Theorem. But now, rather than coordinating degrees of belief described by a single probability function, Bayes' Theorem coordinates different probability functions that emerge from supposing different sets of propositions.

Most primers and encyclopedia entries on Bayesian inference describe Bayes' Theorem as "a simple mathematical formula used for calculating conditional probabilities" (Joyce 2003), that is, as a practical tool for performing Bayesian inference and computing posterior probabilities. On the suppositional analysis, it is more than that: it is an epistemic coordination principle with genuine philosophical significance. In fact, the argument for Bayes' Theorem in the context of hypothetical degrees of belief is very similar to the argument from the introductory chapter that conditional degrees of belief should obey Ratio Analysis and be coordinated with unconditional degrees of belief. The Dutch Book argument for the equality $p(\mathrm{E}|\mathrm{H}) = p(\mathrm{E} \wedge \mathrm{H})/p(\mathrm{H})$ can likewise be used for demonstrating why a rational Bayesian reasoner should coordinate the different probability functions that express her conditional degrees of belief. See pages 17–21 in the introductory chapter, and in particular Table T.5.

Note that Equation (12.1) is a synchronic, not a diachronic, constraint: it describes the conditional degree of belief in H, given E and $\mathcal{M}$, not the degree of belief in H, given $\mathcal{M}$ and after learning E. Only when it is conjoined with Conditionalization $p^{\mathrm{E}}_{\mathcal{M}}(\mathrm{H}) = p_{\mathcal{M}}(\mathrm{H}\,|\,\mathrm{E})$ do we obtain a rule for describing how our degrees of belief should change after a learning experience. Conditionalization emerges as an inference which connects two different modes of reasoning: learning E and supposing E. Exploring this connection is a fascinating topic for future research.

In sum, Bayesian inference should be construed as a model-relative activity where we reason with hypothetical, not with actual, degrees of belief. As we have argued, this impairs neither the functionality of Bayesian reasoning nor its normative pull. On the contrary, we obtain a more unified picture that resolves some persistent puzzles such as the interpretation of subjective prior degrees of belief within statistical models, and the role of chance–credence coordination in scientific inference.

All this suggests that conditional degrees of belief might be a more fruitful primitive notion in Bayesian reasoning than unconditional degrees of belief. This diagnosis resonates well with Hájek's (2003) analysis of conditional probability, and also with the Popper–Rényi axioms, where conditional probability is taken to be primitive (see Popper 1959/2002, Rényi 1970 and the explanations in the introductory chapter). Unconditional probability can then be obtained as a limiting case of conditional probability. It is up to ongoing and future work (e.g., Gyenis, Hofer-Szabó and Rédei 2017; Fitelson and Hájek 2017) to determine the most promising road for spelling out the details of this relationship.

12.5 Conclusion

This chapter set out to explain how we can have positive degrees of belief in highly idealized models when doing Bayesian inference. Quickly this question led us to the task of justifying THE EQUALITY: why do conditional degrees of belief in an observation, given a statistical hypothesis, track the corresponding probability densities? In other words, why is $p(\mathrm{E}\,|\,\mathrm{H}) = \rho_{\mathrm{H}}(\mathrm{E})$? We have argued that the suppositional analysis of conditional degrees of belief provides the key to answering this question, and that it squares well with our intuitive handling of conditional probabilities. On top of this, it explains the role of chance–credence coordination in scientific reasoning, the interpretation of priors and posteriors in highly idealized models, and the epistemic function of Bayes' Theorem.

Moreover, the suppositional account applies naturally to central debates in confirmation theory and philosophy of probability, such as the Problem of Old Evidence. For example, Howson (1984, 1985) and Sprenger (2015) propose to solve that problem by means of counterfactually interpreted conditional probabilities (see Variation 5). The results in this chapter can be invoked and extended for supporting their approach. Another promising application connects conditional probabilities to theories of objective chance (e.g., Suárez 2018) and a more profound discussion of how chance–credence coordination supports scientific inference. Finally, probabilistic semantics of indicative conditionals evaluate the truth or acceptability of a conditional by supposing the antecedent and evaluating the probability of the consequent in the light of this supposition (e.g., Adams 1975; Edgington 1995). It is therefore worthwhile to elaborate the links between conditional degrees of belief and probabilistic accounts of conditionals in more detail (see also Douven 2016).

Finally, we summarize our main results. The form of our synopsis—a list of seven theses—is perhaps more typical of medieval disputations than of twenty-first-century philosophy, but we think that it is excellently suited to leave the reader with a coherent and unified picture.

1. Conditional degrees of belief of the type $p(E|H)$, with observation E and statistical hypothesis H, should be embedded into the context of a general statistical model $\mathcal{M}$ and be interpreted in the suppositional (and possibly counterfactual) way anticipated by Ramsey: we suppose that $\mathcal{M}$ and H are true and reason on this basis about the probability of E. The relevant class of possible worlds for evaluating conditional degree of belief in E is defined by an equivalence relation on the set of possible worlds, namely by assigning the same probabilistic law to S.

2. The suppositional analysis explains why such conditional degrees of belief track the corresponding (objective) probability densities, and why many conditional probability statements appear to be analytic.

3. The suppositional analysis transfers the epistemic function of chance–credence coordination principles, such as the Principal Principle or the Principle of Direct Inference, from actual to counterfactual worlds.

4. By justifying THE EQUALITY, the suppositional analysis ensures agreement on the value of Bayesian measures of evidential support such as the Bayes factor and defends the rationality of Bayesian inference.

5. Ultimately, all probabilities in Bayesian inference are conditional degrees of belief: they are conditional on assuming a general statistical model. Thereby our approach resolves the trilemma from page 311 that we should never assign positive degrees of belief to a hypothesis we know to be wrong.

6. When it comes to transferring model predictions to the real world, the suppositional analysis makes Bayesian inference analogous to other model-based reasoning strategies in science.

7. Bayes' Theorem does not express a mathematical triviality but an epistemic coordination principle for various probability functions that describe conditional degrees of belief.

With these conclusions, we have come full circle: the foundational justification of the subjective interpretation of (conditional) degrees of belief and Bayesian inference goes hand in hand with the practice of Bayesian modeling in statistics and science. It is now time to look back on our achievements.

Conclusion: The Theme Revisited

This book is an attempt to analyze and elucidate scientific reasoning by means of subjective Bayesian inference. Subjective Bayesians assign probabilities to scientific hypotheses and they interpret their values as (conditional) degrees of belief. By the principle of Bayesian Conditionalization, or one of its generalizations (see the introductory chapter and Variation 4), these degrees of belief are changed in the light of incoming evidence.

While scientific reasoning is, of course, much broader and richer than what degrees of belief can express, subjective probability helps us understand it better in at least three important respects. First, subjective probability figures crucially in rational reconstructions of prominent scientific argument patterns. Second, we can give fruitful Bayesian explications of several important concepts in science, such as confirmation, explanatory power and causal strength. Third, Bayesian models can help us to understand important cognitive values in scientific (and in particular, statistical) inference, such as simplicity, objectivity and severe hypothesis testing.

The book brings together and unifies these Bayesian models in philosophy of science. The division into chapters—variations on the Bayesian theme $p(\mathrm{E}\,|\,\mathrm{H}) = p(\mathrm{H})\,p(\mathrm{E}\,|\,\mathrm{H})/p(\mathrm{E})$—corresponds to the rough thematic division sketched above. The first set of Variations investigates **confirmatory arguments in science:** Variation 1 presents and evaluates different proposals for quantifying degree of confirmation. Variation 2 on the No Alternatives Argument shows how the failure to find alternatives to a theory can confirm a theory even in the absence of genuine empirical evidence. The argument also suggests how Inference to the Best Explanation can be justified within a confirmation-theoretic perspective. Variation 3 frames the famous No Miracles Argument for scientific realism in Bayesian terms and investigates its scope and limits according to different ways of framing and modeling the argument. Variation 4 provides a framework for learning conditional evidence (e.g., scientific hypotheses of the form "if A, then B") in a Bayesian framework. Variation 5 provides novel solutions for the Problem of Old Evidence—the problem of describing how explanatory successes of a theory with respect to already known data can boost confidence in a theory.

Taken together, these Variations demonstrate that Bayesian confirmation theory extends beyond the standard case of evaluating predictions of a scientific hypothesis: it suits a remarkable variety of modes of scientific reasoning.

The second set of Variations applies the Bayesian modeling apparatus to **explicating central scientific concepts** such as causal strength, explanatory power and intertheoretic reduction. In all three cases, the use of causal Bayes nets for identifying the relata of the causation/explanation/reduction relation is crucial. Variations 6 and 7 propose axiomatic characterizations of causal strength and explanatory power in terms of sets of adequacy conditions. These conditions are then evaluated from a normative point of view in order to advance or dismiss particular explications. Variation 8 demonstrates how the establishment of intertheoretic reductive relations can raise the probability of a scientific theory at the fundamental level.

The third set of Variations is motivated by the wish to answer open questions in **statistical inference.** Variation 9 closes a lacuna in the methodology of hypothesis testing by developing a probabilistic measure of corroboration—a project which allows one to assign meaning to non-significant results and is therefore of the utmost importance for scientific practice. Variation 10 investigates how Bayesians can account for the value of simplicity when choosing between various statistical models. Variation 11 takes up various challenges to the objectivity of Bayesian inference and demonstrates that it is no less objective than its frequentist competitors. Variation 12, finally, develops a suppositional interpretation of conditional degrees of belief that explains the use of chance–credence coordination in statistical inference and rationalizes the use of highly idealized Bayesian models in science.

These final Variations also demonstrate the limits of Bayesian modeling. For example, Variation 9 argues that there cannot be a purely Bayesian, confirmation-theoretic explication of corroboration. Bayesian and non-Bayesian approaches need to be combined to measure the degree to which a hypothesis has stood up to severe tests. Variation 10 demonstrates that popular "Bayesian" explications of the weight of simplicity in model selection fail to have a strictly Bayesian justification. Rather, they should be seen as being motivated by a Bayesian *heuristics*. Variation 11, despite its overall optimistic tone, also points out limits of objectivity in Bayesian reasoning, and Variation 12 demonstrates the limits of the textbook interpretation of probabilities in Bayesian inference.

It is also notable that there is a high degree of similarity between certain chapters, despite the divergent nature of their explicanda. Variations 6, 7 and 9 transfer techniques from the axiomatic characterization of confirmation measures (Variation 1) to causal strength, explanatory power and corroboration. This leads to a strong methodological unification since in all four explications, probabilistic difference-making is a central concept. Variations 4 and 5 both deal with the topic of learning conditional evidence and logical implications of a scientific hypothesis. Finally, Variations 2 and 3 model the assessment of a scientific theory by means of including an additional variable: the number of available alternatives.

We do not want to convince the reader that Bayesian modeling is a panacea for all research problems in philosophy of science. This standpoint would be exposed to justified and potentially devastating criticism—in particular, to any criticisms directed against purely formal accounts of inductive inference (Norton 2003, 2011). What we hope to have demonstrated is much less ambitious, but still very promising for the future: that Bayesian inference is more than a simple and appealing theory for representing and updating degrees of belief—it is home to powerful models that can be applied to a surprising variety of problems in scientific reasoning.

The use of Bayesian inference as a model for explicating central aspects of scientific reasoning is also characteristic of our general methodology. Indeed our approach can be characterized as **scientific philosophy,** but this should not be understood in the sense of the Logical Empiricists (e.g., Carnap 1935) or of naturalists such as W. V. O. Quine (1969). The Logical Empiricists conceived scientific philosophy primarily as logical and linguistic analysis that would, in the end, lay the foundations for a language of science. Naturalists like Quine saw philosophy as a mere branch of science—for instance, epistemology was supposed to reduce to cognitive psychology. Recently, Maddy (2009) and Ladyman and Ross (2009) have tried to revive this style of philosophical theorizing. We do believe, however, that the epistemic problems of science are genuinely philosophical problems that cannot be reduced to purely scientific questions (above all, because of the normative character of many questions). Our understanding of scientific philosophy is much closer to the views articulated by Hans Reichenbach (1951) and Hannes Leitgeb (2013). Like Leitgeb, we believe that such questions can be addressed by means of scientific tools, that is, formal modeling, case studies, experimentation and computer-based simulations, which can be fruitfully combined with conceptual analysis as a core method of philosophical analysis (see also Hartmann and Sprenger 2012).

All these methods have a history in philosophy, some longer, some shorter. **Conceptual analysis**—breaking complex concepts into their constitutive parts, such as a set of necessary and sufficient conditions—goes back to the classics of philosophy (of science), such as Plato's famous analysis of knowledge as justified true belief and Hume's equally classic analysis of causation (namely as contiguity, succession and constant conjunction). Since then, conceptual analysis has been an indispensable method of philosophical research, and we, too, have amply used it in our book. **Mathematical and logical analysis** has a similarly rich history, going back to Aristotle's logic and the medieval logicians. Interestingly, mathematics, and probability theory in particular, has been used less frequently than logic for explicating philosophical arguments—Hume's *Dialogues on Natural Religion* and the famous tenth chapter of the *Enquiry on Human Understanding* ("Of Miracles") being among the notable exceptions. **Case studies**—already part of Bacon's *Novum Organon* and Descartes' *Discours de la Méthode* (Part V)—have been popular in philosophy of science since the 1960s and 1970s. They answered the need for calibrating general philosophical models of scientific reasoning, such as those provided by the Logical Empiricists, with the practice of science and have inspired philosophical theorizing ever since. For instance, the mechanistic model of explanation popularized by Machamer, Darden and Craver (2000) and Craver (2007) heavily draws on case studies in cognitive science and biology. **Experimental methods,** which challenge the exclusive reliance on conceptual analysis in philosophical research, have a quite young history (Stich 1988; Knobe and Nichols 2007). Primarily, they are used for testing theoretical accounts of central philosophical concepts such as knowledge and justification (e.g., Weinberg, Nichols and Stich 2001; Alexander and Weinberg 2007; Nagel, San Juan and Mar 2013), reference (e.g., Machery et al. 2004), free will (e.g., Nahmias et al. 2005) or intentionality (e.g., Knobe 2003; Hindriks, Douven and Singmann 2016). Part of these investigations, such as experiments on normative aspects in causal and explanatory judgments (Knobe and Fraser 2008; Hitchcock and Knobe 2009), is directly relevant for philosophy of science—see also Variations 6 and 7. More recently, cognitive scientists and philosophers of science have worked together to highlight the empirical dimension of concepts such as confirmation, causation and explanation, and to establish important links between them (e.g., Crupi, Fitelson and Tentori 2008; Crupi, Chater and Tentori 2013; Colombo, Bucher and Sprenger 2017b). Finally, **computational methods,** and agent-based simulations in particular, have gained ground in philosophy of science over the last years. Often they are used to study the emergence and stability of social norms and contracts (e.g., Alexander 2007; Skyrms 2010; Muldoon et al. 2014), but sometimes they are also applied to modeling scientific progress and the communication structure of epistemic communities (e.g., Zollman 2007; Weisberg and Muldoon 2009; De Langhe and Rubbens 2015; Heesen 2017). Of particular interest are those studies where probabilistic reasoning in science (e.g., hypothesis tests) interacts with rewards and biases in the scientific community (e.g., Romero 2016). Notably, all these methods are rarely combined with each other. Integrating them in the treatment of diverse topics in philosophy of science is perhaps one of the main innovations of this book.

Indeed, most Variations in this book feature a majority of these methods. Conceptual analysis and formal modeling, the core methods of our explicative project, are used in almost any of the twelve Variations. Variation 11, which evaluates the objectivity of Bayesian reasoning, is perhaps the only one that explicitly eschews formal modeling. However, it requires a basic understanding of principles of statistical inference. Case studies play an important role in Variation 2 (particle physics as an application of the NAA), Variation 5 (confirmation by old evidence), Variation 6 (applications of causal strength in science), Variation 8 (reduction in statistical mechanics), Variation 9 (null hypothesis significance testing), Variation 10 (analysis of model selection criteria such as AIC and BIC) and Variation 11 (parapsychological research and statistical analysis of experimental data). These Variations also address methodological problems in specific disciplines (e.g., Variation 2 for particle physics, Variations 6 and 11 for cognitive psychology and medical science, Variations 9 and 10 for statistics). Experimental evidence from psychology and cognitive science is cited in Variation 1 (judgments of confirmation), Variation 4 (learning conditionals), Variation 6 (causal reasoning), Variation 7 (judgments of explanatory power), Variation 9 (scientists' use of null hypothesis significance tests), Variation 11 (the social psychology case study) and Variation 12 (experiments on learning vs. supposing). Finally, computational methods are used—sometimes behind the screens—in Variation 2 (calculating the degree of confirmation of the NAA), Variation 3 (idem for the NMA), Variation 5 (comparing our premises to Jeffrey 1983), Variation 7 (calculating explanatory power vs. posterior probability) and Variation 8 (calculating degree of confirmation for successful reductions). Table C.1 gives a schematic overview.

Variation	Topic	CA	FM	CS	EM	CM
1	Confirmation and Induction	✓	✓		✓	
2	The No Alternatives Argument	✓	✓	✓		✓
3	Scientific Realism and the NMA	✓	✓			✓
4	Conditional Evidence	✓	✓		✓	
5	Old Evidence	✓	✓	✓		✓
6	Causal Strength	✓	✓	✓	✓	
7	Explanatory Power	✓	✓		✓	✓
8	Reduction	✓	✓	✓		✓
9	Hypothesis Tests and Corroboration	✓	✓	✓	✓	
10	Simplicity	✓	✓	✓		
11	Objectivity	✓		✓	✓	
12	Models, Idealizations, Objective Chance	✓	✓		✓	

Table C.1: An overview of the methods used in the book, split up according to the Variations where they occur. CA = Conceptual Analysis, FM = Formal Methods, CS = Case Studies and Discipline-Specific Methodological Problems, EM = Experimental Methods, CM = Computational Methods.

We finish by sketching open research questions and recapitulating three open problems for each chapter or topic of the book:

Variation 1 Confirmation

- Investigating the information-theoretic foundations of confirmation measures
- Applying confirmation theory to the diagnostic value of scientific tests (e.g., in medicine)
- Using confirmation judgments to explain phenomena in psychology of reasoning

Variation 2 The No Alternatives Argument (NAA)

- Reframing Inference to the Best Explanation (IBE) in terms of the NAA
- Finding instances of NAA-based reasoning in diverse scientific fields
- Relating the NAA to eliminative inference and to the use of the TINA argument in the political discourse

Variation 3 Scientific Realism and the No Miracles Argument (NMA)

- Conducting a scientometric analysis of theoretical stability in twentieth-century science in order to evaluate the premises of the NMA
- Studying parallels between the NAA and the NMA
- Extending the NMA toward the full realist thesis

Variation 4 Learning Conditional Evidence

- Evaluating promises and limits of a pluralist position for learning non-extreme conditional evidence ($0 < p'(E|H) < 1$) by means of divergence minimization
- Developing normative constraints for the choice of a specific divergence function
- Transferring our analysis from indicative to subjunctive conditionals

Variation 5 The Problem of Old Evidence (POE)

- Applying the POE solutions to the prediction-vs.-accommodation debate
- Integrating the "counterfactual" solutions of the POE with a theory of belief revision (e.g., AGM theory)
- Solving the POE in terms of learning conditional evidence (see Variation 4)

Variation 6 Causal Strength

- Relating the causal strength measure to statistical effect sizes
- Connecting our analysis of causal strength to probabilistic and/or information-theoretic explications of causal specificity
- Investigating how causal strength spreads and combines in more complicated network structures

Variation 7 Explanatory Power

- Investigating the rationality of explanatory-power-based IBE, including the case of uncertain evidence
- Conducting further experiments on the determinants of explanatory judgments, including ecologically valid scenarios
- Synthesizing theories of explanatory power with an analysis of actual causation, statistical normality and so on

Variation 8 Intertheoretic Reduction

- Analyzing the robustness of the confirmatory value of intertheoretic reduction using different confirmation measures
- Modeling intertheoretic reduction as increasing the (probabilistically measured) coherence of a set of theories
- Investigating the value of reductive relationships when evidence confirms one of the theories and disconfirms the other

Variation 9 Hypothesis Tests and Corroboration

- Extending the proposed corroboration measure to more complicated statistical inference problems (nuisance parameters, hierarchical modeling, etc.)
- Reconciling Bayesian and frequentist hypothesis tests from a corroboration-based perspective
- Finding case studies for corroboration-based reasoning in the history of science

Variation 10 Simplicity

- Exploring and systematizing the context-sensitivity of model selection criteria
- Working out the thesis of instrumental Bayesianism in more detail and transferring it to other areas of statistical inference
- Comparing the respective roles of simplicity in statistical and causal inference (model selection vs. causal Bayesian networks)

Variation 11 Objectivity

- Scrutinizing the objectivity claims of Objective Bayesianism
- Using Bayesian inference for addressing biases in statistical reasoning and for addressing the replication crisis
- Aggregating different (subjective) probability functions in an objective way

Variation 12 Models, Idealizations and Objective Chance

- Applying the suppositional interpretation of degrees of belief to the POE
- Exploring the implications of the suppositional interpretation for philosophical theories of objective chance
- Establishing links between suppositional accounts of conditional degrees of belief and suppositional accounts of (indicative and subjunctive) conditionals

Now we proceed to describing five major research projects that fit into the Bayesian philosophy of science research program, and that bridge different topics discussed in this book.

An obvious direction into which our Bayesian research program could be extended is the **integration of various modes of human reasoning.** The analyses of learning conditionals and of confirmation by old evidence in Variations 4 and 5, respectively, showed that causal considerations often constrain an agent's rational degrees of belief. Moreover, the evaluation of conditional degrees of belief proceeds counterfactually (see Variation 12), and via the notion of possible interventions, counterfactual considerations play an important role in evaluating causal relations. Future work should bring forth an integrated theory of causal induction, Bayesian learning and conditional reasoning (e.g., building on Oaksford and Chater 2000; Pearl 2000; Douven 2016; Over 2016). Apart from theoretical pioneer work, we see a lot of promise in experiments that investigate such connections or that try to measure concepts such as causal strength (see Variation 6). Finally, Bayesian models can be used to study human argumentation and the value we assign to logically valid arguments (e.g., Eva and Hartmann 2018a; Stern and Hartmann 2018).

Second, Bayesian confirmation is intimately related to causal and explanatory considerations (Variations 1, 6 and 7). On a theoretical level, this calls for an **extended analysis of the relationship between measures of confirmation, corroboration, causal strength and explanatory power,** similar to what Schupbach (2018) did for measures of explanatory power and posterior probability. This should lead to a sharper demarcation of these concepts and to a description of the conditions when the one is conducive to the other. On an empirical level, we propose experiments that uncover correlations and differences in judgments of explanatory power, causal strength and probability, in order to reveal the determinants of explanatory judgments and to provide a more nuanced and descriptively appropriate view of explanatory reasoning. Since explanation is a concept which is loaded with causal and probabilistic connotations, such experiments strike us as highly valuable. Colombo, Postma and Sprenger (2016) and Colombo, Bucher and Sprenger (2017b) have already made some steps in this direction, but there is still much work to be done (see also the survey in Sloman and Lagnado 2015). This research could also be related to the role of simplicity in scientific reasoning: our formal analysis of simplicity in model selection (see Variation 10) should at some point be complemented by an empirical investigation of how simplicity considerations affect other aspects of scientific reasoning, similar to what Lombrozo (2007) did for simplicity in explanatory judgments.

Third, the material in this book is an outstanding basis for a detailed investigation of the **scope and rationality of Inference to the Best Explanation.** Not only can one assess IBE on the basis of diverse measures of explanatory power (see Variation 7), it is also possible to relate IBE to other argument patterns that we explicated in this book: the No Alternatives Argument (NAA, see Variation 2) and the No Miracles Argument (NMA, see

Variation 3). As we argued in those Variations, both arguments are essentially abductive in arguing that the empirical adequacy of a theory T is the best explanation for the absence of viable alternatives (NAA) and for the predictive success of T (NMA). Our book may therefore give new impulses for research along this route.

Fourth, this book does not cover **theoretical unification and its role in reduction and explanation** (compare Variations 7 and 8)—mainly because this topic is rarely approached from a Bayesian perspective. Unification is traditionally regarded as an important cognitive value in scientific reasoning (e.g., McMullin 1982; Douglas 2013), as a value that, for most scientists, counts as a reason to accept a theory and to pursue it further. Based on the pioneering work done by Myrvold (2003, 2017) and Schupbach (2005), it seems plausible to explicate unification by means of confirmation-theoretical or information-theoretic models, to explain its role in intertheoretic reductions and explanatory reasoning and to describe unification in important case studies, such as Bayesian cognitive science (Colombo and Hartmann 2017).

Fifth and last, Bayesian methods can provide **better foundations for hypothesis testing in science.** It has been frequently noted that the current method of hypothesis testing, essentially based on *p*-values, is not only at odds with the very principles of Bayesian reasoning, but also a danger for the reliability of scientific inquiry (e.g., Berger and Sellke 1987; Goodman 1999a; Cumming 2012, 2014). It is therefore important to integrate Bayesian reasoning into hypothesis tests and to reconcile both paradigms (Wetzels et al. 2009; Wetzels and Wagenmakers 2012; Lee and Wagenmakers 2014; Morey et al. 2014, 2016). However, doing so is often far from straightforward, due to the different motivations that feed Bayesian confirmation theory (see Variation 1) and hypothesis testing in the tradition of Popper and Fisher. Variation 9 makes an attempt to quantify the degree of corroboration of a hypothesis and Variation 6 axiomatizes various measures of causal strength that could result from randomized controlled trials or case-control studies. These projects need to be expanded in order to obtain a theoretically appealing Bayesian account of hypothesis testing that squares well with scientific practice.

This brings us, finally, to a wider perspective on Bayesian philosophy of science. First, this book has neglected the **social dimension of science.** We have, so far, focused on the perspective of an individual scientist (or a homogeneous research team) who does experiments, analyzes data and assesses theories. Future work could link the issues covered in this book to questions about merging opinions and the role of experts in science (for survey articles, see Dietrich and List 2016; Martini and Sprenger 2017). For yet a different research program in the social epistemology of science that can be tackled by Bayesian models, consider the exploration of epistemic landscapes and the credit reward system in science (e.g., Zollman 2007; Weisberg and Muldoon 2009; Heesen 2017, 2018).

Second, we could tighten the link between Bayesian reasoning in philosophy and **Bayesian reasoning in science** (e.g., Bayesian statistics). In this book, we have only scratched the surface, but there is a fascinating and largely unexplored set of questions on how philosophical insights about Bayesian reasoning and hypothesis testing should translate into practical statistical reasoning (e.g., Gallistel 2009; Bernardo 2012; Sprenger 2013b). The fifth research project listed above, concerned with scientific hypothesis testing and Bayesian Inference, falls into this domain. But there are also more general methodological questions. For example, Gelman and Shalizi (2012, 2013) suggest that Bayesian inference is very convenient at the micro-level of statistical inference within a given class of models, but that the proper task of model testing and evaluation rather follows a hypothetico-deductive rationale. Moreover, there has not yet been a systematic investigation and comparison of the philosophical foundations of the different objective Bayesian approaches and the conceptions of objectivity that they endorse. Given the centrality of claims to objectivity in the evaluation of research findings in modern science, this strikes us as a highly worthwhile endeavor.

Both projects—investigating the practice and the social dimension of Bayesian inference—can be pursued in tandem. The social sciences, psychology in particular, are undergoing a **replication crisis,** that is, difficulties in replicating findings from published experiments (e.g., Galak et al. 2012; Makel, Plucker and Hegarty 2012; Open Science Collaboration 2015; Camerer et al. 2016; Nosek and Errington 2017). How much can we trust scientific method if research results are so often unstable? Bayesian methods may provide an answer to this question: they have been used to explain why the current publication culture promotes unreliable findings, and to point out how such biases can be cured (e.g., Sterne and Davey Smith 2001; Wacholder et al. 2004; Ioannidis 2005b; Ioannidis and Trikalinos 2007). The work in this book, especially that in Variations 1, 9 and 11, provides a starting point for further philosophical investigation of these problems. Further research in this direction would combine Bayesian philosophy of science with the practice of Bayesian statistics and a social perspective on the scientific enterprise.

All in all, there is an exciting and virtually inexhaustible set of unanswered research questions in Bayesian philosophy of science. Therefore we predict a bright and fascinating future for the Bayesian research program—in philosophy of science and beyond. We hope that our book inspires future contributions along these lines.

Bibliography

Adams, Ernest W. (1965). The Logic of Conditionals. *Inquiry 8*, 166–197.

Adams, Ernest W. (1975). *The Logic of Conditionals*. Dordrecht: Reidel.

Akaike, Hirotogu (1973). Information Theory and an Extension of the Maximum Likelihood Principle. In B. N. Petrov and F. Csaki (eds.), *Second International Symposium on Information Theory*, pp. 267–281. Budapest: Akademiai Kiado.

Alchourrón, Carlos E., Peter Gärdenfors, and David Makinson (1985). On the Logic of Theory Change: Partial Meet Contraction and Revision Functions. *Journal of Symbolic Logic 50*, 510–530.

Alexander, Joshua and Jonathan M. Weinberg (2007). Analytic Epistemology and Experimental Philosophy. *Philosophy Compass 2*, 56–80.

Alexander, Jason McKenzie (2007). *The Structural Evolution of Morality*. Cambridge: Cambridge University Press.

Allais, Maurice (1953). Le comportement de l'homme rationnel devant le risque: Critique des postulats et axiomes de l'École Américaine. *Econometrica 21*, 503–546.

Allais, Maurice and Ole Hagen (1979). *Expected Utility Hypotheses and the Allais Paradox*. Dordrecht: Reidel.

Aquinas, Thomas (1945). *Basic Writings of St. Thomas Aquinas*. New York: Random House.

Assen, Marcel A. L. M. van, Robbie C. M. van Aert, Michèle B. Nuijten, and Jelte M. Wicherts (2014). Why Publishing Everything is More Effective than Selective Publishing of Statistically Significant Results. *PLoS One 9*. https://doi.org/10.1371/journal.pone.0084896.

Atkinson, David (2012). Confirmation and Justification: A Commentary on Shogenji's Measure. *Synthese 184*, 49–61.

Baker, Alan (2003). Quantitative Parsimony and Explanation. *British Journal for the Philosophy of Science 54*, 245–259.

Baker, Alan (2016). Simplicity. In Ed Zalta (ed.), *The Stanford Encyclopedia of Philosophy*. Retrieved from https://plato.stanford.edu/archives/sum2018/entries/simplicity/.

Bakker, Marjan, Jelte Wicherts, and Annette van Dijk (2012). The Rules of the Game Called Psychological Science. *Perspectives on Psychological Science 7*, 543–554.

Bandyopadhyay, Prasanta S., Robert J. Boik, and Prasun Basu (1996). The Curve Fitting Problem: A Bayesian Approach. *Philosophy of Science 63*, S264–S272.

Barnes, Eric Christian (2008). *The Paradox of Predictivism*. Cambridge: Cambridge University Press.

Batterman, Robert W. (2002). *The Devil in the Details: Asymptotic Reasoning in Explanation, Reduction, and Emergence*. Oxford: Oxford University Press.

Bayarri, M. J., M. H. De Groot, and J. B. Kadane (1988). What Is the Likelihood Function? In S. S. Gupta and J. O. Berger (eds.), *Statistical Decision Theory and Related Topics*, Volume IV, pp. 1–27. Springer.

Bayes, Thomas (1763). An Essay Towards Solving a Problem in the Doctrine of Chances. *Philosophical Transactions of the Royal Society 53*, 370–418.

Bechtel, William and Adele Abrahamsen (2005). Explanation: A Mechanist Alternative. *Studies in History and Philosophy of Science Part C: Studies in History and Philosophy of Biological and Biomedical Sciences 36*, 421–441.

Beckers, Sander and Joost Vennekens (2018). A Principled Approach to Defining Actual Causation. *Synthese 195*, 835–862.

Begley, C. Glenn and Lee M. Ellis (2012). Drug Development: Raise Standards for Preclinical Cancer Research. *Nature 483*, 531–533.

Bem, Daryl J. (2011). Feeling the Future: Experimental Evidence for Anomalous Retroactive Influences on Cognition and Affect. *Journal of Personality and Social Psychology 100*, 407–425.

Bem, Daryl J., Jessica Utts, and Wesley O. Johnson (2011). Must Psychologists Change the Way they Analyze Their Data? *Journal of Personality and Social Psychology 101*, 716–719.

Benjamin, Daniel J., James O. Berger, Magnus Johannesson, et al. (2018). Redefine statistical significance. *Nature Human Behaviour 2*, 6–10.

Bennett, Jonathan (2003). *A Philosophical Guide to Conditionals*. Oxford: Oxford University Press.

Berger, James O. (2003). Could Fisher, Jeffreys and Neyman have Agreed on Testing? *Statistical Science 18*, 1–32.

Berger, James O., B. Boukai, and Y. Wang (1997). Unified Frequentist and Bayesian Testing of a Precise Hypothesis. *Statistical Science 12*, 133–160.

Berger, James O. and Thomas Sellke (1987). Testing a Point Null Hypothesis: The Irreconcilability of *P* Values and Evidence. *Journal of the American Statistical Association 82*, 112–122.

Berger, James O. and Robert L. Wolpert (1984). *The Likelihood Principle.* Hayward, Calif.: Institute of Mathematical Statistics.

Bernardo, José M. (1979a). Expected Information as Expected Utility. *Annals of Statistics 7*, 686–690.

Bernardo, José M. (1979b). Reference Posterior Distributions for Bayesian Inference. *Journal of the Royal Statistical Society, Series B 41*, 113–147.

Bernardo, José M. (1999). Nested Hypothesis Testing: The Bayesian Reference Criterion. In José M. Bernardo, James O. Berger, A. P. Dawid, and Adrian F. M. Smith (eds.), *Bayesian Statistics 6: Proceedings of the Sixth Valencia Meeting on Bayesian Statistics*, pp. 101–130 (with discussion). Oxford: Oxford University Press.

Bernardo, José M. (2012). Integrated Objective Bayesian Estimation and Hypothesis Testing. In *Bayesian Statistics 9: Proceedings of the Ninth Valencia Meeting*, pp. 1–68 (with discussion). Oxford: Oxford University Press.

Bernardo, José M. and Adrian F. M. Smith (1994). *Bayesian Theory*. New York: Wiley.

Bickle, John (1998). *Psychoneural Reduction: The New Wave*. Cambridge, Mass.: MIT Press.

Birnbaum, Allan (1962). On the Foundations of Statistical Inference. *Journal of the American Statistical Association 57*, 269–306.

Blackwell, David and Lester Dubins (1962). Merging of Opinions with Increasing Information. *Annals of Mathematical Statistics 33*, 882–886.

Boniolo, Giovanni (2003). Kant's Explication and Carnap's Explication: The *Redde Rationem*. *International Philosophical Quarterly 3*, 289–298.

Bornstein, Robert F. (1989). Exposure and Affect: Overview and Meta-Analysis of Research, 1968–1987. *Psychological Bulletin 106*, 265–289.

Bovens, Luc (2010). Judy Benjamin is a Sleeping Beauty. *Analysis 70*, 23–26.

Bovens, Luc and Stephan Hartmann (2003). *Bayesian Epistemology*. New York: Oxford University Press.

Box, George E. P. (1976). Science and Statistics. *Journal of the American Statistical Association 71*, 791–799.

Boyd, Richard (1980). Scientific Realism and Naturalistic Epistemology. In Peter D. Asquith and Ronald N. Giere (eds.), *PSA 1980*, Volume II, pp. 613–662. East Lansing, Mich.: Philosophy of Science Association.

Boyd, Richard (1983). On the Current Status of the Issue of Scientific Realism. *Erkenntnis 19*, 45–90.

Boyd, Richard (1984). The Current Status of Scientific Realism. In Jarrett Leplin (ed.), *Scientific Realism*, pp. 41–82. Berkeley, Calif.: University of California Press.

Bradley, Seamus (2014). Imprecise Probabilities. In Ed Zalta (ed.), *The Stanford Encyclopedia of Philosophy*. Retrieved from https://plato.stanford.edu/archives/sum2018/entries/imprecise-probabilities/.

Braham, Matthew and Martin van Hees (2009). Degrees of Causation. *Erkenntnis 71*, 323–344.

Brier, Glenn W. (1950). Verification of Forecasts Expressed in Terms of Probability. *Monthly Weather Review 78*, 1–3.

Briggs, R.A. (2015). Foundations of Probability. *Journal of Philosophical Logic 44*, 625–640.

Broadbent, Alex (2013). *Philosophy of Epidemiology*. Basingstroke: Palgrave Macmillan.

Bromberger, Sylvain (1965). An Approach to Explanation. In R. J. Butler (ed.), *Analytical Philosophy*, pp. 72–105. Oxford: Oxford University Press.

Brössel, Peter (2013). The Problem of Measure Sensitivity Redux. *Philosophy of Science 80*, 378–397.

Brössel, Peter (forthcoming). *Rethinking Bayesian Confirmation Theory*. Berlin: Springer.

Brössel, Peter and Franz Huber (2015). Bayesian Confirmation: A Means with No End. *British Journal for the Philosophy of Science 66*, 737–749.

Brush, Stephen G. (1989). Prediction and Theory Evaluation: The Case of Light Bending. *Science 246*, 1124–1129.

Buchak, Lara (2013). *Risk and Rationality*. Oxford: Oxford University Press.

Burnham, Kenneth P. and David R. Anderson (2002). *Model Selection and Multimodel Inference: A Practical Information-theoretic Approach*. New York: Springer.

Burnham, Kenneth P. and David R. Anderson (2004). Multimodel Inference Understanding AIC and BIC in Model Selection. *Sociological Methods Research 33*, 261–304.

Buss, David M. and David P. Schmitt (1993). Sexual Strategies Theory: An Evolutionary Perspective on Human Mating. *Psychological Review 100*, 204–232.

Bylander, Tom, Dean Allemang, Michael C. Tanner, and John R. Josephson (1991). The Computational Complexity of Abduction. *Artificial Intelligence 49*, 25–60.

Callender, Craig (2001). Taking Thermodynamics Too Seriously. *Studies in History and Philosophy of Modern Physics 32*, 539–553.

Camerer, Colin F., Anna Dreber, Eskil Forsell, et al. (2016). Evaluating Replicability of Laboratory Experiments in Economics. *Science*. https://doi.org/10.1126/science.aaf0918.

Carnap, Rudolf (1935). *Philosophy and Logical Syntax*. London: Kegan Paul, Trench, Trubner & Co.

Carnap, Rudolf (1947). On the Application of Inductive Logic. *Philosophy and Phenomenological Research 8*, 133–148.

Carnap, Rudolf (1950). *Logical Foundations of Probability*. Chicago: University of Chicago Press.

Cartwright, Nancy (1979). Causal Laws and Effective Strategies. *Noûs 13*, 419–437.

Cartwright, Nancy (1989). *Nature's Capacities and Their Measurement*. Oxford: Clarendon Press.

Casscells, W., A. Schoenberger, and T. B. Graboys (1978). Interpretation by Physicians of Clinical Laboratory Results. *New England Journal of Medicine 299*, 999–1001.

Cevolani, Gustavo and Luca Tambolo (2013). Progress as Approximation to the Truth: A Defence of the Verisimilitudinarian Approach. *Erkenntnis 78*, 921–935.

Chakravartty, Anjan (2017). Scientific Realism. In Ed Zalta (ed.), *The Stanford Encyclopedia of Philosophy*. Retrieved from https://plato.stanford.edu/archives/sum2018/entries/scientific-realism/.

Chase, Warren and Fred Brown (2000). *General Statistics*. New York: Wiley.

Cheng, Patricia W. (1997). From Covariation to Causation: A Causal Power Theory. *Psychological Review 104*, 367–405.

Chihara, Charles S. (1987). Some Problems for Bayesian Confirmation Theory. *British Journal for the Philosophy of Science 38*, 551–560.

Chockler, Hana and Joseph Y. Halpern (2004). Responsibility and Blame: A Structural-Model Approach. *Journal of Artificial Intelligence Research 22*, 93–115.

Christensen, David (1999). Measuring Confirmation. *Journal of Philosophy 96*, 437–461.

Churchland, Paul M. (1979). *Scientific Realism and the Plasticity of Mind*. Cambridge: Cambridge University Press.

Churchland, Paul M. (1985). Reduction, Qualia, and the Direct Introspection of Brain States. *Journal of Philosophy 82*, 8–28.

Cohen, Jacob (1988). *Statistical Power Analysis for the Behavioral Sciences*. Newark, N.J.: Lawrence & Erlbaum.

Cohen, Jacob (1994). The Earth is Round ($p < .05$). *Psychological Review 49*, 997–1001.

Cohen, Michael (2018). Explanatory Justice: The Case of Disjunctive Explanations. *Philosophy of Science 85*, 442–454.

Cohen, Michael P. (2015). On Schupbach and Sprenger's Measures of Explanatory Power. *Philosophy of Science 82*, 97–109.

Cohen, Michael P. (2016). On Three Measures of Explanatory Power with Axiomatic Representations. *British Journal for the Philosophy of Science 67*, 1077–1089.

Colombo, Matteo (2017). Experimental Philosophy of Explanation Rising: The Case for a Plurality of Concepts of *Explanation*. *Cognitive Science 41*, 503–517.

Colombo, Matteo, Leandra Bucher, and Jan Sprenger (2017a). Determinants of Judgments of Explanatory Power: Credibility, Generality, and Statistical Relevance. *Frontiers in Psychology 8*, 1430.

Colombo, Matteo, Leandra Bucher, and Jan Sprenger (2017b). Determinants of Judgments of Explanatory Power: Credibility, Generalizability, and Causal Framing. In *Proceedings of the 39th Annual Conference of the Cognitive Science Society*, pp. 1806–1811. Austin, Tex.: Cognitive Science Society.

Colombo, Matteo and Stephan Hartmann (2017). Bayesian Cognitive Science, Unification and Explanation. *British Journal for the Philosophy of Science 68*, 451–484.

Colombo, Matteo, Marie Postma, and Jan Sprenger (2016). Explanatory Value, Probability and Abductive Inference. In *Proceedings of the 38th Annual Conference of the Cognitive Science Society*, pp. 432–437. Austin, Tex.: Cognitive Science Society.

Colombo, Matteo and Cory Wright (2017). Explanatory Pluralism: An Unrewarding Prediction Error for Free Energy Theorists. *Brain and Cognition 112*, 3–12.

Colyvan, Mark (2001). *The Indispensability of Mathematics*. New York: Oxford University Press.

Cox, David and David Hinkley (1974). *Theoretical Statistics*. London: Chapman & Hall.

Cox, David and Deborah G. Mayo (2010). Objectivity and Conditionality in Frequentist Inference. In Deborah G. Mayo and Aris Spanos (eds.), *Error and Inference: Recent Exchanges on Experimental Reasoning, Reliability and the Objectivity and Rationality of Science*, Chapter 2, pp. 276–304. Cambridge: Cambridge University Press.

Cox, Richard (1946). Probability, Frequency and Reasonable Expectation. *American Journal of Physics 14*, 1–10.

Craver, Carl F. (2007). *Explaining the Brain*. Oxford: Oxford University Press.

Crupi, Vincenzo (2012). An Argument for Not Equating Confirmation and Explanatory Power. *The Reasoner 6*, 39–40.

Crupi, Vincenzo (2015). Confirmation. In Ed Zalta (ed.), *The Stanford Encyclopedia of Philosophy*. Retrieved from https://plato.stanford.edu/archives/sum2018/entries/confirmation/.

Crupi, Vincenzo, Nick Chater, and Katya Tentori (2013). New Axioms for Probability and Likelihood Ratio Measures. *British Journal for the Philosophy of Science 64*, 189–204.

Crupi, Vincenzo, Branden Fitelson, and Katya Tentori (2008). Probability, Confirmation, and the Conjunction Fallacy. *Thinking & Reasoning 14*, 182–199.

Crupi, Vincenzo and Katya Tentori (2012). A Second Look at the Logic of Explanatory Power (with Two Novel Representation Theorems). *Philosophy of Science 79*, 365–385.

Crupi, Vincenzo and Katya Tentori (2013). Confirmation as Partial Entailment: A Representation Theorem in Inductive Logic. *Journal of Applied Logic 11*, 364–372.

Crupi, Vincenzo and Katya Tentori (2014). Measuring Information and Confirmation. *Studies in the History and Philosophy of Science 47*, 81–90.

Crupi, Vincenzo, Katya Tentori, and Michel González (2007). On Bayesian Measures of Evidential Support: Theoretical and Empirical Issues. *Philosophy of Science 74*, 229–252.

Crupi, Vincenzo, Katya Tentori, and Luigi Lombardi (2009). Pseudodiagnosticity Revisited. *Psychological Review 116*, 971–985.

Cumming, Geoff (2012). *Understanding the New Statistics*. New York: Routledge.

Cumming, Geoff (2014). The New Statistics: Why and How. *Psychological Science 25*, 7–29.

Cziszár, Imre (1967). Information Type Measures of Difference of Probability Distributions and Indirect Observations. *Studia Scientiarum Mathematicarum Hungarica 2*, 299–318.

Cziszár, Imre (1975). *I*-Divergence Geometry of Probability Distributions of Minimization Problems. *Annals of Probability 3*, 146–158.

Dawid, Richard (2006). Underdetermination and Theory Succession from the Perspective of String Theory. *Philosophy of Science 73*, 298–322.

Dawid, Richard (2009). On the Conflicting Assessments of the Current Status of String Theory. *Philosophy of Science 76*, 984–996.

Dawid, Richard and Stephan Hartmann (2018). The No Miracles Argument Without the Base Rate Fallacy. *Synthese 195*, 4063–4079.

Dawid, Richard, Stephan Hartmann, and Jan Sprenger (2015). The No Alternatives Argument. *British Journal for the Philosophy of Science 66*, 213–234.

De Bona, Glauber and Julia Staffel (2018). Why Be (Approximately) Coherent? *Analysis 78*, 405–415.

de Finetti, Bruno (1931). Sul significato soggettivo della probabilità. *Fundamenta Mathematicae 17*, 298–329.

de Finetti, Bruno (1936). La logique de la probabilité. In *Actes du Congrès International de Philosophie Scientifique*, Volume IV: Induction et Probabilité. Paris: Hermann & Cie.

de Finetti, Bruno (1937). La prévision: Ses lois logiques, ses sources subjectives. In *Annales de l'Institut Henri Poincaré*.

de Finetti, Bruno (1972). *Probability, Induction and Statistics: The Art of Guessing*. New York: John Wiley & Sons.

de Finetti, Bruno (1974). *Theory of Probability*. New York: John Wiley & Sons.

de Finetti, Bruno (2008). *Philosophical Lectures on Probability*. Berlin: Springer.

De Langhe, Rogier and Peter Rubbens (2015). From Theory Choice to Theory Search: The Essential Tension between Exploration and Exploitation in Science. In William J. Devlin and Alisa Bokulich (eds.), *Kuhn's* Structure of Scientific Revolutions*—50 Years On*, pp. 105–114. New York: Springer.

Diaconis, Persi and Sandy L. Zabell (1982). Updating Subjective Probability. *Journal of the American Statistical Association 77*, 822–830.

Dietrich, Franz and Christian List (2016). Probabilistic Opinion Pooling. In Alan Hájek and Christopher Hitchcock (eds.), *Oxford Handbook of Probability and Philosophy*, pp. 179–207. Oxford: Oxford University Press.

Dietrich, Franz and Luca Moretti (2005). On Coherent Sets and the Transmission of Confirmation. *Philosophy of Science 72*, 403–424.

Díez, José (2011). On Popper's Strong Inductivism (or Strongly Inconsistent Anti-Inductivism). *Studies in History and Philosophy of Science A 42*, 105–116.

Dizadji-Bahmani, Foad, Roman Frigg, and Stephan Hartmann (2010). Who's Afraid of Nagelian Reduction? *Erkenntnis 73*, 393–412.

Dizadji-Bahmani, Foad, Roman Frigg, and Stephan Hartmann (2011). Confirmation and Reduction: A Bayesian Account. *Synthese 179*, 321–338.

Douglas, Heather (2000). Inductive Risk and Values in Science. *Philosophy of Science 67*, 559–579.

Douglas, Heather (2004). The Irreducible Complexity of Objectivity. *Synthese 138*, 453–473.

Douglas, Heather (2009a). Reintroducing Prediction to Explanation. *Philosophy of Science 76*, 444–463.

Douglas, Heather (2009b). *Science, Policy, and the Value-Free Ideal*. Pittsburgh: Pittsburgh University Press.

Douglas, Heather (2013). The Value of Cognitive Values. *Philosophy of Science 80*, 796–806.

Douven, Igor (2012). Learning Conditional Information. *Mind & Language 27*, 239–263.

Douven, Igor (2016). *The Epistemology of Indicative Conditionals*. Cambridge: Cambridge University Press.

Douven, Igor (2017). Abduction. In Ed Zalta (ed.), *The Stanford Encyclopedia of Philosophy*. Retrieved from http://plato.stanford.edu/archives/sum2018/entries/abduction/.

Douven, Igor and Richard Dietz (2011). A Puzzle about Stalnaker's Hypothesis. *Topoi 30*, 31–37.

Douven, Igor and Jan-Willem Romeijn (2011). A New Resolution of the Judy Benjamin Problem. *Mind 120*, 637–670.

Douven, Igor and Jonah N. Schupbach (2015a). Probabilistic Alternatives to Bayesianism: The Case of Explanationism. *Frontiers in Psychology 6*. https://doi.org/10.3389/fpsyg.2015.00459.

Douven, Igor and Jonah N. Schupbach (2015b). The Role of Explanatory Considerations in Updating. *Cognition 142*, 299–311.

Dowe, David L. (2011). MML, Hybrid Bayesian Network Graphical Models, Statistical Consistency, Invariance and Uniqueness. In Prasanta S. Bandyopadhyay and Malcolm Forster (eds.), *Philosophy of Statistics*, Handbook of the Philosophy of Science, Volume VII, pp. 901–982. Dordrecht: Elsevier.

Dowe, David L., Steve Gardner, and Graham Oppy (2007). Bayes not Bust! Why Simplicity is no Problem for Bayesians. *British Journal for the Philosophy of Science 58*, 709–754.

Dowe, Phil (2000). *Physical Causation*. Cambridge: Cambridge University Press.

Doya, Kenji, Shin Ishii, Alexandre Pouget, and Rajesh P. N. Rao (eds.) (2007). *Bayesian Brain: Probabilistic Approaches to Neural Coding*. Cambridge, Mass.: MIT Press.

Duhem, Pierre (1914). *La théorie physique: Son objet, sa structure*. Paris: Vrin.

Dupré, John (1984). Probabilistic Causality Emancipated. *Midwest Studies in Philosophy 9*, 169–175.

Eagle, Antony (2004). Twenty-one Arguments Against Propensity Analyses of Probability. *Erkenntnis 60*, 371–416.

Earman, John (1992). *Bayes or Bust? A Critical Examination of Bayesian Confirmation Theory*. Cambridge, Mass.: MIT Press.

Easwaran, Kenny (2011a). Bayesianism I: Introduction and Arguments in Favor. *Philosophy Compass 6*, 312–320.

Easwaran, Kenny (2011b). Bayesianism II: Applications and Criticisms. *Philosophy 6*, 321–332.

Easwaran, Kenny (2011c). The Varieties of Conditional Probability. In Prasanta S. Bandyopadhyay and Malcolm Forster (eds.), *Philosophy of Statistics*, Handbook of the Philosophy of Science, Volume VII, pp. 137–148. Amsterdam: Elsevier.

Easwaran, Kenny (2016). Dr. Truthlove, or: How I Learned to Stop Worrying and Love Bayesian Probabilities. *Noûs 50*, 816–853.

Easwaran, Kenny and Branden Fitelson (2016). Accuracy, Coherence and Evidence. In Tamar Szabó Gendler and John Hawthorne (eds.), *Oxford Studies in Epistemology*, Volume 5, pp. 61–96. New York: Oxford University Press.

Edgington, Dorothy (1995). On Conditionals. *Mind 104*, 235–329.

Edwards, A. W. F. (1972). *Likelihood*. Cambridge: Cambridge University Press.

Eells, Ellery (1985). Problems of Old Evidence. *Pacific Philosophical Quarterly 66*, 283.

Eells, Ellery (1990). Bayesian Problems of Old Evidence. In C. W. Savage (ed.), *Scientific Theories*, pp. 205–223. Minneapolis: University of Minnesota Press.

Eells, Ellery (1991). *Probabilistic Causality*. Cambridge: Cambridge University Press.

Eells, Ellery and Branden Fitelson (2000). Measuring Confirmation and Evidence. *Journal of Philosophy 97*, 663–672.

Eells, Ellery and Branden Fitelson (2002). Symmetries and Asymmetries in Evidential Support. *Philosophical Studies 107*, 129–142.

Efron, Bradley (1986). Why Isn't Everyone a Bayesian? *American Statistician 40*, 1–11 (with discussion).

Eiter, Thomas and Georg Gottlob (1995). The Complexity of Logic-based Abduction. *Journal of the ACM 42*, 3–42.

Elga, Adam (2010). Subjective Probabilities Should Be Sharp. *Philosophical Imprints 10*, 1–11. Retrieved from http://hdl.handle.net/2027/spo.3521354.0010.005.

Ellsberg, Daniel (1961). Risk, Ambiguity and the Savage Axioms. *Quarterly Journal of Economics 75*, 643–669.

Ellsberg, Daniel (2001). *Risk, Ambiguity and Decision*. New York: Routledge.

Eriksson, Lina and Alan Hájek (2007). What Are Degrees of Belief? *Studia Logica 86*, 185–215.

Eva, Benjamin and Stephan Hartmann (2018a). Bayesian Argumentation and the Value of Logical Validity. *Psychological Review 125*, 806–821.

Eva, Benjamin and Stephan Hartmann (2018b). When No Reason For Is a Reason Against. *Analysis 78*, 426–431.

Eva, Benjamin, Stephan Hartmann, and Soroush Rafiee Rad (forthcoming). Updating on Conditionals. *Mind*. Retrieved from https://www.stephanhartmann.org on March 29, 2019.

Eva, Benjamin and Reuben Stern (forthcoming). Causal Explanatory Power. *British Journal for the Philosophy of Science*. https://doi.org/10.1093/bjps/axx033.

Fahrbach, Ludwig (2009). Pessimistic Meta-Induction and the Exponential Growth of Science. In Alexander Hieke and Hannes Leitgeb (eds.), *Reduction–Abstraction–Analysis*, pp. 95–111. Heusenstamm: Ontos.

Fahrbach, Ludwig (2011). How the Growth of Science Ends Theory Change. *Synthese 108*, 139–155.

Fanelli, Daniele (2016). Set Up a 'Self-Retraction' System for Honest Errors. *Nature 531*, 415.

Festa, Roberto (2012). *"For Unto Every One that Hath Shall be Given"*: Matthew Properties for Incremental Confirmation. *Synthese 184*, 89–100.

Festa, Roberto, Atocha Aliseda, and Jeanne Peijnenburg (eds.) (2005). *Confirmation, Empirical Progress and Truth Approximation: Essays in Debate with Theo Kuipers*. Amsterdam: Rodopi.

Festa, Roberto and Gustavo Cevolani (2017). Unfolding the grammar of Bayesian confirmation: likelihood and anti-likelihood principles. *Philosophy of Science 84*, 56–81.

Feyerabend, Paul (1975). *Against Method*. London: Verso.

Fisher, R. A. (1935). The Fiducial Argument in Statistical Inference. *Annals of Human Genetics 6*, 391–398.

Fisher, R. A. (1935/74). *The Design of Experiments*. New York: Hafner Press. Reprint of the ninth edition from 1971. Originally published in 1935 (Edinburgh: Oliver & Boyd).

Fisher, R. A. (1956). *Statistical Methods and Scientific Inference*. New York: Hafner.

Fitelson, Branden (1999). The Plurality of Bayesian Measures of Confirmation and the Problem of Measure Sensitivity. *Philosophy of Science 66*, S362–S378.

Fitelson, Branden (2001a). A Bayesian Account of Independent Evidence with Applications. *Philosophy of Science 68*, S123–S140.

Fitelson, Branden (2001b). *Studies in Bayesian Confirmation Theory*. Ph.D. thesis, University of Wisconsin–Madison.

Fitelson, Branden (2003). A Probabilistic Theory of Coherence. *Analysis 63*, 194–199.

Fitelson, Branden (2004). Earman on Old Evidence and Measures of Confirmation. Unpublished manuscript, retrieved from http://fitelson.org/oe_old.pdf on August 10, 2018.

Fitelson, Branden (2006). The Paradox of Confirmation. *Philosophy Compass 1*, 95–113.

Fitelson, Branden (2008a). A Decision Procedure for Probability Calculus with Applications. *Review of Symbolic Logic 1*, 111–125.

Fitelson, Branden (2008b). Goodman's 'New Riddle'. *Journal of Philosophical Logic 37*, 613–643.

Fitelson, Branden (2018). Coherence. Unpublished manuscript, retrieved from http://fitelson.org/coherence/coherence_duke.pdf on August 10, 2018.

Fitelson, Branden and Alan Hájek (2017). Declarations of Independence. *Synthese 194*, 3979–3995.

Fitelson, Branden and Stephan Hartmann (2015). A New Garber-Style Solution to the Problem of Old Evidence. *Philosophy of Science 82*, 712–717.

Fitelson, Branden and James Hawthorne (2011). How Bayesian Confirmation Theory Handles the Paradox of the Ravens. In James H. Fetzer and Ellery Eells (eds.), *The Place of Probability in Science*, pp. 247–275. New York: Springer.

Fitelson, Branden and Christopher Hitchcock (2011). Probabilistic Measures of Causal Strength. In Phyllis McKay Illari, Federica Russo, and Jon Williamson (eds.), *Causality in the Sciences*, pp. 600–627. Oxford: Oxford University Press.

Forber, Patrick (2011). Reconceiving Eliminative Inference. *Philosophy of Science 78*, 185–208.

Forster, Malcolm (1995). Bayes or Bust: Simplicity as a Problem for a Probabilist's Approach to Confirmation. *British Journal for the Philosophy of Science 46*, 399–424.

Forster, Malcolm (1999). Model Selection in Science: The Problem of Language Variance. *British Journal for the Philosophy of Science 50*, 83–102.

Forster, Malcolm (2000). Key Concepts in Model Selection: Performance and Generalizability. *Journal of Mathematical Psychology 44*, 205–231.

Forster, Malcolm (2002). Predictive Accuracy as an Achievable Goal of Science. *Philosophy of Science 69*, S124–S134.

Forster, Malcolm, Garvesh Raskutti, Reuben Stern, and Naftali Weinberger (2018). The Frugal Inference of Causal Relations. *British Journal for the Philosophy of Science 69*, 821–848.

Forster, Malcolm and Elliott Sober (1994). How to Tell When Simpler, More Unified, or Less *Ad Hoc* Theories will Provide More Accurate Predictions. *British Journal for the Philosophy of Science 45*, 1–35.

Forster, Malcolm and Elliott Sober (2010). AIC Scores as Evidence: A Bayesian Interpretation. In Prasanta S. Bandyopadhyay and Malcolm Forster (eds.), *Philosophy of Statistics*, Handbook of the Philosophy of Science, Volume VII, pp. 535–549. Dordrecht: Elsevier.

Francis, Greg (2012). Publication Bias and the Failure of Replication in Experimental Psychology. *Psychonomic Bulletin & Review 19*, 975–991.

Francis, Gregory (2014). The Frequency of Excess Success for Articles in Psychological Science. *Psychonomic Bulletin & Review 21*, 1180–1187.

Friedman, Michael (1974). Explanation and Scientific Understanding. *Journal of Philosophy 71*, 5–19.

Frigg, Roman (2008). A Field Guide to Recent Work on the Foundations of Statistical Mechanics. In Dean Rickles (ed.), *The Ashgate Companion to Contemporary Philosophy of Physics*, pp. 99–196. London: Ashgate.

Frigg, Roman and Stephan Hartmann (2012). Models in Science. In Ed Zalta (ed.), *The Stanford Encyclopedia of Philosophy*. Retrieved from https://plato.stanford.edu/archives/sum2018/entries/models-science/.

Frigg, Roman and Carl Hoefer (2015). The Best Humean System for Statistical Mechanics. *Erkenntnis 80*, 551–574.

Gaifman, Haim and Marc Snir (1982). Probabilities over Rich Languages, Testing and Randomness. *Journal of Symbolic Logic 47*, 495–548.

Galak, Jeff, Robyn A. LeBoeuf, Leif D. Nelson, and Joseph P. Simmons (2012). Correcting the Past: Failures to Replicate Psi. *Journal of Personality and Social Psychology 103*, 933–948.

Galavotti, Maria Carla (1989). Anti-Realism in the Philosophy of Probability: Bruno de Finetti's Subjectivism. *Erkenntnis 31*, 239–261.

Gallistel, C. R. (2009). The Importance of Proving the Null. *Psychological Review 116*, 439–453.

Garber, Daniel (1983). Old Evidence and Logical Omniscience in Bayesian Confirmation Theory. In John Earman (ed.), *Testing Scientific Theories*, pp. 99–132. Minneapolis: University of Minnesota Press.

Gelman, Andrew and Christian Hennig (2017). Beyond Objective and Subjective in Statistics. *Journal of the Royal Statistical Society, Series A 180*, 967–1033 (with discussion).

Gelman, Andrew and Cosma Shalizi (2012). Philosophy and the Practice of Bayesian Statistics in the Social Sciences. In Harold Kincaid (ed.), *Oxford Handbook of the Philosophy of the Social Sciences*, pp. 259–273. Oxford: Oxford University Press.

Gelman, Andrew and Cosma Shalizi (2013). Philosophy and the Practice of Bayesian Statistics. *British Journal of Mathematical and Statistical Psychology 66*, 8–38.

Gemes, Ken (1993). Hypothetico-Deductivism, Content and the Natural Axiomatisation of Theories. *Philosophy of Science 60*, 477–487.

Gemes, Ken (1998). Hypothetico-Deductivism: The Current State of Play; the Criterion of Empirical Significance: Endgame. *Erkenntnis 49*, 1–20.

Genest, Christian and James V. Zidek (1986). Combining Probability Distributions: A Critique and an Annotated Bibliography. *Statistical Science 1*, 114–135.

Gigerenzer, Gerd (2004). Mindless Statistics. *Journal of Socio-Economics 33*, 587–606.

Gigerenzer, Gerd and Julian N. Marewski (2015). Surrogate Science: The Idol of a Universal Method for Scientific Inference. *Journal of Management 41*, 421–440.

Glymour, Clark (1980). *Theory and Evidence*. Princeton, N.J.: Princeton University Press.

Glymour, Clark (2015). Probability and the Explanatory Virtues. *British Journal for the Philosophy of Science 66*, 591–604.

Good, I. J. (1952). Rational Decisions. *Journal of the Royal Statistical Society, Series B 14*, 107–114.

Good, I. J. (1960). Weight of Evidence, Corroboration, Explanatory Power, Information and the Utility of Experiments. *Journal of the Royal Statistical Society, Series B 22*, 319–331.

Good, I. J. (1961a). A Causal Calculus (I). *British Journal for the Philosophy of Science 11*, 305–318.

Good, I. J. (1961b). A Causal Calculus (II). *British Journal for the Philosophy of Science 12*, 43–51.

Good, I. J. (1967). The White Shoe is a Red Herring. *British Journal for the Philosophy of Science 17*, 322.

Good, I. J. (1968a). Corrigendum: Weight of Evidence, Corroboration, Explanatory Power, Information and the Utility of Experiments. *Journal of the Royal Statistical Society, Series B 30*, 203.

Good, I. J. (1968b). Corroboration, Explanation, Evolving Probability, Simplicity and a Sharpened Razor. *British Journal for the Philosophy of Science 19*, 123–143.

Good, I. J. (1975). Explicativity, Corroboration, and the Relative Odds of Hypotheses. *Synthese 30*, 39–73.

Good, I. J. (2009). *Good Thinking*. Mineola, N.Y.: Dover. Originally published in 1983 (Minneapolis: University of Minnesota Press).

Goodman, Nelson (1955). *Fact, Fiction and Forecast*. Cambridge, Mass.: Harvard University Press.

Goodman, Stephen N. (1999a). Toward Evidence-Based Medical Statistics 1: The *P* value Fallacy. *Annals of Internal Medicine 130*, 995–1004.

Goodman, Stephen N. (1999b). Toward Evidence-Based Medical Statistics 2: The Bayes Factor. *Annals of Internal Medicine 130*, 1005–1013.

Gould, Stephen J. and Richard C. Lewontin (1979). The Spandrels of San Marco and the Panglossian Paradigm: A Critique of the Adaptationist Programme. *Proceedings of the Royal Society of London, Series B (Biological Sciences) 205*, 581–598.

Greaves, Hilary and David Wallace (2006). Justifying Conditionalization: Conditionalization Maximizes Expected Epistemic Utility. *Mind 115*, 607–632.

Griffiths, Paul, Arnaud Pocheville, Brett Calcott, Karola Stotz, Hyunju Kim, and Rob Knight (2015). Measuring Causal Specificity. *Philosophy of Science 82*, 529–555.

Grünwald, Peter (2005). A Tutorial Introduction to the Minimum Description Length Principle. In P. Grünwald, I. J. Myung, and M. Pitt (eds.), *Advances in Minimum Description Length: Theory and Applications*. Cambridge, Mass.: MIT Press.

Gyenis, Zalán, Gábor Hofer-Szabó, and Miklós Rédei (2017). Conditioning Using Conditional Expectations: The Borel–Kolmogorov Paradox. *Synthese 194*, 2595–2630.

Hacking, Ian (1965). *Logic of Statistical Inference*. Cambridge: Cambridge University Press.

Haenni, Rolf, Jan-Willem Romeijn, Gregory Wheeler, and Jon Williamson (2011). *Probabilistic Logic and Probabilistic Networks*. Berlin: Springer.

Hahn, Ulrike and Mike Oaksford (2007). The Rationality of Informal Argumentation: A Bayesian Approach to Reasoning Fallacies. *Psychological Review 114*, 704–732.

Hailperin, Theodore (1984). Probability Logic. *Notre Dame Journal of Formal Logic 25*, 198–212.

Hailperin, Theodore (1996). *Sentential Probability Logic: Origins, Development, Current Status, and Technical Applications*. Bethlehem, Pa.: Lehigh University Press.

Hájek, Alan (2003). What Conditional Probability Could Not Be. *Synthese 137*, 273–323.

Hájek, Alan (2008). Arguments For—or Against—Probabilism? *British Journal for the Philosophy of Science 59*, 793–819.

Hájek, Alan (2011). Interpretations of Probability. In Ed Zalta (ed.), *The Stanford Encyclopedia of Philosophy*. Retrieved from http://plato.stanford.edu/archives/sum2018/entries/probability-interpret/.

Hájek, Alan and Stephan Hartmann (2010). Bayesian Epistemology. In Jonathan Dancy, Ernest Sosa, and Matthias Steup (eds.), *A Companion to Epistemology* (2nd ed.)., pp. 93–106. Chichester: Blackwell.

Hall, Ned (2004). The Intrinsic Character of Causation. In John Collins, Ned Hall, and Laurie Paul (eds.), *Oxford Studies in Metaphysics*, pp. 225–276. Oxford: Oxford University Press.

Halpern, Joseph Y. and Christopher Hitchcock (2015). Graded Causation and Defaults. *British Journal for the Philosophy of Science 66*, 413–457.

Halpern, Joseph Y. and Judea Pearl (2005a). Causes and Explanations: A Structural-Model Approach. Part I: Causes. *British Journal for the Philosophy of Science 56*, 843–887.

Halpern, Joseph Y. and Judea Pearl (2005b). Causes and Explanations: A Structural-Model Approach. Part II: Explanations. *British Journal for the Philosophy of Science 56*, 889–911.

Harding, Sandra (1991). *Whose Science? Whose Knowledge? Thinking from Women's Lives*. Ithaca, N.Y.: Cornell University Press.

Harman, Gilbert (1965). The Inference to the Best Explanation. *Philosophical Review 74*, 88–95.

Hart, H. L. A. and Tony Honoré (1985). *Causation in the Law*. Oxford: Oxford University Press.

Hartmann, Stephan (1999). Models and Stories in Hadron Physics. In Mary Morgan and Margaret Morrison (eds.), *Models as Mediators: Perspectives on Natural and Social Science*, pp. 326–346. Cambridge: Cambridge University Press.

Hartmann, Stephan and Jan Sprenger (2010). Bayesian Epistemology. In Duncan Pritchard (ed.), *Routledge Companion to Epistemology*, pp. 609–620. London: Routledge.

Hartmann, Stephan and Jan Sprenger (2012). The Future of Philosophy of Science: Introduction. *European Journal for Philosophy of Science 2*, 157–159.

Hawthorne, James (2005). Degree-of-Belief and Degree-of-Support: Why Bayesians Need Both Notions. *Mind 114*, 277–320.

Hawthorne, James and Branden Fitelson (2004). Re-Solving Irrelevant Conjunction with Probabilistic Independence. *Philosophy of Science 71*, 505–514.

Heckerman, David (1988). An Axiomatic Framework for Belief Updates. In J. F. Lemmer and L. N. Kanal (eds.), *Uncertainty in Artificial Intelligence 2*, pp. 11–22. Amsterdam: North-Holland.

Heesen, Remco (2017). Communism and the Incentive to Share in Science. *Philosophy of Science 84*, 698–716.

Heesen, Remco (2018). When Journal Editors Play Favorites. *Philosophical Studies 175*, 831–858.

Hempel, Carl G. (1945a). Studies in the Logic of Confirmation I. *Mind 54*, 1–26.

Hempel, Carl G. (1945b). Studies in the Logic of Confirmation II. *Mind 54*, 97–121.

Hempel, Carl G. (1960). Inductive Inconsistencies. *Synthese 12*, 439–469.

Hempel, Carl G. (1965a). *Aspects of Scientific Explanation and Other Essays in the Philosophy of Science*. New York: Free Press.

Hempel, Carl G. (1965b). Science and Human Values. In *Aspects of Scientific Explanation*, pp. 81–96. New York: Free Press.

Hempel, Carl G. and Paul Oppenheim (1948). Studies in the Logic of Explanation. *Philosophy of Science 15*, 135–175.

Henderson, Leah, Noah D. Goodman, Joshua B. Tenenbaum, and James F. Woodward (2010). The Structure and Dynamics of Scientific Theories: A Hierarchical Bayesian Perspective. *Philosophy of Science 77*, 172–200.

Herfeld, Catherine and Malte Doehne (forthcoming). The Diffusion of Scientific Innovations: A Role Typology. *Studies in the History and Philosophy of Science, Part A*. https://doi.org/10.1016/j.shpsa.2017.12.001.

Hindriks, Frank, Igor Douven, and Henrik Singmann (2016). A New Angle on the Knobe Effect: Intentionality Correlates with Blame, Not with Praise. *Mind & Language 31*, 204–220.

Hitchcock, Christopher and Joshua Knobe (2009). Cause and Norm. *Journal of Philosophy 106*, 587–612.

Hitchcock, Christopher and Elliott Sober (2004). Prediction versus Accommodation and the Risk of Overfitting. *British Journal for the Philosophy of Science 55*, 1–34.

Hitchcock, Christopher and James Woodward (2003). Explanatory Generalizations, Part II: Plumbing Explanatory Depth. *Noûs 37*, 181–199.

Hobbs, Jerry R., Mark Stickel, Paul Martin, and Douglas Edwards (1988). Interpretation as Abduction. In *Proceedings of the 26th Annual Meeting of the Association for Computational Linguistics*, pp. 95–103. Buffalo, N.Y.

Hoefer, Carl (2007). The Third Way on Objective Probability: A Sceptic's Guide to Objective Chance. *Mind 116*, 549–596.

Holland, Paul W. (1986). Statistics and Causal Inference. *Journal of the American Statistical Association 81*, 945–960.

Howson, Colin (1984). Bayesianism and Support by Novel Facts. *British Journal for the Philosophy of Science 35*, 245–251.

Howson, Colin (1985). Some Recent Objections to the Bayesian Theory of Support. *British Journal for the Philosophy of Science 36*, 305–309.

Howson, Colin (1991). The 'Old Evidence' Problem. *British Journal for the Philosophy of Science 42*, 547–555.

Howson, Colin (2000). *Hume's Problem: Induction and the Justification of Belief.* Oxford: Oxford University Press.

Howson, Colin (2008). De Finetti, Countable Additivity, Consistency and Coherence. *British Journal for the Philosophy of Science 59*, 1–23.

Howson, Colin (2013). Exhuming the No-Miracles Argument. *Analysis 73*, 205–211.

Howson, Colin (2017). Putting on the Garber Style? Better Not. *Philosophy of Science 84*, 659–676.

Howson, Colin and Peter Urbach (2006). *Scientific Reasoning: The Bayesian Approach* (3rd ed.). La Salle, Ill.: Open Court.

Huber, Franz (2005a). Subjective Probabilities as Basis for Scientific Reasoning. *British Journal for the Philosophy of Science 56*, 101–116.

Huber, Franz (2005b). What Is the Point of Confirmation? *Philosophy of Science 72*, 1146–1159.

Huber, Franz (2006). Ranking Functions and Rankings on Languages. *Artificial Intelligence 170*, 462–471.

Huber, Franz (2016). Formal representations of belief. In Ed Zalta (ed.), *Stanford Encyclopedia of Philosophy*. Retrieved from https://plato.stanford.edu/archives/sum2018/entries/formal-belief/.

Huber, Franz (2019). *A Logical Introduction to Probability and Induction*. New York: Oxford University Press.

Huber, Peter J. (2009). *Robust Statistics* (2nd ed.). New York: Wiley.

Hume, David (1739). *A Treatise of Human Nature.* Oxford: Clarendon Press.

Hume, David (1748). *Enquiry Concerning Human Understanding.* Oxford: Clarendon Press.

Hyman, Ray and Charles Honorton (1986). A Joint Communiqué: The Psi Ganzfeld Controversy. *Journal of Parapsychology 50*, 351–364.

Icard, Thomas F., Jonathan F. Kominsky, and Joshua Knobe (2017). Normality and Actual Causal Strength. *Cognition 161*, 80–93.

Ioannidis, John P. A. (2005a). Contradicted and Initially Stronger Effects in Highly Cited Clinical Research. *Journal of the Americal Medical Association 294*, 218–228.

Ioannidis, John P. A. (2005b). Why Most Published Research Findings Are False. *PLoS Medicine* 2. https://doi.org/10.1371/journal.pmed.0020124.

Ioannidis, John P. A. and Thomas A. Trikalinos (2007). An Exploratory Test for an Excess of Significant Findings. *Clinical Trials 4*, 245–253.

Jackson, Frank (1975). Grue. *Journal of Philosophy 72*, 113–131.

Jaynes, Edwin T. (1968). Prior Probabilities. In *IEEE Transactions on Systems Science and Cybernetics (SSC-4)*, pp. 227–241.

Jaynes, Edwin T. (2003). *Probability Theory: The Logic of Science*. Cambridge: Cambridge University Press.

Jeffrey, Richard C. (1965). *The Logic of Decision*. Chicago and London: University of Chicago Press. Second edition 1983.

Jeffrey, Richard C. (1983). Bayesianism with a Human Face. In John Earman (ed.), *Testing Scientific Theories*, pp. 133–156. Minneapolis: University of Minnesota Press.

Jeffrey, Richard C. (2004). *Subjective Probability: The Real Thing*. Cambridge: Cambridge University Press.

Jeffreys, Harold (1961). *Theory of Probability* (3rd ed.). Oxford: Oxford University Press.

Joyce, James (1998). A Nonpragmatic Vindication of Probabilism. *Philosophy of Science 65*, 575–603.

Joyce, James (2003). Bayes' Theorem. In Ed Zalta (ed.), *Stanford Encyclopedia of Philosophy*. Retrieved from https://plato.stanford.edu/archives/sum2018/entries/bayes-theorem/.

Joyce, James (2009). Accuracy and Coherence: Prospects for an Alethic Epistemology of Partial Belief. In Franz Huber and Christoph Schmidt-Petri (eds.), *Degrees of Belief*, pp. 263–297. Berlin: Springer.

Kadane, Joseph B., Mark J. Schervish, and Teddy Seidenfeld (1999). *Rethinking the Foundations of Statistics*. Cambridge: Cambridge University Press.

Kahneman, Daniel, Paul Slovic, and Amos Tversky (eds.) (1982). *Judgment under Uncertainty: Heuristics and Biases*. Cambridge: Cambridge University Press.

Kahneman, Daniel and Amos Tversky (1973). On the Psychology of Prediction. *Psychological Review 80*, 237–251.

Kaiserman, Alexander (2016). Causal Contribution. *Proceedings of the Aristotelian Society 116*, 387–394.

Kaiserman, Alexander (2017). Partial Liability. *Legal Theory 23*, 1–26.

Kass, Robert E. and Adrian E. Raftery (1995). Bayes Factors. *Journal of the American Statistical Association 90*, 773–795.

Kelly, Kevin (1996). *The Logic of Reliable Inquiry*. Oxford: Oxford University Press.

Kemeny, John G. (1955). Fair Bets and Inductive Probability. *Journal of Symbolic Logic 20*, 263–273.

Kemeny, John G. and Paul Oppenheim (1952). Degree of Factual Support. *Philosophy of Science 19*, 307–324.

Kieseppä, Ilkka A. (1997). Akaike Information Criterion, Curve-Fitting, and the Philosophical Problem of Simplicity. *British Journal for the Philosophy of Science 48*, 21–48.

Kitcher, Philip (1981). Explanatory Unification. *Philosophy of Science 48*, 507–531.

Knobe, Joshua (2003). Intentional Action and Side Effects in Ordinary Language. *Analysis 63*, 190–194.

Knobe, Joshua and Ben Fraser (2008). Causal Judgment and Moral Judgment: Two Experiments. *Moral Psychology 2*, 441–448.

Knobe, Joshua and Shaun Nichols (2007). An Experimental Philosophy Manifesto. In Joshua Knobe and Shaun Nichols (eds.), *Experimental Philosophy*, pp. 3–14. London: Oxford University Press.

Kolmogorov, Andrey (1933). *Grundbegriffe der Wahrscheinlichkeitsrechnung*. Berlin: Springer.

Kominsky, Jonathan F., Jonathan Phillips, Tobias Gerstenberg, David Lagnado, and Joshua Knobe (2015). Causal Superseding. *Cognition 137*, 196–209.

Koole, Sander L. and Daniel Lakens (2012). Rewarding Replications. *Perspectives on Psychological Science 7*, 608–614.

Koopman, Bernard O. (1940). The Axioms and Algebra of Intuitive Probability. *Annals of Mathematics 41*, 269–292.

Korb, Kevin B., Lucas Hope, and Erik P. Nyberg (2009). Information-Theoretic Causal Power. In F. Emmert-Streib and M. Dehmer (eds.), *Information Theory and Statistical Learning*, pp. 231–265. Berlin: Springer.

Korb, Kevin B., Erik P. Nyberg, and Lucas Hope (2011). A New Causal Power Theory. In Phyllis Illari, Federica Russo, and Jon Williamson (eds.), *Causality in the Sciences*, pp. 628–652. Oxford: Oxford University Press.

Kuhn, Thomas S. (1977a). Objectivity, Value Judgment, and Theory Choice. In *The Essential Tension*, pp. 320–339. Chicago: University of Chicago Press.

Kuhn, Thomas S. (1977b). *The Essential Tension: Selected Studies in Scientific Tradition and Change*. Chicago: Chicago University Press.

Kuipers, Theo A. F. (2000). *From Instrumentalism to Constructive Realism*. Berlin: Springer.

Kullback, Solomon and Richard A. Leibler (1951). On Information and Sufficiency. *Annals of Mathematical Statistics 22*, 79–86.

Kyburg, Henry E. (1961). *Probability and the Logic of Rational Belief*. Middletown, Conn.: Wesleyan University Press.

Kyburg, Henry E. (1974). *The Logical Foundations of Statistical Inference*. Dordrecht: Reidel.

Kyburg, Henry E. and Choh Man Teng (2001). *Uncertain Inference*. Cambridge: Cambridge University Press.

Lacey, Hugh (1999). *Is Science Value Free? Values and Scientific Understanding*. London: Routledge.

Ladyman, James and Don Ross (2009). *Every Thing Must Go: Metaphysics Naturalized*. Oxford: Oxford University Press.

Laudan, Larry (1981). A Confutation of Convergent Realism. *Philosophy of Science 48*, 19–48.

Lee, Michael D. and Eric-Jan Wagenmakers (2014). *Bayesian Cognitive Modeling: A Practical Course*. Cambridge: Cambridge University Press.

Lehrer, Keith and Carl Wagner (1981). *Rational Consensus in Science and Society*. Dordrecht: Reidel.

Leitgeb, Hannes (2013). Scientific Philosophy, Mathematical Philosophy, and All That. *Metaphilosophy 44*, 267–275.

Leitgeb, Hannes (2014). The Stability Theory of Belief. *Philosophical Review 123*, 131–171.

Leitgeb, Hannes (2017). *The Stability of Belief*. Oxford: Oxford University Press.

Leitgeb, Hannes and Richard Pettigrew (2010a). An Objective Justification of Bayesianism I: Measuring Inaccuracy. *Philosophy of Science 77*, 201–235.

Leitgeb, Hannes and Richard Pettigrew (2010b). An Objective Justification of Bayesianism II: The Consequences of Minimizing Inaccuracy. *Philosophy of Science 77*, 236–272.

Levi, Isaac (1960). Must the Scientist Make Value Judgments? *Journal of Philosophy 11*, 345–357.

Levi, Isaac (1963). Corroboration and Rules of Acceptance. *British Journal for the Philosophy of Science 13*, 307–313.

Levi, Isaac (1974). On Indeterminate Probabilities. *Journal of Philosophy 71*, 391—418.

Levi, Isaac (1977). Direct Inference. *Journal of Philosophy 74*, 5–29.

Levi, Isaac (1980). *The Enterprise of Knowledge*. Cambridge, Mass.: MIT Press.

Levi, Isaac (1996). *For the Sake of the Argument*. Cambridge: Cambridge University Press.

Lewis, David (1973). Causation. *Journal of Philosophy 70*, 556–567.

Lewis, David (1976). Probabilities of Conditionals and Conditional Probabilities. *Philosophical Review 85*, 297–315.

Lewis, David (1980). A Subjectivist's Guide to Objective Chance. In Richard C. Jeffrey (ed.), *Studies in Inductive Logic and Probability*, pp. 263–293. Berkeley, Calif.: University of California Press.

Lewis, David (1986). Postscript to "Causation". In *Philosophical Papers*, Volume 2, pp. 172–213. Cambridge: Cambridge University Press.

Lewis, David (1999). *Papers in Metaphysics and Epistemology*. Cambridge: Cambridge University Press.

Lindley, Dennis (1983). Reconciliation of Probability Distributions. *Operations Research 31*, 866–880.

Lipton, Peter (2000). Inference to the Best Explanation. In W. H. Newton-Smith (ed.), *A Companion to the Philosophy of Science*, pp. 184–193. Oxford: Blackwell.

Lipton, Peter (2001). Is Explanation a Guide to Inference? A Reply to Wesley C. Salmon. In Giora Hon and Sam S. Rakover (eds.), *Explanation: Theoretical Approaches and Applications*, pp. 93–120. Dordrecht: Kluwer Academic.

Lipton, Peter (2004). *Inference to the Best Explanation* (2nd ed.). New York: Routledge.

Lombrozo, Tania (2006). The Structure and Function of Explanations. *Trends in Cognitive Sciences 10*, 464–470.

Lombrozo, Tania (2007). Simplicity and Probability in Causal Explanation. *Cognitive Psychology 55*, 232–257.

Lombrozo, Tania (2009). Explanation and Categorization: How 'Why?' Informs 'What?'. *Cognition 110*, 248–253.

Lombrozo, Tania (2011). The Instrumental Value of Explanations. *Philosophy Compass 6*, 539–551.

Lombrozo, Tania (2012). Explanation and Abductive Inference. In K. J. Holyoak and R. G. Morrison (eds.), *The Oxford Handbook of Thinking and Reasoning*, pp. 260–276. Oxford: Oxford University Press.

Longino, Helen (1990). *Science as Social Knowledge: Values and Objectivity in Scientific Inquiry*. Princeton, N.J.: Princeton University Press.

Machamer, Peter, Lindley Darden, and Carl F. Craver (2000). Thinking about Mechanisms. *Philosophy of Science 67*, 1–25.

Machery, Édouard (2012). Power and Negative Results. *Philosophy of Science 79*, 808–820.

Machery, Édouard, Ron Mallon, Shaun Nichols, and Stephen Stich (2004). Semantics, Cross-Cultural Style. *Cognition 92*, 1–12.

Mackie, John L. (1974). *The Cement of the Universe: A Study in Causation*. Oxford: Clarendon Press.

Maddy, Penelope (2009). *Second Philosophy: A Naturalistic Method*. Oxford: Oxford University Press.

Magnani, Lorenzo (2001). *Abduction, Reason and Science*. New York: Springer.

Magnus, P. D. and Craig Callender (2004). Realist Ennui and the Base Rate Fallacy. *Philosophy of Science 71*, 320–338.

Maher, Patrick (1993). *Betting on Theories*. Cambridge: Cambridge University Press.

Maher, Patrick (1999). Inductive Logic and the Ravens Paradox. *Philosophy of Science 66*, 50–70.

Maher, Patrick (2002). Joyce's Argument for Probabilism. *Philosophy of Science 69*, 73–81.

Maher, Patrick (2004). Probability Captures the Logic of Scientific Confirmation. In Christopher Hitchcock (ed.), *Contemporary Debates in Philosophy of Science*, pp. 69–93. London: Blackwell.

Maher, Patrick (2007). Explication Defended. *Studia Logica 86*, 331–341.

Maher, Patrick (2010). What Is Probability? Unfinished book manuscript, retrieved from http://patrick.maher1.net/preprints/pop.pdf on August 10, 2018.

Makel, Matthew C., Jonathan A. Plucker, and Boyd Hegarty (2012). Replications in Psychology Research: How Often Do They Really Occur? *Perspectives on Psychological Science 7*, 537–542.

Makinson, David (1965). The Paradox of the Preface. *Analysis 25*, 205–207.

Martini, Carlo and Jan Sprenger (2017). Opinion Aggregation and Individual Expertise. In Thomas Boyer-Kassem, Conor Mayo-Wilson, and Michael Weisberg (eds.), *Scientific Collaboration and Collective Knowledge*, pp. 180–201. New York: Oxford University Press.

Mayo, Deborah G. (1996). *Error and the Growth of Experimental Knowledge.* Chicago: University of Chicago Press.

Mayo, Deborah G. (2018). *Statistical Inference as Severe Testing: How to Get Beyond the Science Wars.* Cambridge: Cambridge University Press.

Mayo, Deborah G. and Aris Spanos (2006). Severe Testing as a Basic Concept in a Neyman–Pearson Philosophy of Induction. *British Journal for the Philosophy of Science* 57, 323–357.

McCullagh, Peter (2002). What Is a Statistical Model? *Annals of Statistics* 30, 1225–1310.

McGrew, Timothy (2003). Confirmation, Heuristics, and Explanatory Reasoning. *British Journal for the Philosophy of Science* 54, 553–567.

McMullin, Ernan (1982). Values in Science. In *Proceedings of the Biennal Meeting of the Philosophy of Science Association*, pp. 3–28.

McMullin, Ernan (2008). The Virtues of a Good Theory. In Martin Curd and Stathis Psillos (eds.), *The Routledge Companion to Philosophy of Science*, pp. 498–508. London: Routledge.

Meadows, A. J. (1974). *Communication in Science.* London: Butterworths.

Meek, Christopher and Clark Glymour (1994). Conditioning and Intervening. *British Journal for the Philosophy of Science* 45, 1001–1021.

Megill, Alan (ed.) (1994). *Rethinking Objectivity.* Durham, N.C., and London: Duke University Press.

Meijs, Wouter (2005). *Probabilistic Measures of Coherence.* Ph.D. thesis, Erasmus University Rotterdam.

Menzies, Peter (2014). Counterfactual Theories of Causation. In Ed Zalta (ed.), *Stanford Encyclopedia of Philosophy.* Retrieved from https://plato.stanford.edu/archives/sum2018/entries/causation-counterfactual/.

Mikkelson, Gregory M. (2006). Realism vs. Instrumentalism in a New Statistical Framework. *Philosophy of Science* 73, 440–447.

Miller, David (1974). Popper's Qualitative Theory of Verisimilitude. *British Journal for the Philosophy of Science* 25, 166–177.

Milne, Peter (1996). $\log\big[P(h/eb)/P(h/b)\big]$ Is the One True Measure of Confirmation. *Philosophy of Science* 63, 21–26.

Monton, Bradley and Chad Mohler (2017). Constructive Empiricism. In Ed Zalta (ed.), *The Stanford Encyclopedia of Philosophy.*

Retrieved from https://plato.stanford.edu/archives/sum2018/entries/constructive-empiricism/.

Moretti, Luca (2007). Ways in Which Coherence Is Confirmation Conducive. *Synthese 157*, 309–319.

Morey, Richard D., Rink Hoekstra, Jeffrey N. Rouder, Michael D. Lee, and Eric-Jan Wagenmakers (2016). The Fallacy of Placing Confidence in Confidence Intervals. *Psychonomic Bulletin & Review 23*, 103–123.

Morey, Richard D., Jeffrey N. Rouder, Josine Verhagen, and Eric-Jan Wagenmakers (2014). Why Hypothesis Tests are Essential for Psychological Science: A Comment on Cumming (2014). *Psychological Science 25*, 1289–1290.

Moss, Sarah (2018). *Probabilistic Knowledge*. Oxford: Oxford University Press.

Moyé, Lemuel A. (2008). Bayesians in Clinical Trials: Asleep at the Switch. *Statistics in Medicine 27*, 469–482.

Muldoon, Ryan, Chiara Lisciandra, Cristina Bicchieri, Stephan Hartmann, and Jan Sprenger (2014). On the Emergence of Descriptive Norms. *Politics, Philosophy and Economics 13*, 3–22.

Munafò, Marcus R., Brian Nosek, Dorothy V. M. Bishop, et al. (2017). A manifesto for reproducible science. *Nature Human Behaviour 1*, 0021.

Myrvold, Wayne C. (2003). A Bayesian Account of the Virtue of Unification. *Philosophy of Science 70*, 399–423.

Myrvold, Wayne C. (2015). You Can't Always Get What You Want: Some Considerations Regarding Conditional Probabilities. *Erkenntnis 80*, 573–603.

Myrvold, Wayne C. (2017). On the Evidential Import of Unification. *Philosophy of Science 84*, 92–114.

Nagel, Ernest (1961). *The Structure of Science*. London: Routledge.

Nagel, Ernest (1979). *Teleology Revisited and Other Essays in the Philosophy of Science*. New York: Columbia University Press.

Nagel, Jennifer, Valerie San Juan, and Raymond A. Mar (2013). Lay Denial of Knowledge for Justified True Beliefs. *Cognition 129*, 652–661.

Nahmias, Eddy, Stephen Morris, Thomas Nadelhoffer, and Jason Turner (2005). Surveying Freedom: Folk Intuitions about Free Will and Moral Responsibility. *Philosophical Psychology 18*, 561–584.

Nardini, Cecilia and Jan Sprenger (2013). Bias and Conditioning in Sequential Medical Trials. *Philosophy of Science 80*, 1053–1064.

Neyman, Jerzy and Egon S. Pearson (1933). On the Problem of the Most Efficient Tests of Statistical Hypotheses. *Philosophical Transactions of the Royal Society A 231*, 289–337.

Neyman, Jerzy and Egon S. Pearson (1967). *Joint Statistical Papers*. Berkeley, Calif.: University of California Press.

Nicod, Jean (1925/61). *Le problème logique de l'induction*. Paris: Presses Universitaires de France. Originally published in 1925 (Paris: Alcan).

Niiniluoto, Ilkka (1983). Novel Facts and Bayesianism. *British Journal for the Philosophy of Science 34*, 375–379.

Niiniluoto, Ilkka (1999). *Critical Scientific Realism*. Oxford: Oxford University Press.

Niiniluoto, Ilkka (2011). Revising Beliefs Towards the Truth. *Erkenntnis 75*, 165–181.

Nolan, Daniel (1997). Quantitative Parsimony. *British Journal for the Philosophy of Science 48*, 329–343.

Norton, John D. (2003). A Material Theory of Induction. *Philosophy of Science 70*, 647–670.

Norton, John D. (2011). Challenges to Bayesian Confirmation Theory. In Prasanta S. Bandyopadhyay and Malcolm Forster (eds.), *Philosophy of Statistics*, Handbook of the Philosophy of Science, Volume VII, pp. 391–439. Amsterdam: Elsevier.

Norton, John D. (forthcoming). A Demonstration of the Incompleteness of Calculi of Inductive Inference. *British Journal for the Philosophy of Science*. https://doi.org/10.1093/bjps/axx004.

Nosek, Brian A. and Timothy M. Errington (2017). Reproducibility in cancer biology: Making sense of replications. *eLife 6*, e23383.

Oaksford, Mike and Nick Chater (2000). *Bayesian Rationality*. Oxford: Oxford University Press.

Oddie, Graham (1986). *Likeness to Truth*. Dordrecht: Reidel.

Oddie, Graham (2014). Truthlikeness. In Ed Zalta (ed.), *Stanford Encyclopedia of Philosophy*. Retrieved from http://plato.stanford.edu/archives/sum2018/entries/truthlikeness.

O'Hagan, Tony (2012). Posting on the Statistical Methods Used in the Discovery of the Higgs Boson. Email list of the International Society for Bayesian Analysis (ISBA). Retrieved from www.isba.org on July 27, 2017.

Okasha, Samir (2000). Van Fraassen's Critique of Inference to the Best Explanation. *Studies in the History and Philosophy of Science 31*, 691–710.

Okasha, Samir (2007). What Does Goodman's 'Grue' Problem Really Show? *Philosophical Papers 36*, 483–502.

Olsson, Erik J. (2011). A Simulation Approach to Veritistic Social Epistemology. *Episteme 8*, 127–143.

Open Science Collaboration (2015). Estimating the Reproducibility of Psychological Science. *Science* 349. Retrieved from http://science.sciencemag.org/content/349/6251/aac4716.full.pdf.

Over, David (2016). Causation and the Probability of Causal Conditionals. In Michael Waldmann (ed.), *Oxford Handbook of Causal Reasoning*, pp. 307–325. Oxford: Oxford University Press.

Oxburgh, Ron, Huw Davies, Kerry Emanuel, et al. (2010). Report of the International Panel Set Up by the University of East Anglia to Examine the Research of the Climatic Research Unit. Technical report, University of East Anglia. Downloaded on July 28, 2017, at http://www.uea.ac.uk/documents/3154295/7847337/SAP.pdf/a6f591fc-fc6e-4a70-9648-8b943d84782b.

Paris, Jeff B. and Alina Vencovská (1989). On the Applicability of Maximum Entropy to Inexact Reasoning. *International Journal of Approximate Reasoning* 3, 1–34.

Paris, Jeff B., Alina Vencovská, and G. M. Wilmers (1994). A Natural Prior Probability Distribution Derived from the Propositional Calculus. *Annals of Pure and Applied Logic* 70, 243–285.

Pearl, Judea (2000). *Causality*. Cambridge: Cambridge University Press.

Pearl, Judea (2001). Direct and Indirect Effects. In Jack Breese and Daphne Koller (eds.), *Proceedings of the Seventeenth Conference on Uncertainty in Artificial Intelligence*, pp. 411–420.

Pearl, Judea (2011). The Structural Theory of Causation. In Phyllis McKay Illari, Frederic Russo, and Jon Williamson (eds.), *Causality in the Sciences*, pp. 697–727. Oxford: Oxford University Press.

Peirce, Charles Sanders (1931–1935). *The Collected Papers of Charles Sanders Peirce*, Volume I–VI. Cambridge, Mass.: Harvard University Press.

Pettigrew, Richard (2015). Accuracy and the Credence–Belief Connection. *Philosophical Imprints*.

Pettigrew, Richard (2016). *Accuracy and the Laws of Credence*. Oxford: Oxford University Press.

Poole, Charles (2010). On the Origin of Risk Relativism. *Epidemiology* 21, 3–9.

Popper, Karl R. (1954). Degree of Confirmation. *British Journal for the Philosophy of Science* 5, 143–149.

Popper, Karl R. (1957). A Second Note on Degree of Confirmation. *British Journal for the Philosophy of Science* 7, 350–353.

Popper, Karl R. (1958). A Third Note on Degree of Corroboration or Confirmation. *British Journal for the Philosophy of Science* 8, 294–302.

Popper, Karl R. (1959/2002). *The Logic of Scientific Discovery*. London: Routledge. Reprint of the revised English 1959 edition. Originally published in German in 1934 as "Logik der Forschung".

Popper, Karl R. (1963). *Conjectures and Refutations: The Growth of Scientific Knowledge*. London: Routledge.

Popper, Karl R. (1979). *Objective Knowledge: An Evolutionary Approach*. Oxford: Clarendon Press.

Popper, Karl R. (1983). *Realism and the Aim of Science*. Totowa, N.J.: Rowman & Littlefield.

Popper, Karl R. and David Miller (1983). A Proof of the Impossibility of Inductive Probability. *Nature 302*, 687–688.

Porter, Theodore (1996). *Trust in Numbers: The Pursuit of Objectivity in Science and Public Life*. Princeton, N.J.: Princeton University Press.

Predd, Joel B., Robert Seiringer, Elliott H. Lieb, Daniel N. Osherson, Vincent H. Poor, and Sanjeev R. Kulkarni (2009). Probabilistic Coherence and Proper Scoring Rules. In *IEEE Transactions on Information Theory*, Volume 55, pp. 4786–4792.

Prinz, Florian, Thomas Schlange, and Khusru Asadullah (2011). Believe It or Not: How Much Can We Rely on Published Data on Potential Drug Targets? *Nature Reviews Drug Discovery 10*, 712.

Psillos, Stathis (1999). *Scientific Realism: How Science Tracks Truth*. London: Routledge.

Psillos, Stathis (2009). *Knowing the Structure of Nature: Essays on Realism and Explanation*. London: Palgrave Macmillan.

Putnam, Hilary (1975). *Mathematics, Matter, and Method*, Volume I of *Philosophical Papers*. Cambridge: Cambridge University Press.

Quine, W. V. O. (1951). Two Dogmas of Empiricism. *Philosophical Review 60*, 20–43.

Quine, W. V. O. (1969). Epistemology Naturalized. In *Ontological Relativity and Other Essays*, pp. 69–90. New York: Columbia University Press.

Quine, W. V. O. (1992). *Pursuit of Truth*. Cambridge, Mass.: Harvard University Press.

Quintana, Daniel S. (2015). From pre-registration to publication: a non-technical primer for conducting a meta-analysis to synthesize correlational data. *Frontiers in Psychology 6*, 1549.

Raftery, Adrian E. (1995). Bayesian Model Selection in Social Research. *Sociological Methodology 25*, 111–163.

Ramsey, Frank P. (1926). Truth and Probability. In D. H. Mellor (ed.), *Philosophical Papers*, pp. 52–94. Cambridge: Cambridge University Press.

Rawls, John (1971). *A Theory of Justice*. Cambridge, Mass.: Harvard University Press.

Rédei, Miklós and Zalán Gyenis (2016). Measure-theoretic Analysis of Consistency of the Principal Principle. *Philosophy of Science 83*, 972–987.

Reichenbach, Hans (1949). *The Theory of Probability*. Berkeley, Calif.: University of California Press.

Reichenbach, Hans (1951). *The Rise of Scientific Philosophy*. Berkeley, Calif.: University of California Press.

Reichenbach, Hans (1956). *The Direction of Time*. Berkeley, Calif., and Los Angeles: University of California Press.

Reiss, Julian and Jan Sprenger (2014). Scientific Objectivity. In Ed Zalta (ed.), *The Stanford Encyclopedia of Philosophy*. Retrieved from https://plato.stanford.edu/archives/sum2018/entries/scientific-objectivity/.

Rényi, Alfred (1970). *Foundations of Probability*. San Francisco: Holden-Day.

Reutlinger, Alexander (2016). Is There a Monist Theory of Causal and Non-Causal Explanations? The Counterfactual Theory of Scientific Explanation. *Philosophy of Science 83*, 733–745.

Reutlinger, Alexander (2017). Explanation Beyond Causation? New Directions in the Philosophy of Scientific Explanation. *Philosophy Compass 12*. https://doi.org/10.1111/phc3.12395.

Richard, F. D., Charles F. Bond, Jr., and Juli J. Stokes-Zoota (2003). One Hundred Years of Social Psychology Quantitatively Described. *Review of General Psychology 7*, 331–363.

Rips, Lance J. (1989). Similarity, Typicality, and Categorization. In S. Vosniadou and A. Ortony (eds.), *Similarity and Analogical Reasoning*, pp. 21–59. Cambridge: Cambridge University Press.

Rizzo, Mario J. and Frank S. Arnold (1980). Causal Apportionment in Tort Law: An Economic Theory. *Columbia Law Review 85*, 1399–1429.

Roche, William (2014). A Note on Confirmation and Matthew Properties. *Logic and Philosophy of Science 12*, 91–101.

Romeijn, Jan-Willem (2014). Philosophy of Statistics. In E. Zalta (ed.), *The Stanford Encyclopedia of Philosophy*. Retrieved from https://plato.stanford.edu/archives/sum2018/entries/statistics/.

Romeijn, Jan-Willem (2017). Inherent Complexity: A Problem for Statistical Model Evaluation. *Philosophy of Science 84*, 797–809.

Romeijn, Jan-Willem, Rens van de Schoot, and Herbert Hoijtink (2012). One Size does Not Fit All: Derivation of a Prior-Adapted BIC. In Dennis Dieks, Wenceslao J. Gonzáles, Stephan Hartmann, Michael Stöltzner, and Marcel Weber (eds.), *Probabilities, Laws, and Structures*, pp. 87–106. Berlin: Springer.

Romero, Felipe (2016). Can the Behavioral Sciences Self-Correct? A Socio-Epistemic Assessment. *Studies in the History and Philosophy of Science 60*, 55–69.

Romero, Felipe (2017). Novelty vs. Replicability: Virtues and Vices in the Reward System of Science. *Philosophy of Science 84*, 1031–1043.

Romero, Felipe (2018). Who Should Do Replication Labor? *Advances in Methods and Practices in Psychological Science 1*, 516–537.

Rosenbaum, Paul R. and Donald B. Rubin (1983). The Central Role of the Propensity Score in Observational Studies for Causal Effects. *Biometrika 70*, 41–55.

Rosenkrantz, Roger (1981). *Foundations and Applications of Inductive Probability*. Atascadero, Calif.: Ridgeview Press.

Rosenthal, Jacob (2004). *Wahrscheinlichkeiten als Tendenzen*. Paderborn: Mentis.

Rosenthal, Robert (1979). The File Drawer Problem and Tolerance for Null Results. *Psychological Bulletin 86*, 638–641.

Rouder, Jeffrey N. and Richard D. Morey (2011). A Bayes Factor Meta-Analysis of Bem's ESP Claim. *Psychonomic Bulletin & Review 18*, 682–689.

Rouder, Jeffrey N., Paul L. Speckman, Dongchu Sun, Richard D. Morey, and Geoffrey Iverson (2009). Bayesian t Tests for Accepting and Rejecting the Null Hypothesis. *Psychonomic Bulletin & Review 16*, 225–237.

Rowbottom, Darrell P. (2008). The Big Test of Corroboration. *International Studies in the Philosophy of Science 22*, 293–302.

Rowbottom, Darrell P. (2011). *Popper's Critical Rationalism: A Philosophical Investigation*. London: Routledge.

Rowbottom, Darrell P. (2012). Popper's Measure of Corroboration and $P(h|b)$. *British Journal for the Philosophy of Science 64*, 739–745.

Royall, Richard (1997). *Statistical Evidence: A Likelihood Paradigm*. London: Chapman & Hall.

Rubin, Donald B. (1974). Estimating Causal Effects of Treatments in Randomized and Nonrandomized Studies. *Journal of Educational Psychology 66*, 688–701.

Rudner, Richard (1953). The Scientist *qua* Scientist Makes Value Judgments. *Philosophy of Science 20*, 1–6.

Sakamoto, Y., M. Ishiguro, and G. Kitagawa (1986). *Akaike Information Criterion Statistics*. Dordrecht: Reidel.

Salmon, Wesley C. (1971). Statistical Explanation and Statistical Relevance. In R. G. Colodny (ed.), *The Nature and Function of Scientific Theories*, pp. 173–231. Pittsburgh: Pittsburgh University Press.

Salmon, Wesley C. (1984). *Scientific Explanation and the Causal Structure of the World*. Princeton, N.J.: Princeton University Press.

Salmon, Wesley C. (2001). Explanation and Confirmation: A Bayesian Critique of Inference to the Best Explanation. In Giora Hon and Sam S. Rakover (eds.), *Explanation: Theoretical Approaches and Applications*, pp. 61–91. Dordrecht: Kluwer Academic.

Savage, Leonard J. (1972). *The Foundations of Statistics* (2nd ed.). New York: Wiley. Originally published in 1954.

Schaffner, Kenneth F. (1967). Approaches to Reduction. *Philosophy of Science* 34, 137–147.

Schaffner, Kenneth F. (1969). The Watson–Crick Model and Reductionism. *British Journal for the Philosophy of Science* 20, 325–348.

Schaffner, Kenneth F. (1976). Reductionism in Biology: Prospects and Problems. In *PSA 1974 Special Edition—Boston Studies in the Philosophy of Science*, Volume 32, pp. 613–632. New York: Springer.

Schaffner, Kenneth F. (1977). Reduction, Reductionism, Values, and Progress in the Biomedical Sciences. In R. G. Colodny (ed.), *Logic, Laws and Life*, pp. 143–171. Pittsburgh: Pittsburgh University Press.

Schaffner, Kenneth F. (1993). *Discovery and Explanation in Biology and Medicine*. Chicago: Chicago University Press.

Schippers, Michael (2017). A Representation Theorem for Absolute Confirmation. *Philosophy of Science* 84, 82–91.

Schupbach, Jonah N. (2005). On a Bayesian Analysis of the Virtue of Unification. *Philosophy of Science* 72, 594–607.

Schupbach, Jonah N. (2011a). Comparing Probabilistic Measures of Explanatory Power. *Philosophy of Science* 78, 813–829.

Schupbach, Jonah N. (2011b). *Inference to the Best Explanation, Cleaned Up and Made Respectable*. Ph.D. thesis, University of Pittsburgh.

Schupbach, Jonah N. (2018). Inference to the Best Explanation, Cleaned Up and Made Respectable. In Kevin McCain and Ted Poston (eds.), *Best Explanations: New Essays on Inference to the Best Explanation*, pp. 39–61. Oxford: Oxford University Press.

Schupbach, Jonah N. and Jan Sprenger (2011). The Logic of Explanatory Power. *Philosophy of Science* 78, 105–127.

Schurz, Gerhard (1991). Relevant Deduction. *Erkenntnis* 35, 391–437.

Schwan, Ben and Reuben Stern (2016). A Causal Understanding of When and When Not to Jeffrey Conditionalize. *Philosophers' Imprint* 17, 1–21. Retrieved from http://hdl.handle.net/2027/spo.3521354.0017.008.

Schwarz, Gideon (1978). Estimating the Dimension of a Model. *Annals of Statistics 6*, 461–464.

Scott, Dana (1964). Measurement Structures and Linear Inequalities. *Journal of Mathematical Psychology 1*, 233–247.

Seidenfeld, Teddy (1979a). *Philosophical Problems of Statistical Inference: Learning from R. A. Fisher*. New York: Springer.

Seidenfeld, Teddy (1979b). Why I Am Not an Objective Bayesian: Some Reflections Prompted by Rosenkrantz. *Theory and Decision 11*, 413–440.

Seidenfeld, Teddy (1986). Entropy and Uncertainty. *Philosophy of Science 53*, 467–491.

Senn, Stephen (2011). You May Believe You Are a Bayesian but You Are Probably Wrong. *Rationality, Markets and Morals 2*, 48–66.

Shafer, Glenn (1976). *A Mathematical Theory of Evidence*. Princeton, N.J.: Princeton University Press.

Shannon, Claude (1949). Communication Theory of Secrecy Systems. *Bell Systems Technical Journal 28*, 656–715.

Shogenji, Tomoji (2012). The Degree of Epistemic Justification and the Conjunction Fallacy. *Synthese 184*, 29–48.

Simmons, Joseph P., Leif D. Nelson, and Uri Simonsohn (2011). False-positive psychology: Undisclosed flexibility in data collection and analysis allows presenting anything as significant. *Psychological Science 22*, 1359–1366.

Simonsohn, Uri, Leif D. Nelson, and Joseph P. Simmons (2014a). *P*-curve: A Key to the File Drawer. *Journal of Experimental Psychology: General 143*, 534–547.

Simonsohn, Uri, Leif D. Nelson, and Joseph P. Simmons (2014b). *P*-curve and Effect Size. *Perspectives on Psychological Science 9*, 666–681.

Skyrms, Brian (2000). *Choice & Chance* (4th ed.). Belmont, Calif.: Wadsworth.

Skyrms, Brian (2010). *Signals: Evolution, Learning, and Information*. Oxford: Oxford University Press.

Sloman, Steven A. and David Lagnado (2015). Causality in Thought. *Annual Review of Psychology 66*, 223–247.

Smith, Adrian F. M. (1986). Comment [on "Why Isn't Everyone a Bayesian?", Efron 1986]. *American Statistician 40*, 10.

Sober, Elliott (2002). Instrumentalism, Parsimony, and the Akaike Framework. *Philosophy of Science 69*, S112–S123.

Sober, Elliott (2008). *Evidence and Evolution: The Logic Behind the Science*. Cambridge: Cambridge University Press.

Sober, Elliott (2009). Absence of Evidence and Evidence of Absence: Evidential Transitivity in Connection with Fossils, Fishing, Fine-Tuning, and Firing Squads. *Philosophical Studies* 143, 63–90.

Spiegelhalter, David J., Nicola G. Best, Bradley P. Carlin, and Angelika van der Linde (2002). Bayesian Measures of Model Complexity and Fit. *Journal of the Royal Statistical Society, Series B* 64, 583–639.

Spiegelhalter, David J. and Adrian F. M. Smith (1980). Bayes Factors and Choice Criteria for Linear Models. *Journal of the Royal Statistical Society, Series B* 42, 213–220.

Spielman, Stephen (1974). The Logic of Tests of Significance. *Philosophy of Science* 41, 211–226.

Spirtes, Peter, Clark Glymour, and Richard Scheines (2000). *Causation, Prediction, and Search* (2nd ed.). New York: Springer.

Spohn, Wolfgang (1988). Ordinal Conditional Functions: A Dynamic Theory of Epistemic States. In W. L. Harper and B. Skyrms (eds.), *Causation in Decision, Belief Change, and Statistics*, Volume 2, pp. 105–134. Dordrecht: Kluwer.

Spohn, Wolfgang (2012). *The Laws of Belief: Ranking Theory and Its Philosophical Applications*. Oxford: Oxford University Press.

Sprenger, Jan (2009a). Evidence and Experimental Design in Sequential Trials. *Philosophy of Science* 76, 637–649.

Sprenger, Jan (2009b). Statistics between Inductive Logic and Empirical Science. *Journal of Applied Logic* 7, 239–250.

Sprenger, Jan (2010a). Hempel and the Paradoxes of Confirmation. In Dov Gabbay, Stephan Hartmann, and John Woods (eds.), *Handbook of the History of Logic*, Volume 10, pp. 235–263. Amsterdam: North-Holland.

Sprenger, Jan (2010b). Statistical Inference Without Frequentist Justifications. In *EPSA Epistemology and Methodology of Science: Launch of the European Philosophy of Science Association*, pp. 289–297. Berlin: Springer.

Sprenger, Jan (2011). Hypothetico-Deductive Confirmation. *Philosophy Compass* 6, 497–508.

Sprenger, Jan (2013a). A Synthesis of Hempelian and Hypothetico-Deductive Confirmation. *Erkenntnis* 78, 727–738.

Sprenger, Jan (2013b). Testing a Precise Null Hypothesis: The Case of Lindley's Paradox. *Philosophy of Science* 80, 733–744.

Sprenger, Jan (2013c). The Role of Bayesian Philosophy within Bayesian Model Selection. *European Journal for Philosophy of Science* 2, 101–114.

Sprenger, Jan (2015). A Novel Solution of the Problem of Old Evidence. *Philosophy of Science* 82, 383–401.

Sprenger, Jan (2016a). Bayesianism and Frequentism in Statistical Inference. In Alan Hájek and Christopher Hitchcock (eds.), *Handbook of Philosophy of Probability*, pp. 382–405. Oxford: Oxford University Press.

Sprenger, Jan (2016b). Bayésianisme versus fréquentisme en inférence statistique. In Isabelle Drouet (ed.), *Le Bayésianisme aujourd'hui*, pp. 167–192. Paris: Éditions matériologiques.

Sprenger, Jan (2016c). Confirmation and Induction. In Paul W Humphreys (ed.), *Handbook of Philosophy of Probability*, pp. 185–209. Oxford: Oxford University Press.

Sprenger, Jan (2016d). The Probabilistic No Miracles Argument. *European Journal for Philosophy of Science 6*, 173–189.

Sprenger, Jan (2018a). Conditional Degree of Belief. Unpublished manuscript, retrieved on from http://philsci-archive.pitt.edu/13515/ on August 10, 2018.

Sprenger, Jan (2018b). Foundations for a Probabilistic Theory of Causal Strength. *Philosophical Review 127*, 371–398.

Sprenger, Jan (2018c). The Objectivity of Subjective Bayesianism. *European Journal for Philosophy of Science 8*, 539–558.

Sprenger, Jan (2018d). Two Impossibility Results for Popperian Corroboration. *British Journal for the Philosophy of Science 69*, 139–159.

Sprenger, Jan and Jacob Stegenga (2017). Three Arguments for Absolute Outcome Measures. *Philosophy of Science 84*, 840–852.

Stainforth, D. A., M. R. Allen, E. R. Tredger, and L. A. Smith (2007). Confidence, Uncertainty and Decision-Support Relevance in Climate Predictions. *Philosophical Transactions of the Royal Society, Series A 365*, 2145–2161.

Staley, Kent (2012). Strategies for Securing Evidence through Model Criticism. *European Journal for Philosophy of Science 2*, 21–43.

Stalnaker, Robert (1968). A Theory of Conditionals. In Nicholas Rescher (ed.), *Studies in Logical Theory*, Number 2 in American Philosophical Quarterly Monograph Series, pp. 98–112. Oxford: Blackwell.

Stalnaker, Robert (1970). Probability and Conditionals. *Philosophy of Science 37*, 64–80.

Stalnaker, Robert (1975). Indicative Conditionals. *Philosophia 5*, 269–286.

Stanford, P. Kyle (2006). *Exceeding Our Grasp: Science, History, and the Problem of Unconceived Alternatives*. Oxford: Oxford University Press.

Stegenga, Jacob (2015). Measuring Effectiveness. *Studies in History and Philosophy of Biological and Biomedical Sciences 54*, 62–71.

Stern, Reuben and Stephan Hartmann (2018). Two Sides of Modus Ponens. *Journal of Philosophy 115*, 605–621.

Sterne, Jonathan A. C. and George Davey Smith (2001). Sifting the Evidence–What's Wrong with Significance Tests? *British Medical Journal* 322, 226.

Stich, Stephen (1988). Reflective Equilibrium, Analytic Epistemology and the Problem of Cognitive Diversity. *Synthese* 74, 391–413.

Storm, Lance, Patrizio E. Tressoldi, and Lorenzo Di Risio (2010). Meta-Analysis of Free Response Studies 1992–2008: Assessing the Noise Reduction Model in Parapsychology. *Psychological Bulletin* 136, 471–485.

Strawson, P. F. (1963). Carnap's View on Constructed Systems versus Natural Languages in Analytic Philosophy. In Paul Arthur Schilpp (ed.), *The Philosophy of Rudolf Carnap*, pp. 503–518. La Salle, Ill.: Open Court.

Strevens, Michael (1999). Objective Probability as a Guide to the World. *Philosophical Studies* 95, 243–275.

Strevens, Michael (2009). *Depth*. Cambridge, Mass.: Harvard University Press.

Suárez, Mauricio (2016). Representations in Science. In P. Humphreys (ed.), *The Oxford Handbook of Philosophy of Science*, pp. 440–459. New York: Oxford University Press.

Suárez, Mauricio (2018). The Chances of Propensities. *British Journal for the Philosophy of Science* 69, 1155–1177.

Suppes, Patrick (1970). *A Probabilistic Theory of Causality*. Amsterdam: North-Holland.

Taper, Mark and Subhash Lele (eds.) (2004). *The Nature of Scientific Evidence*. Chicago: University of Chicago Press.

Tebaldi, Claudia and Reto Knutti (2007). The Use of the Multi-Model Ensemble in Probabilistic Climate Projections. *Philosophical Transactions of the Royal Society, Series A* 365, 2053–2075.

Teller, Paul (1973). Conditionalization and Observation. *Synthese* 26, 218–258.

Tentori, Katya, Vincenzo Crupi, Nicolao Bonini, and Daniel Osherson (2007). Comparison of Confirmation Measures. *Cognition* 103, 107–119.

Tentori, Katya, Vincenzo Crupi, and Daniel Osherson (2007). Determinants of Confirmation. *Psychonomic Bulletin & Review* 14, 877–883.

Thagard, Paul (1989). Explanatory Coherence. *Behavioral and Brain Sciences* 12, 435–502.

Tichý, Pavel (1974). On Popper's Definitions of Verisimilitude. *British Journal for the Philosophy of Science* 25, 155–160.

Titelbaum, Michael G. (2013). *Quitting Certainties: A Bayesian Framework Modeling Degrees of Belief*. Oxford: Oxford University Press.

Titelbaum, Michael G. (forthcoming). *Fundamentals of Bayesian Epistemology*. Oxford: Oxford University Press.

Trafimow, David and Michael Marks (2015). Editorial. *Basic and Applied Social Psychology* 37, 1–2.

Trpin, Borut and Max Pellert (forthcoming). Inference to the Best Explanation in Uncertain Evidential Situations. *British Journal for the Philosophy of Science*. https://doi.org/10.1093/bjps/axy027.

US Food and Drug Administration (2010). Guidance for the Use of Bayesian Statistics in Medical Device Clinical Trials.

Utts, Jessica (1991). Replication and Meta-Analysis in Parapsychology. *Statistical Science* 6, 363–403 (with discussion).

van Fraassen, Bas (1980). *The Scientific Image*. New York: Oxford University Press.

van Fraassen, Bas (1981). A Problem for Relative Information Minimizers in Probability Kinematics. *British Journal for the Philosophy of Science* 32, 375–379.

van Fraassen, Bas (1989). *Laws and Symmetry*. New York: Oxford University Press.

van Riel, Raphael and Robert Van Gulick (2014). Scientific Reduction. In Ed Zalta (ed.), *The Stanford Encyclopedia of Philosophy*. Retrieved from https://plato.stanford.edu/archives/sum2018/entries/scientific-reduction/.

Vassend, Olav Benjamin (2018). Goals and the Informativeness of Prior Probabilities. *Erkenntnis* 83, 647–670.

Vassend, Olav Benjamin (forthcoming). New Semantics for Bayesian Inference: The Interpretive Problem and Its Solutions. *Philosophy of Science*. Retrieved from https://sites.google.com/site/olavbvassend/research on March 29, 2019.

Villegas, C. (1964). On Qualitative Probability Sigma-Algebras. *Annals of Mathematical Statistics* 35, 1787–1796.

Vineberg, Susan (2016). Dutch Book Arguments. In Ed Zalta (ed.), *The Stanford Encyclopedia of Philosophy*. Retrieved from https://plato.stanford.edu/archives/sum2018/entries/dutch-book/.

Wacholder, Sholom, Stephen Chanock, Montserrat Garcia-Closas, Laure El Ghormli, and Nathaniel Rothman (2004). Assessing the probability that a positive report is false: An approach for molecular epidemiology studies. *JNCI: Journal of the National Cancer Institute* 96, 434–442.

Wagenmakers, Eric-Jan, Ruud Wetzels, Denny Borsboom, and Han L. J. van der Maas (2011a). Why Psychologists Must Change the Way They Analyze Their Data: The Case of Psi: Comment on Bem (2011). *Journal of Personality and Social Psychology* 100, 426–432.

Wagenmakers, Eric-Jan, Ruud Wetzels, Denny Borsboom, and Han L. J. van der Maas (2011b). Yes, Psychologists Must Change the Way They

Analyze Their Data: Clarifications for Bem, Utts and Johnson (2011). Unpublished manuscript, retrieved from http://www.ejwagenmakers.com/2011/ClarificationsForBemUttsJohnson.pdf on August 10, 2018.

Wallace, Chris (2005). *Statistical and Inductive Inference by Minimum Message Length*. New York: Springer.

Walton, Douglas (1995). *Arguments from Ignorance*. Philadelphia, Pa.: Penn State University Press.

Wasserman, Larry (2004). *All of Statistics*. New York: Springer.

Waters, C. Kenneth (2007). Causes That Make a Difference. *Journal of Philosophy 104*, 551–579.

Weber, Marcel (2006). The Central Dogma as a Thesis of Causal Specificity. *History and Philosophy of the Life Sciences 28*, 595–609.

Weinberg, Jonathan M., Shaun Nichols, and Stephen Stich (2001). Normativity and Epistemic Intuitions. *Philosophical Topics 29*, 429–460. Page references are to the version reprinted in Knobe and Nichols 2007, pp. 16–45.

Weisberg, Jonathan (2009). Varieties of Bayesianism. In Dov Gabbay, Stephan Hartmann, and John Woods (eds.), *Handbook of the History of Logic*, Volume 10: Inductive Logic, pp. 477–551. Amsterdam: North-Holland.

Weisberg, Michael (2007). Who is a Modeler? *British Journal for the Philosophy of Science 58*, 207–233.

Weisberg, Michael (2012). *Simulations and Similarity: Using Models to Understand the World*. Oxford: Oxford University Press.

Weisberg, Michael and Ryan Muldoon (2009). Epistemic Landscapes and the Division of Cognitive Labor. *Philosophy of Science 76*, 225–252.

Wenmackers, Sylvia and Jan-Willem Romeijn (2016). A New Theory about Old Evidence. *Synthese 193*, 1225–1250.

Wetzels, Ruud, Jeroen G. W. Raaijmakers, Emöke Jakab, and Eric-Jan Wagenmakers (2009). How to Quantify Support For and Against the Null Hypothesis: A Flexible WinBUGS Implementation of a Default Bayesian *t* Test. *Psychonomic Bulletin & Review 16*, 752–760.

Wetzels, Ruud and Eric-Jan Wagenmakers (2012). A Default Bayesian Hypothesis Test for Correlations and Partial Correlations. *Psychonomic Bulletin & Review 19*, 1057–1064.

Whewell, William (1847). *Philosophy of the Inductive Sciences, Founded Upon Their History*. London: Parker.

Williams, J. Robert G. (2012). Generalized Probabilism: Dutch Books and Accuracy Domination. *Journal of Philosophical Logic 41*, 811–840.

Williamson, Jon (2007). Motivating Objective Bayesianism: From Empirical Constraints to Objective Probabilities. In William Harper and Gregory Wheeler (eds.), *Probability and Inference: Essays in Honour of Henry E. Kyburg Jr.*, pp. 151–179. London: College Publications.

Williamson, Jon (2010). *In Defence of Objective Bayesianism*. Oxford: Oxford University Press.

Williamson, Jon (2013). Why Frequentists and Bayesians Need Each Other. *Erkenntnis 78*, 293–318.

Williamson, Jon (2017). *Lectures on Inductive Logic*. Oxford: Oxford University Press.

Woodward, James (2003). *Making Things Happen: A Theory of Causal Explanation*. Oxford: Oxford University Press.

Woodward, James (2016). Causation and Manipulability. In Ed Zalta (ed.), *The Stanford Encyclopedia of Philosophy*. Retrieved from https://plato.stanford.edu/archives/sum2018/entries/causation-mani/.

Worrall, John (1989). Structural Realism: The Best of Both Worlds? *Dialectica 43*, 99–124.

Yang, Yuhong (2005). Can the Strengths of AIC and BIC be Shared? A Conflict between Model Identification and Regression Estimation. *Biometrika 92*, 937–950.

Zamora Bonilla, Jesús Pedro (2000). Truthlikeness, Rationality and Scientific Method. *Synthese 122*, 321–335.

Zhao, Jiaying, Vincenzo Crupi, Katya Tentori, Branden Fitelson, and Daniel Osherson (2012). Updating: Learning versus Supposing. *Cognition 124*, 373–378.

Zhao, Jiaying, Anuj Shah, and Daniel Osherson (2009). On the Provenance of Judgments of Conditional Probability. *Cognition 113*, 26–36.

Ziliak, Stephen T. and Deirdre N. McCloskey (2008). *The Cult of Statistical Significance: How the Standard Error Costs Us Jobs, Justice, and Lives*. Ann Arbor, Mich.: University of Michigan Press.

Zollman, Kevin J. S. (2007). The Communication Structure of Epistemic Communities. *Philosophy of Science 74*, 574–587.

Index

The manufacturer's authorised representative in the EU for product safety is Oxford University Press España S.A. of el Parque Empresarial San Fernando de Henares, Avenida de Castilla, 2 – 28830 Madrid (www.oup.es/en or product.safety@oup.com). OUP España S.A. also acts as importer into Spain of products made by the manufacturer.

www.ingramcontent.com/pod-product-compliance
Lightning Source LLC
Chambersburg PA
CBHW051255200926
51612CB00001B/54
9780199672110